Empire of AI

Empire of AI

Dreams and Nightmares
in Sam Altman's OpenAI

KAREN HAO

PENGUIN PRESS   NEW YORK   2025

PENGUIN PRESS
An imprint of Penguin Random House LLC
1745 Broadway, New York, NY 10019
penguinrandomhouse.com

Copyright © 2025 by Karen Hao
Penguin Random House values and supports copyright. Copyright fuels creativity, encourages diverse voices, promotes free speech, and creates a vibrant culture. Thank you for buying an authorized edition of this book and for complying with copyright laws by not reproducing, scanning, or distributing any part of it in any form without permission. You are supporting writers and allowing Penguin Random House to continue to publish books for every reader. Please note that no part of this book may be used or reproduced in any manner for the purpose of training artificial intelligence technologies or systems.

Library of Congress record available at https://lccn.loc.gov/2025935502

ISBN 9780593657508 (hardcover)
ISBN 9798217060481 (international edition)
ISBN 9780593657515 (ebook)

Printed in the United States of America
5th Printing

Book design by Daniel Lagin

The authorized representative in the EU for product safety and compliance is Penguin Random House Ireland, Morrison Chambers, 32 Nassau Street, Dublin D02 YH68, Ireland, https://eu-contact.penguin.ie.

To my family,
past, present, and future.

To the movements
around the world
who refuse dispossession
in the name of abundance.

It is said that to explain is to explain away. This maxim is nowhere so well fulfilled as in the area of computer programming, especially in what is called heuristic programming and artificial intelligence. For in those realms machines are made to behave in wondrous ways, often sufficient to dazzle even the most experienced observer. But once a particular program is unmasked, once its inner workings are explained in language sufficiently plain to induce understanding, its magic crumbles away; it stands revealed as a mere collection of procedures, each quite comprehensible. The observer says to himself "I could have written that." With that thought he moves the program in question from the shelf marked "intelligent," to that reserved for curios, fit to be discussed only with people less enlightened than he.

**—JOSEPH WEIZENBAUM, MIT PROFESSOR
AND INVENTOR OF THE FIRST CHATBOT, ELIZA, 1966**

"Successful people create companies. More successful people create countries. The most successful people create religions."

I heard this from Qi Lu; I'm not sure what the source is. It got me thinking, though—the most successful founders do not set out to create companies. They are on a mission to create something closer to a religion, and at some point it turns out that forming a company is the easiest way to do so.

—SAM ALTMAN, 2013

CONTENTS

AUTHOR'S NOTE — xi

PROLOGUE A Run for the Throne — 1

I

1 Divine Right — 23
2 A Civilizing Mission — 46
3 Nerve Center — 73
4 Dreams of Modernity — 88
5 Scale of Ambition — 117

II

6 Ascension — 141
7 Science in Captivity — 158
8 Dawn of Commerce — 175
9 Disaster Capitalism — 189

III

10	Gods and Demons	227
11	Apex	256
12	Plundered Earth	271
13	The Two Prophets	301
14	Deliverance	326

IV

15	The Gambit	343
16	Cloak-and-Dagger	361
17	Reckoning	377
18	A Formula for Empire	399

EPILOGUE How the Empire Falls — 409

ACKNOWLEDGMENTS 423

NOTES 427

INDEX 469

AUTHOR'S NOTE

This book is based on over 300 interviews with around 260 people and an extensive trove of correspondence and documents. Most of the interviews were conducted for this book. Some were drawn from my last seven years of reporting on OpenAI, the AI industry, and its global impacts for *MIT Technology Review*, *The Wall Street Journal*, and *The Atlantic*. Over 150 of the interviews were with more than 90 current or former OpenAI executives and employees, and a handful of contractors who had access to detailed documentation of parts of OpenAI's model development practices. Another share of the interviews was with some 40 current and former executives and employees at Microsoft, Anthropic, Meta, Google, DeepMind, and Scale, as well as people close to Sam Altman.

Any quoted emails, documents, or Slack messages come from copies or screenshots of those documents and correspondences or are exactly as they appear in lawsuits. In cases where I do not have a copy, I paraphrase the text without quotes. There is one exception, which I mark in the endnotes. All dialogue is reconstructed from people's memories, from contemporaneous notes, or, when marked in the endnotes, pulled from an audio recording or transcript. In most cases, I or my fact-checking team asked those recalling quotes to repeat or confirm them again several months apart to test their stability. Every scene, every number, every name and code name, and every technical detail about OpenAI's models, such as the composition of their training data or the number of chips they were trained on, is corroborated by at least two people, with contemporaneous notes and documentation, or, in a few cases that I mark in the endnotes, with other media reporting. The same is true for most every other detail about OpenAI in the book. If I named someone, it does not mean I spoke to them directly. When I reference anyone's thoughts or feelings, it is because they described that thought or feeling, either to me, to someone I spoke to, in an email or recording I obtained, or in a public interview.

This book is not a corporate book. While it tells the inside story of OpenAI, that story is meant to be a prism through which to see far beyond this one company. It is a profile of a scientific ambition turned into an aggressive ideological, money-fueled quest; an examination of its multifaceted and expansive footprint; a meditation on power. To that end, in the course of my reporting, I spent significant time embedding with communities on the ground in countries around the world to understand their histories, cultures, lives, and experiences grappling with the visceral impacts of AI. My hope is that their stories shine through in these pages as much as the stories within the walls of one of Silicon Valley's most secretive organizations.

I reached out to all of the key figures and companies that are described in this book to seek interviews and comment. OpenAI and Sam Altman chose not to cooperate.

Empire of AI

Prologue

A Run for the Throne

On Friday, November 17, 2023, around noon Pacific time, Sam Altman, CEO of OpenAI, Silicon Valley's golden boy, avatar of the generative AI revolution, logged on to a Google Meet to see four of his five board members staring at him.

From his video square, board member Ilya Sutskever, OpenAI's chief scientist, was brief: Altman was being fired. The announcement would go out momentarily.

Altman was in his room at a luxury hotel in Las Vegas to attend the city's first Formula One race in a generation, a star-studded affair with guests from Rihanna to David Beckham. The trip was a short reprieve in the middle of the punishing travel schedule he had maintained ever since the company released ChatGPT about a year earlier. For a moment, he was too stunned to speak. He looked away as he sought to regain his composure. As the conversation continued, he tried in his characteristic way to smooth things over.

"How can I help?" he asked.

The board told him to support the interim chief executive they had selected, Mira Murati, who had been serving as his chief technology officer. Altman, still confused and wondering whether this was a bad dream, acquiesced.

Minutes later, Sutskever sent another Google Meet link to Greg Brockman, OpenAI's president and a close ally to Altman who had been the only board member missing from the previous meeting. Sutskever told Brockman he would no longer be on the board but would retain his role at the company.

The public announcement went up soon thereafter. "Mr. Altman's departure follows a deliberative review process by the board, which concluded that he was not consistently candid in his communications with the board, hindering its ability to exercise its responsibilities. The board no longer has confidence in his ability to continue leading OpenAI."

On the face of it, OpenAI had been at the height of its power. Ever since the launch of ChatGPT in November 2022, it had become Silicon Valley's most spectacular success story. ChatGPT was the fastest-growing consumer app in history. The startup's valuation was on the kind of meteoric ascent that made investors salivate and top talent clamor to join the rocket-ship company. Just weeks before, it had been valued at up to $90 billion as part of a tender offer it was in the middle of finalizing that would allow employees to sell their shares to said eager investors. A few days before, it had held a highly anticipated and highly celebrated event to launch its most aggressive slate of products.

Altman was, as far as the public was concerned, the man who had made it all happen. He had spent the spring and summer touring the world, reaching a level of celebrity that was leading the media to compare him to Taylor Swift. He had wowed just about everyone with his unassuming small frame, bold declarations, and apparent sincerity.

Before Vegas, he had once again been globe-trotting, sitting on a panel at the APEC CEO Summit, delivering lines with his usual dazzling effect.

"Why are you devoting your life to this work?" Laurene Powell Jobs, founder and president of the Emerson Collective and Steve Jobs's widow, had asked him.

"I think this will be the most transformative and beneficial technology humanity has yet invented," he said. "Four times now in the history of OpenAI—the most recent time was just in the last couple of weeks—I have gotten to be in the room, when we sort of push the veil of ignorance back and the frontier of discovery forward, and getting to do that is, like, the professional honor of a lifetime."

Shocked employees learned about Altman's firing just as everyone else did, the link to the public announcement zipping from one phone to the next across the company. It was the chasm between the news and Altman's glowing reputation that startled them the most. The company was by now pushing eight hundred people. These days, employees had fewer opportunities to meet and interact with their CEO in person. But his charming demeanor on global stages was not unlike how he behaved during all-hands meetings, at company functions, and, when he wasn't traveling, around the office.

As the rumor mill kicked into a frenzy and employees doomscrolled X, formerly Twitter, for any shreds of information, someone in the office latched on to what they saw as the most logical explanation and shouted, "Altman's running for president!" It created a momentary release of tension, before people realized this was not the case, and speculation started anew with fresh intensity and dread. Had Altman done something illegal? Maybe it was related to his sister, employees wondered. She had alleged in tweets that had gone viral a month before that her brother had abused her. Maybe it wasn't something illegal but ethically untoward, they speculated, perhaps related to Altman's other investments or his fundraising with Saudi investors for a new AI chip venture.

Sutskever posted an announcement in OpenAI's Slack. In two hours, he would hold a virtual all-hands meeting to answer employee questions. "That was the longest two hours ever," an employee remembers.

Sutskever, Murati, and OpenAI's remaining executives came onto the screen side by side, stiff and unrehearsed, as the all-hands streamed to employees in the office and working from home.

Sutskever looked solemn. He was known among employees as a deep thinker and a mystic, regularly speaking in spiritual terms with a force of sincerity that could be endearing to some and off-putting to others. He was also a goofball and gentlehearted. He wore shirts with animals on them to the office and loved to paint them as well—a cuddly cat, cuddly alpacas, a cuddly fire-breathing dragon—alongside abstract faces and everyday objects. Some of his amateur paintings hung around the office, including a trio of flowers blossoming in the shape of OpenAI's logo, a symbol of what he always urged employees to build: "A plurality of humanity-loving AGIs."

Now, he attempted to project a sense of certainty to anxious employees submitting rapid-fire questions via an online document. But Sutskever was an imperfect messenger; he was not one that excelled at landing messages with his audience.

"Was there a specific incident that led to this?" Murati read aloud first from the list of employee questions.

"Many of the questions in the document will be about the details," Sutskever responded. "What, when, how, who, exactly. I wish I could go into the details. But I can't." Anyone curious should read the press release, he added. "It actually says a lot of stuff. Read it maybe a few times."

The response baffled employees. They had just received cataclysmic news. Surely, as the people most directly affected by the situation, they deserved more specifics than the general public.

Murati read off a few more questions. How did this affect the relationship with Microsoft? Microsoft, OpenAI's biggest backer and exclusive licensee of its technologies, was the sole supplier of its computing infrastructure. Without it, all the startup's work—performing research, training AI models, launching products—would grind to a halt. Murati responded that she didn't expect it to be affected. They had just had a call with Microsoft's chief executive Satya Nadella and chief technology officer Kevin Scott. "They're all very committed to our work," she said.

What about OpenAI's tender offer? Employees with a certain tenure had been given the option to sell what could amount to millions of dollars' worth of their equity. The tender was so soon that many had made plans to buy property, or already had. "The tender—we're, um, we're going to see," Brad Lightcap, the chief operating officer, waffled. "I am in touch with investors leading the tender and some of our largest investors already on the cap table. All have committed their steadfast support to the company."

After several more questions were met with vague responses, another employee tried again to ascertain what Sam had done. Was this related to his role at the company? Or did it involve his personal life? Sutskever once again directed people to the press release. "The answer is actually there," he said.

Murati read on from the document. "Will questions about details be answered at some point or never?"

Sutskever responded: "Keep your expectations low."

As the all-hands continued and Sutskever's answers seemed to grow more and more out of touch, employee unease quickly turned into anger.

"When a group of people grow through a difficult experience, they often end up being more united and closer to each other," Sutskever said. "This difficult experience will make us even closer as a team and therefore more productive."

"How do you reconcile the desire to grow together through crisis with a frustrating lack of transparency?" an employee wrote in. "Typically truth is a necessary condition for reconciliation."

"I mean, fair enough," Sutskever replied. "The situation isn't perfect."

Murati tried to quell the rising tension. "The mission is so much bigger than any of us," she said.

Lightcap echoed her message: OpenAI's partners, customers, and investors had all stressed that they continued to resonate with the mission. "If anything, we have a greater duty now, I think, to push hard on that mission."

Sutskever again attempted to be reassuring. "We have all the ingredients, all of them: The computer, the research, the breakthroughs are astounding," he said. "When you feel uncertain, when you feel scared, remember those things. Visualize the size of the cluster in your mind's eye. Just imagine all those GPUs working together."

An employee submitted a new question. "Are we worried about the hostile takeover via coercive influence of the existing board members?"

"Hostile takeover?" Sutskever repeated, a new edge in his voice. "The OpenAI nonprofit board has acted entirely in accordance to its objective. It is not a hostile takeover. Not at all. I disagree with this question."

That night, several employees gathered at a colleague's house for a party that had been planned before Altman's firing. There were guests from other AI companies as well, including Google DeepMind and Anthropic.

Right before the event, an alert went out to all attendees. "We are adding a second themed room for tonight: 'The no-OpenAI talk room.' See you all!" In the end, few people stayed long in the room. Most people wanted to talk about OpenAI.

Brockman had announced that afternoon that he was quitting in protest. Microsoft's Nadella, who had been furious about being told about Altman's firing only minutes before it happened, had put out a carefully crafted tweet: "We have a long-term agreement with OpenAI with full access to everything we need to deliver on our innovation agenda and an exciting product roadmap; and remain committed to our partnership, and to Mira and the team."

As rumors continued to proliferate, word arrived that three more senior researchers had quit the company: Jakub Pachocki and Szymon Sidor, early employees who had among the longest tenures at OpenAI, and Aleksander Mądry, an MIT professor on leave who had joined recently. Their departures further alarmed some OpenAI employees, a signal of a bleeding out of leadership and talent that could spook investors and halt the tender offer or, worse, ruin the company. At the party, employees grew more and more despondent and agitated. A dissolution of the tender offer would snatch away a significant financial upside to all their hard labor, to say nothing of a dissolution of the company, which would squander so much promise and hard work.

Also that night, the board and the remaining leadership at the company were holding a series of increasingly hostile meetings. After the all-hands, the false projection of unity between Sutskever and the other leaders had collapsed. Many of the executives who had sat next to Sutskever during the livestream had been nearly as blindsided as the rest of the staff, having learned of Altman's dismissal moments before it was announced. Riled up by Sutskever's poor performance, they had demanded to meet with the rest of the board. Roughly a dozen executives, including Murati and Lightcap, had gathered in a conference room at the office.

Sutskever was dialed in virtually along with the three independent directors: Adam D'Angelo, the cofounder and CEO of the question-and-answer site Quora; Tasha McCauley, an entrepreneur and adjunct senior management scientist at the policy think tank RAND; and Helen Toner, an Australian-born researcher at another think tank, Georgetown University's CSET, or Center for Security and Emerging Technology.

Under an onslaught of questions, the four board members repeatedly evaded making further disclosures, citing their legal responsibilities to protect confidentiality. Several leaders grew visibly enraged. "You're saying that Sam is untrustworthy," Anna Makanju, the vice president of global affairs, who had often accompanied Altman on his global charm offensive, said furiously. "That's just not our experience with him at all."

The gathered leadership pressed the board to resign and hand their seats to three employees, threatening to all quit if the board didn't comply immediately. Jason Kwon, the chief strategy officer, a lawyer who had previously served as OpenAI's general counsel, upped the ante. It was in fact *illegal* for the board not to resign, he said, because if the company fell apart, this would be a breach of the board members' fiduciary duties.

The board members disagreed. They maintained that they had carefully consulted lawyers in making the decision to fire Altman and had acted in accordance with their delineated responsibilities. OpenAI was not like a normal company, its board not like a normal board. It had a unique structure that Altman had designed himself, giving the board broad authority to act in the best interest not of OpenAI's shareholders but of its mission: to ensure that AGI, or artificial general intelligence, benefits humanity. Altman had long touted the board's ability to fire him as its most important governance mechanism. Toner underscored the point: "If this action destroys the company, it could in fact be consistent with the mission."

The leadership relayed her words back to employees in real time: Toner didn't care if she destroyed the company. Perhaps, many employees began to conclude, that was even her intention. At the thought of losing all of their equity, a person at the party began to cry.

The next day, Saturday, November 18, dozens of people, including OpenAI employees, gathered together at Altman's $27 million mansion to await more news.

The three senior researchers who had quit, Pachocki, Sidor, and Mądry, had met with Altman and Brockman to talk about re-forming the company and continuing their work. To some, word of their discussions increased employee anxiety: A new OpenAI competitor could intensify the instability at the company. To others it offered hope: If Altman indeed founded a new venture, they would leave to go with him.

OpenAI's remaining leadership gave the board a deadline of 5 p.m. Pacific time that day: Reinstate Altman and resign, or risk a mass employee exodus from the company. The board members refused. Through the weekend, they frantically made calls, sometimes in the middle of the night, to anyone on their roster of connections who would pick up. In the face of mounting ire from employees and investors over Altman's firing, Murati was no longer willing to serve as interim CEO. They needed to replace her with someone who could help restore stability, or find new board members who could hold their own against Altman if he actually came back.

That night, after the deadline came and went, Jason Kwon sent a memo to employees. "We are still working towards a resolution and we remain optimistic," he wrote. "By resolution, we mean bringing back Sam, Greg, Jakub, Szymon, Aleksander."

Altman tweeted in his signature lowercase style. "i love the openai team so much."

Dozens of other employees began retweeting it with a heart emoji.

On Sunday, Altman and Brockman arrived back at the office to negotiate their return. Over the course of the day, more and more employees joined them to wait in suspense. By then, most employees, leadership, and the board had barely slept in more than thirty-six hours; everything was beginning to blur together. Altman tweeted a selfie, lips pursed, brows furrowed, displaying a guest badge in his hand. "first and last time i ever wear one of these," he added as the caption. Leadership set another 5 p.m. deadline for the board to reinstate Altman and to resign.

The pressure was now piling on from all directions. Microsoft, OpenAI's other investors, and heavyweights across Silicon Valley were publicly siding with Altman. A source relayed the playbook to the media: Not only would employees leave en masse if the decision were not reversed, but Microsoft would withhold access to its computing infrastructure, and investors would file lawsuits. The combination would make an OpenAI without Altman untenable.

Still, the board continued to resist. Nearing 9 p.m., once again well past the latest deadline, Sutskever posted a long message on Slack on behalf of the board. Altman was not returning; Emmett Shear, the former CEO of Twitch, was now the new interim head of OpenAI. He and Shear would arrive at the office in five minutes to give a speech about the company's new vision.

"The board firmly stands by its decision as the only path to advance and defend the mission of OpenAI," he wrote. "Put simply, Sam's behavior and lack of transparency in his interactions with the board undermined the board's ability to effectively supervise the company in the manner it was mandated to do."

The Slack instantly lit up with dozens of angry replies from employees.

"You and what fucking army"

"you're delusional"

"Emmett will be the CEO of nothing"

Roughly two hundred employees paraded out of the office to boycott the talk. Murati rushed the executives out the building. By the time

Shear arrived with Sutskever, only a dozen or so people were in the audience.

Anna Brockman, Greg's wife, approached Sutskever, who four years earlier had officiated the couple's civil ceremony. Through tears, she flung her arms around him and pleaded with him to reconsider his position.

Many of the employees who had left the office gathered at a few colleagues' houses to weather the night; hundreds joined a Signal group for updates. Late that evening, Nadella announced that he was hiring Altman and Brockman to lead a new AI division. Word spread rapidly: Anyone who wanted to join Altman would have a guaranteed job at Microsoft.

The news flipped the mood from fear to defiance. With the perception of a backup option in hand, employees had new leverage to speak out against the board and Shear. At one employee's house, overflowing with well over a hundred OpenAI colleagues, executives and senior researchers wrote an open letter to amp up the pressure, reiterating the leadership team's threats with greater force: Without Altman's reinstatement and the board's resignation, they could all quit immediately and join Microsoft.

The group worked to circulate the letter as far and wide as possible, posting it on various private channels and phoning employees who were not present to sign it. As it reached a critical mass of signatures, many more employees rushed to join in, under pressure to avoid raising questions about their absence. Within twenty-four hours, the letter had reached more than 700 signatories of the roughly 770 employees. Dozens of employees sent identical emails to the board in rapid succession. In droves, they took to X to post the same message: "OpenAI is nothing without its people."

Then, in the middle of the night, employees saw Sutskever's name appear on the open letter.

Sutskever soon addressed it publicly. "I deeply regret my participation in the board's actions," he tweeted in the early hours of Monday morning. "I never intended to harm OpenAI. I love everything we've built together and I will do everything I can to reunite the company."

On Tuesday, November 21, leadership dialed the board from Altman's house. Five days in, everyone was caving from the lack of sleep and exhaustion. Thanksgiving was around the corner, and desperation on both sides had finally opened a pathway toward a resolution.

Through the day, everyone began to consider different configurations in earnest.

Altman and Brockman, originally adamant about returning to OpenAI with board seats, finally acquiesced to no longer having them. The board, seeing no path for preserving the company without Altman, finally acquiesced to his return.

Late that night, they agreed on the three independent board seats. D'Angelo would stay; Toner and McCauley would step down; Bret Taylor, a former co-CEO of Salesforce and former CTO of Facebook, and Larry Summers, a former treasury secretary and former president of Harvard, would fill their vacancies. As part of the deal, the new board would eventually add more members. Altman would submit to an investigation.

What was most important for the company now, they also agreed, was to project unity, stability, and reconciliation. Two days later, Altman would tweet a staged message: "just spent a really nice few hours with @adamdangelo. happy thanksgiving from our families to yours🦃" Ten days later, Brockman would tweet a photo: him and Sutskever, arms around each other, smiling widely. In the office, the company's artist in residence would hook up OpenAI's image generator DALL-E to a color printer to create tiny kaleidoscopic heart-shaped stickers. Next to the printer would be a giant pink heart emblazoned with the line "OpenAI is nothing without its people."

In December, Altman would describe the experience to Trevor Noah in a podcast as the second worst moment of his life, surpassed only by his dad's death. The following month, in January, the tender would close, valuing OpenAI at $86 billion.

But that was all to come. Tuesday night, November 21, was just about celebrating. With the announcement of Altman's return and the new agreement, employees came flooding back to the office to hug, to cry, to blast music. At some point, someone turned on a smoke machine. It set off the fire alarm. Everyone kept partying.

Brockman snapped a group selfie with the crowd, a picture bursting with the ecstatic, slaphappy delirium of surviving a crisis. He tweeted it with a caption: "we are so back."

The news of Altman's ouster broke as I was in the middle of an interview for this book. I had silenced my phone, blissfully unaware of the chaotic week about to unfold. Twenty minutes later, I tapped my screen to check the time and saw a slew of missed notifications. So began a hazy, adrenaline-fueled series of days as I raced to understand what was going on.

In the weeks that followed, friends, family, and media would ask me dozens of times: What did all this mean, if anything? Was the back-and-forth just an entertaining distraction? Or would it have consequences for the rest of us? I had by then been following OpenAI for five years. In 2019, I was the first journalist to gain extensive access to the company and to write its first profile. To me, these events were not just some frivolous Silicon Valley power moves. The drama highlighted one of the most urgent questions of our generation: How do we govern artificial intelligence?

AI is one of the most consequential technologies of this era. In a little over a decade, it has reformed the backbone of the internet, becoming a ubiquitous mediator of digital activities. In even less time, it is now on track to rewire a great many other critical functions in society, from health care to education, from law to finance, from journalism to government. The future of AI—the shape that this technology takes—is inextricably tied to our future. The question of how to govern AI, then, is really a question about how to ensure we make our future better, not worse.

From the beginning, OpenAI had presented itself as a bold experiment in answering this question. It was founded by a group including Elon Musk and Sam Altman, with other billionaire backers like Peter Thiel, to be more than just a research lab or a company. The founders asserted a radical commitment to develop so-called artificial general intelligence, what they described as the most powerful form of AI anyone had ever seen, not for the financial gains of shareholders but for the benefit of humanity. To that end, Musk and Altman had set it up as a nonprofit and pledged $1 billion for its operation. It would not work on commercial products; instead it would be dedicated fully to research, driven by only the purest intentions of ushering in a form of AGI that would unlock global utopia, and not its opposite. Musk and Altman also pledged to share as much of its research as possible along the way and to collaborate widely with other institutions. If the goal was to do good by the world, openness—hence *Open*AI—and democratic participation in the technology's development were key. A few years later, leadership went even further, making a promise to self-sacrifice if necessary. "We are concerned about late-stage AGI development becoming a competitive race without time for adequate safety precautions," they wrote. If another attempt to create beneficial AGI surpassed OpenAI's progress, "we commit to stop competing with and start assisting this project."

But by the time I began to profile OpenAI, its commitment to these ideals were fast eroding. Merely a year and a half in, OpenAI's executives realized that the path they wanted to take in AI development would demand extraordinary amounts of money. Musk and Altman, who had until then both taken more hands-off approaches as cochairmen, each tried to install himself as CEO. Altman won out. Musk left the organization in early 2018 and took his money with him. In hindsight, the rift was the first major sign that OpenAI was not in fact an altruistic project but rather one of ego.

The loss of its primary backer pushed OpenAI into financial uncertainty. To plug the hole, Altman reformulated OpenAI's legal structure. Nested within the nonprofit, he created a for-profit arm, OpenAI LP, to raise capital, commercialize products, and provide returns to investors much like any other company. Four months later, in July 2019, OpenAI announced a new $1 billion funder: software giant and cloud services provider Microsoft.

I arrived at OpenAI's offices for the first time shortly thereafter, in

August 2019. After three days embedded among employees and dozens of interviews, I could see that the experiment in idealistic governance was unraveling. OpenAI had grown competitive, secretive, and insular, even fearful of the outside world under the intoxicating power of controlling such a paramount technology. Gone were notions of transparency and democracy, of self-sacrifice and collaboration. OpenAI executives had a singular obsession: to be the first to reach artificial general intelligence, to make it in their own image.

Over the next four years, OpenAI became everything that it said it would not be. It turned into a nonprofit in name only, aggressively commercializing products like ChatGPT and seeking unheard-of valuations. It grew even more secretive, not only cutting off access to its own research but shifting norms across the industry to bar a significant share of AI development from public scrutiny. It triggered the very race to the bottom that it had warned about, massively accelerating the technology's commercialization and deployment without shoring up its harmful flaws or the dangerous ways that it could amplify and exploit the fault lines in our society. Along the way, clashes between leaders and employees grew ever more fierce, as different groups inside the company sought to seize control and re-form OpenAI around their vision.

The ouster and reinstatement of Altman in November 2023 was final proof that the governance experiment had failed. Not simply because OpenAI's nonprofit board buckled under moneyed interests, dissolving the last remnant of the organization's altruistic facade. It illustrated in the clearest terms just how much a power struggle among a tiny handful of Silicon Valley elites is shaping the future of AI. Even if events had gone a different way and the board had succeeded in replacing Altman, nothing would have changed about the fact that such a consequential decision was made behind closed doors. Beyond a small group of ultrarich techno-optimists, their fiercest ideological rivals, and a multibillion-dollar tech giant, even OpenAI's own employees found themselves largely in the dark about which way their fates would fall.

I began reporting on artificial intelligence long before OpenAI and

ChatGPT became synonymous with the technology. I watched it evolve through the messy process of science and innovation as researchers trialed new ideas, presented their best successes at packed conferences, and brought them to bear on commercial products at the world's biggest companies, including Google and Facebook, Alibaba and Baidu. I read hundreds of research papers and interviewed scientists, engineers, and executives to understand their worldviews and their decisions—and how those left fingerprints on the technology's design and application.* As AI's footprint sprawled out globally, I tracked the subtle and dramatic ways it changed lives and communities. I traveled to five continents to hear from people about these experiences. In Colombia and Kenya, I met people who in the face of economic crisis turned to annotating data for the AI industry, only to find themselves working under conditions that resembled indentured servitude. In Arizona and Chile, I met with local politicians and activists worried about the growing shadow metropolis of data centers guzzling their homes' precious water resources.

Through my reporting, I've come to understand two things: Artificial intelligence is a technology that takes many forms. It is in fact a multitude of technologies that shape-shift and evolve, not merely based on technical merit but with the ideological drives of the people who create them and the winds of hype and commercialization. While ChatGPT and other so-called large language models or generative AI applications have now taken the limelight, they are but one manifestation of AI, a manifestation that embodies a particular and remarkably narrow view about the way the world is and the way it should be. Nothing about this form of AI coming to the fore or even existing at all was inevitable; it was the culmination of thousands of subjective choices, made by the people who had

---

* A note on AI research paper conventions and peer review: In the AI field, researchers often post their papers directly online to a free and open repository called arXiv (pronounced "archive") and either go through a peer-review process with a conference or publication many months or years later, or do not bother to get peer reviewed at all. This practice has become so normalized that many people in the field cite papers based on impact rather than whether they have passed peer review. In this book, I will do the same. The endnotes denote which papers did not get peer-reviewed as preprints.

the power to be in the decision-making room. In the same way, future generations of AI technologies are not predetermined. But the question of governance returns: Who will get to shape them?

The other thing I've learned: This current manifestation of AI, and the trajectory of its development, is headed in an alarming direction. On the surface, generative AI is thrilling: a creative aid for instantly brainstorming ideas and generating writing; a companion to chat with late into the night to ward off loneliness; a tool that could perhaps one day be so effective at boosting productivity that it will increase top-line economic activity. But in the same way we once thought Facebook was merely a place for posting vacation pictures and connecting with long-lost elementary school friends, or for sparking positive and transformative social movements, there is more to the sleek, entrancing exterior than meets the eye. Under the hood, generative AI models are monstrosities, built from consuming previously unfathomable amounts of data, labor, computing power, and natural resources. GPT-4, the successor to the first ChatGPT, is, by one measure, reportedly over fifteen thousand times larger than its first generation, GPT-1, released five years earlier. The exploding human and material costs are settling onto wide swaths of society, especially the most vulnerable, people I met around the world, whether workers and rural residents in the Global North or impoverished communities in the Global South, all suffering new degrees of precarity. Rarely have they seen any "trickle-down" gains of this so-called technological revolution; the benefits of generative AI mostly accrue upward.

Over the years, I've found only one metaphor that encapsulates the nature of what these AI power players are: empires. During the long era of European colonialism, empires seized and extracted resources that were not their own and exploited the labor of the people they subjugated to mine, cultivate, and refine those resources for the empires' enrichment. They projected racist, dehumanizing ideas of their own superiority and modernity to justify—and even entice the conquered into accepting—the invasion of sovereignty, the theft, and the subjugation. They justified their quest for power by the need to compete with other empires: In an arms race, all bets are off. All this ultimately served to entrench each empire's power and to drive its expansion and progress. In the simplest terms, empires amassed extraordinary riches across space and time, through imposing a colonial world order, at great expense to everyone else.

The empires of AI are not engaged in the same overt violence and brutality that marked this history. But they, too, seize and extract precious resources to feed their vision of artificial intelligence: the work of artists and writers; the data of countless individuals posting about their experiences and observations online; the land, energy, and water required to house and run massive data centers and supercomputers. So too do the new empires exploit the labor of people globally to clean, tabulate, and prepare that data for spinning into lucrative AI technologies. They project tantalizing ideas of modernity and posture aggressively about the need to defeat other empires to provide cover for, and to fuel, invasions of privacy, theft, and the cataclysmic automation of large swaths of meaningful economic opportunities.

OpenAI is now leading our acceleration toward this modern-day colonial world order. In the pursuit of an amorphous vision of progress, its aggressive push on the limits of scale have set the rules for a new era of AI development. Now every tech giant is racing to out-scale one another, spending sums so astronomical that even they have scrambled to redistribute and consolidate their resources. Around the time Microsoft invested $10 billion in OpenAI, it laid off ten thousand workers to cut costs. After Google watched OpenAI outpace it, it centralized its AI labs into Google DeepMind. As Baidu raced to develop its ChatGPT equivalent, employees working to advance AI technologies for drug discovery had to suspend their research and cede their computer chips to develop the chatbot instead. The current AI paradigm is also choking off alternative paths to AI development. The number of independent researchers not affiliated with or receiving funding from the tech industry has rapidly dwindled, diminishing the diversity of ideas in the field not tied to short-term commercial benefit. Companies themselves, which once invested in sprawling exploratory research, can no longer afford to do so under the weight of the generative AI development bill. Younger generations of scientists are falling in line with the new status quo to make themselves more employable. What was once unprecedented has become the norm.

Today, the empires have never been richer. As I finished writing this book in January 2025, OpenAI topped a $157 billion valuation. Anthropic, a competitor, was nearing a deal that would value it at $60 billion. After striking its partnership with OpenAI, Microsoft tripled its market capitalization to over $3 trillion. Since ChatGPT, the six largest tech giants together have seen their market caps increase $8 trillion. At the same time, more and more doubts have risen about the true economic value of generative AI. In June 2024, a Goldman Sachs report noted spending on the technology's development was projected to hit $1 trillion in a few years with so far "little to show for it." The following month, a survey from The Upwork Research Institute of 2,500 workers globally found that while 96 percent of C-suite leaders expected generative AI to boost productivity, 77 percent of the employees actually using the tools reported them instead adding to their workload; this was in part due to the amount of time spent reviewing AI-generated content, in part due to growing demands from superiors to do more work. In a November *Bloomberg* article reviewing the financial tally of generative AI impacts, staff writers Parmy Olson and Carolyn Silverman summarized it succinctly—the data "raises an uncomfortable prospect: that this supposedly revolutionary technology might never deliver on its promise of broad economic transformation, but instead just concentrate more wealth at the top."

Meanwhile, the rest of the world is beginning to collapse under the weight of the exploding human and material costs of this new era. Workers in Kenya earned starvation wages to filter out violence and hate speech from OpenAI's technologies, including ChatGPT. Artists are being replaced by the very AI models that were built from their work without their consent or compensation. The journalism industry is atrophying as generative AI technologies spawn heightened volumes of misinformation. Before our eyes, we're seeing an ancient story repeat itself—and this is only the beginning.

OpenAI is not slowing down. It is continuing to chase even greater scales with unparalleled resources, and the rest of the industry is following. To quell the rising concerns about generative AI's present-day performance, Altman has trumpeted the future benefits of AGI ever louder. In a September 2024 blog post, he declared that the "Intelligence Age," characterized by "massive prosperity," would soon be upon us, with superintelligence perhaps arriving as soon as in "a few thousand days." "I believe the future is going to be so bright that no one can do it justice by trying to write about it now," he wrote. "Although it will happen incrementally, astounding triumphs—fixing the climate, establishing a space colony, and the discovery of all of physics—will eventually become commonplace." At this point, AGI is largely rhetorical—a fantastical, all-purpose excuse for OpenAI to continue pushing for ever more wealth and power. Few others have the comparable capital to invest in alternative options. OpenAI and its small handful of competitors will have an oligopoly on the technology they're selling us as the key to the future; anyone—whether company or government—who wants a piece of that vision will have to rely on the empires to provide it.

There is a different way forward. Artificial intelligence doesn't have to be what it is today. We don't need to accept the logic of unprecedented scale and consumption to achieve advancement and progress. So much of what our society actually needs—better health care and education, clean air and clean water, a faster transition away from fossil fuels—can be assisted and advanced with, and sometimes even necessitates, significantly smaller AI models and a diversity of other approaches. AI alone won't be enough, either: We'll also need more social cohesion and global cooperation, some of the very things being challenged by the existing vision of AI development.

But the empires of AI won't give up their power easily. The rest of us will need to wrest back control of this technology's future. And we're at a pivotal moment when that's still possible. Just as empires of old eventually fell to more inclusive forms of governance, we, too, can shape the future of AI together. Policymakers can implement strong data privacy and transparency rules and update intellectual property protections to return people's agency over their data and work. Human rights organizations can advance international labor norms and laws to give data labelers guaranteed wage minimums and humane working conditions as well as to shore up labor rights and guarantee access to dignified economic opportunities across all sectors and industries. Funding agencies can foster renewed diversity in AI research to develop fundamentally new manifestations of what this technology could be. Finally, we can all resist the narratives that OpenAI and the AI industry have told us to hide the mounting social and environmental costs of this technology behind an elusive vision of progress.

I

Chapter 1

Divine Right

Everyone else had arrived, but Elon Musk was late as usual.

It was the summer of 2015, and a group of men had gathered for a private dinner at Sam Altman's invitation to discuss the future of AI and humanity.

Musk had met Altman, fourteen years his junior, a while earlier and had formed a good impression. President of the famed Silicon Valley startup accelerator Y Combinator, Altman's reputation preceded him. After starting his first company at age nineteen, he had rapidly established himself within Silicon Valley as a brilliant strategist and dealmaker with grand ambitions, even for the land of big-thinking founders. Musk found him to be smart, driven, and, most important, someone who espoused like-minded views on the need to carefully develop and govern artificial intelligence. It was as if, Musk would describe in a lawsuit years later, Altman had mirrored everything Musk had ever said about the subject to win his trust.

For Altman's part, he often said that Musk had been a childhood hero. After the older entrepreneur had shown him around the sprawling SpaceX factory in Hawthorne, California, that admiration had only deepened. "The thing that sticks in memory was the look of absolute certainty on his face when he talked about sending large rockets to Mars," Altman wrote later of the experience. "I left thinking 'huh, so that's the benchmark for what conviction looks like.'"

Musk had been deeply concerned about AI for some time. In 2012, he'd met Demis Hassabis, the professorial CEO of the London-based AI lab DeepMind Technologies. Shortly thereafter, Hassabis had also paid Musk a visit at his SpaceX factory. As the two men sat in the canteen, surrounded by the sounds of massive rocket parts being transported and assembled, Hassabis raised the possibility that more advanced AI, of the kind that might one day exceed human intelligence, could pose a threat to humanity. What's more, Musk's fail-safe of colonizing Mars to escape would not work in this scenario. Superintelligence, Hassabis said with amusement, would simply follow humans into the galaxy. Musk, decidedly less amused, invested $5 million in DeepMind to keep tabs on the company.

Later, at his 2013 birthday party in the lush wine-growing landscapes of Napa Valley, Musk had gotten into a heated and emotional debate with his longtime friend and Google cofounder Larry Page over whether AI surpassing human intelligence was in fact a problem. Page didn't think so, calling it the next stage of evolution. When Musk balked, Page accused him of being a "specist," discriminating against nonhuman species.

After that, Musk began to speak incessantly about the existential risk of AI. At an MIT symposium, he described AI as probably the "biggest existential threat" to humanity and its development as "summoning the demon." He met with publishers in New York, gripped by the thought of writing his own book about extinction-level threats, including AI. Later, at a recurring AI Salon event at Stanford, a young researcher named Timnit Gebru would come up to him after a talk and ask him why he was so obsessed with AI when the threat of climate change was more clearly existential. "Climate change is bad, but it's not going to kill everyone," he said. "AI could render humanity extinct."

In late 2013, when Musk learned that Google would acquire DeepMind, he was convinced that such a union would end very badly. Publicly, he warned that if Google gave a hypothetical AGI an objective to maximize profits, the software could seek to take out the company's competitors at any cost. "Murdering all competing A.I. researchers as its first move strikes me as a bit of a character flaw," Musk told *The New Yorker*. Over an hour-long Skype call in a closet upstairs at a house party in Los Angeles, he urged Hassabis to reconsider the deal. "The future of AI," said Musk, "should not be controlled by Larry." But although Musk didn't know it, Google had already dispatched a team of AI researchers via private jet to DeepMind's offices to vet the acquisition. As part of the evaluation, Jeff Dean, one of the earliest and most senior Googlers, had reviewed a sample of the company's codebase personally and given the deal his approval. In January 2014, Google confirmed the acquisition. It had reportedly gone through for between $400 million and $650 million.

Musk began hosting his own dinners to discuss ways of countering Google. In early 2015, he also met with US president Barack Obama to explain the dangers of AI, how to make it safer, and how to regulate it. Around the same time, Musk would see Hassabis again at SpaceX, this time for the first meeting of the Google DeepMind AI Ethics Board, a governance structure that Page and Hassabis had proposed to help oversee the responsible development of DeepMind's technologies. The meeting convinced Musk that the board was a fraud and inflamed his concerns into an all-consuming obsession to counter Hassabis's vision.

For years afterward, Musk would regularly characterize Hassabis as a supervillain who needed to be stopped. Musk would make unequivocally clear that OpenAI was the good to DeepMind's evil. In the summer of 2016, not long after OpenAI was founded, several employees met Hassabis and reported back to the office: DeepMind did intend to take over the world; Musk's characterization seemed correct. The following year, Musk hosted an off-site meeting for OpenAI employees at his SpaceX factory and launched into a rant about Hassabis. Before founding DeepMind, Hassabis had spent seven years running a video game design studio he'd founded. "He literally made a video game where an evil genius tries to create AI to take over the world," Musk shouted, referring to Hassabis's 2004 title *Evil Genius*, "and fucking people don't see it. Fucking people don't see it! And Larry? Larry thinks he controls Demis but he's too busy fucking windsurfing to realize that Demis is gathering all the power."

Musk's paranoia about Hassabis would become a source of entertainment for DeepMind employees. Hassabis was incredibly ambitious and could be intense, certainly, but he was also kind and measured. "The creation of OpenAI felt like this semi-hysterical reaction to a fairly mild-mannered man," recalls a former DeepMind researcher. "It seemed a little absurd."

On Musk's list of recommended books was *Superintelligence: Paths, Dangers, Strategies*, in which Oxford philosopher Nick Bostrom argues that if AI ever became smarter than humans, it would be difficult to control and could cause an existential catastrophe. Given a simple objective like producing paper clips, this superior AI could determine that humans pose a threat to its paper clip–producing objective because they take up paper clip–producing resources. Bostrom then proposed a solution: It could be possible to avert the superintelligence control problem by "aligning" AI with human values—giving it the ability to extrapolate beyond explicit instructions to achieve its objectives without harming humans. This idea formed the basis of the AI alignment research discipline, which OpenAI would come to champion. To his far-reaching Twitter following, Musk called the book "worth reading."

In January 2023, the resurfacing of an email Bostrom wrote to a LISTSERV in the midnineties would make people question his own human values. "I have always liked the uncompromisingly objective way of thinking and speaking," he had written. "Take for example the following sentence: Blacks are more stupid than whites. I like that sentence and think it is true." Bostrom would apologize, calling the email "disgusting" and an inaccurate representation of his views.

To Musk, Altman seemed like a fellow traveler, someone who harbored his own streak for hedging against catastrophe. In 2016, Altman would tell longtime California chronicler Tad Friend at *The New Yorker* that in the event of a doomsday scenario, he planned to escape to New Zealand with his close friend and mentor, billionaire investor Peter Thiel.

Thiel would describe Altman in the same article as "culturally very Jewish—an optimist yet a survivalist, with a sense that things can always go deeply wrong." Two years later Altman would tell *Bloomberg* that he had been joking but still had a go bag at the ready. He was particularly concerned about novel biological viruses and had packed gas masks alongside antibiotics, water, batteries, a tent, and a gun. But on his blog in February 2015, he agreed with Musk that superintelligence was "probably the greatest threat to the continued existence of humanity." Even though a devastating engineered virus was more likely to happen, he said, it was "unlikely to destroy every human in the universe." "Incidentally," he wrote in a parenthetical, "Nick Bostrom's excellent book 'Superintelligence' is the best thing I've seen on this topic. It is well worth a read."

A few months later, in May 2015, Altman emailed Musk. "Been thinking a lot about whether it's possible to stop humanity from developing AI," Altman wrote. "I think the answer is almost definitely not. If it's going to happen anyway, it seems like it would be good for someone other than Google to do it first." He proposed for Y Combinator, or YC as it was known, to start a "Manhattan Project for AI," structured "so that the tech belongs to the world via some sort of nonprofit." "Obviously we'd comply with/aggressively support all regulation," he added, nodding to Musk's recent pushes for government oversight.

"Probably worth a conversation," Musk replied.

In June, Altman emailed again with more details. "The mission would be to create the first general AI and use it for individual empowerment—ie, the distributed version of the future that seems the safest. More generally, safety should be a first-class requirement." He then proposed a governance structure that would defer to him and Musk. The two of them would sit on the board and invite three others to join them. "The technology would be owned by the foundation and used 'for the good of the world,' and in cases where it's not obvious how that should be applied the 5 of us would decide," Altman said.

If Musk could also commit to meeting the team around once a month, Altman continued, it would help with "getting the best people to be part of it." If Musk didn't have time, his public endorsement "would still probably be really helpful for recruiting."

"Agree on all," Musk responded.

Altman proceeded to invite Musk to the private dinner on the future of AI and humanity to meet a group of top engineers and AI researchers that he hoped to get on board the project. With Musk's confirmation of attendance, the dinner venue upgraded to a restaurant at one of the SpaceX founder's go-to spots: the upscale sixteen-acre, $1,000-a-night Rosewood Hotel, nestled between dozens of venture-capital firms along the picturesque, tree-lined Sand Hill Road, which slices through Silicon Valley. The private dining room they gathered in opened to a balcony that overlooked a beautiful pool rimmed with Italian cypress trees and garden roses. As Musk walked in over an hour late, the rest of the men were eagerly waiting. Among them: Altman, Greg Brockman, Dario Amodei, and Ilya Sutskever.

The group would soon become the key leaders of the nonprofit. To capture the spirit of their shared mission, Musk would name it OpenAI. Over time, nearly all of the men would depart the organization after clashing with Altman and his vision of artificial intelligence.

Once Altman and Musk were no longer on speaking terms, and Altman had replaced Musk as the new Silicon Valley "it guy," Altman would change the public record on his beliefs about the dangers of what he was building. "I am now very much in the AI-will-be-a-tool camp," he told *Business Insider* in 2023, "though I do think future humans and human society will be extremely different and we have a chance to be thoughtful about how to design that future."

Musk would come to feel like Altman had used him to catapult to prominence.

It was an echo of an observation that has followed Altman throughout his life. "You could parachute him into an island full of cannibals and come back in 5 years and he'd be the king," his mentor, Paul Graham, once famously said. Graham reinforced the point again years later: "Sam is extremely good at becoming powerful."

Samuel Harris Gibstine Altman was born April 22, 1985, the first son of Jewish parents, in Chicago, Illinois.

His mother, Connie Gibstine, is a doctor. Her father, Marvin Gibstine, had also been a doctor, a pediatrician who, as a US Army physician, was dispatched with his new wife to Germany after World War II. Connie received both medical and law degrees, defying the gender norms of her generation. She specialized in dermatology, a profession with a stable paycheck and flexible hours, allowing her to come home to cook dinner and be there for her children.

It was before law school at Loyola University Chicago that Connie met Jerold Altman, a handsome man three years her senior. Jerry, the son of a shoe manufacturer and businessman, had been married once before in his late twenties after attending the University of Pennsylvania's Wharton School and becoming a consultant in Boston. His former wife had retained her maiden name. When Connie married Jerry, she did as well. A few years later, they moved from Chicago back to their hometown of St. Louis.

Jerry went into real estate and property management, for a time serving as chief counsel and vice president of the Roberts Companies, a St. Louis developer. Jerry was a people person. He had a passion for affordable housing and worked on several commercial and residential projects that sought to foster community and revitalize St. Louis. Sam would later repeat one of the biggest lessons his father taught him: "You always help people—even if you don't think you have time, you figure it out."

Connie and Jerry had three boys in rapid succession: After Sam, there was Max, then Jack. Five years later—nine years after Sam—Connie gave birth to Annie, delighted to finally have a daughter. Connie referred to herself as an atheist but culturally Jewish; Jerry was more religious. He attended services during Jewish high holidays like Passover and insisted on all four children having bat and bar mitzvahs, Jewish coming-of-age ceremonies. Connie's rationality and discipline and Jerry's spirituality and focus on service would each manifest in their children in various ways.

From a young age, Sam was driven and intensely curious. At two, he learned how to operate the family VCR; by three, he was fixing it. When his parents gifted him a Mac computer five years later, he quickly learned how to program and disassemble it. He settled well into the role of oldest brother, at times bossing around his younger siblings, at times playing their caretaker. He was extremely competitive, always insistent on winning board games.

As much as Sam was a sore loser, he also had a zest for victory. When his grandmother gifted each of her grandchildren some stock, he picked Apple; Jack picked Applebee's. It became a running joke in the family. Over twenty years, Jack's stock barely grew; Sam's shot up. "Your Apple has gone up—I don't even want to think about it," Jack later said, recounting the story, "hundreds and hundreds of times."

"Yes, it's been a lot," Sam said smugly.

As Sam got older, Connie gave him a choice that she would give to all of her children: whether or not to transfer to a local private school, John Burroughs, known for its rigorous academics and impressive roster of famed alumni. Sam made the switch, Max switched but didn't stay, Jack declined, and Annie followed her oldest brother. At Burroughs, Sam thrived. He excelled academically and socially with his extroverted personality and goofy humor. He was drawn not just to STEM but to writing and a variety of extracurriculars. He was head of the yearbook, captain of the water polo team, and did Model UN, a program that brings students together in events around the world to simulate the United Nations and debate public policy. "I remember thinking—and this is an embarrassing confession—'I hope he doesn't go into technology. He's so creative and such a good writer,'" Andy Abbott, his English teacher who would become the head of Burroughs, would recall. "I hoped he would be an author or something like that."

Even then, Altman was charismatic and a natural leader. He loved to push the boundaries of what was politically acceptable at his more conservative school, once getting in trouble for leading his water polo team in a striptease down to their Speedos at an annual pep rally. It was during those years that he came out to his parents and classmates as gay. While it surprised his mother, she accepted it, as did the rest of the family. A group of Christian students at his school did not. On National Coming Out Day, they boycotted an assembly that he led about sexuality. Altman, seventeen, decided to confront them in a speech to the student body that his college counselor would credit for opening up the school's culture. "Either you have tolerance to open community or you don't, and you don't get to pick and choose," he later said, recalling his last line.

Behind the confident facade, Sam was also sensitive. He worried about what people thought of him. He often grappled with anxiety, a trait that would carry over into his adult life. As his star rose in Silicon Valley, he'd sometimes call his mom with a headache, having convinced himself that he actually had meningitis or lymphoma. He would grow so panicked once while negotiating a deal that he'd have to lie down on the ground, bare chested, arms splayed, to calm himself.

It was these two parts of him—his ambition and his sensitivity—that would come to mark the shape of his career. After spending many hours with Altman to profile him in 2016, *The New Yorker*'s Tad Friend would note this duality: On any given issue, Altman seemed as driven by a relentless desire to push ahead as he was attuned to the countervailing need for caution. Reach AGI as fast as possible; also: Don't destroy humanity.

Upon graduating from Burroughs in 2003, Altman left the Midwest for Stanford University, drawn in by its proximity to the tech industry. He didn't settle on getting into tech immediately, however. As his teacher Andy Abbott had hoped, he did in fact consider being a writer. He also ever so briefly entertained the idea of being an investment banker. In the end, he leaned into his fascination with programming and computers. "I realized that the world does not need or value the seven-millionth novel," he later said. "That was not where I could make the best contribution, and, in cases like that, it also is generally harder to make a lot of money or even enough money."

Altman majored in computer science and took a particular interest in AI and security. He dug deep into assignments, once disemboweling a piece of software he was supposed to use for his homework to its low-level code, a classmate remembered, and finding a bug in the assignment itself. As a sophomore, he became interested in mobile technology. After learning that phones would soon all be equipped with GPS, he went to a campus entrepreneur event and stepped onstage holding a flip phone. He made an open call for people to join him in building something that took advantage of the location-tracking feature.

Around that time, he met Paul Graham, an entrepreneur and influential tech blogger who was beginning a new startup incubator called Y Combinator with his girlfriend Jessica Livingston. Altman joined YC's first batch of companies in 2005 as the founder of his new startup, Loopt, and spent the summer in Cambridge, Massachusetts, where the incubator initially started. Loopt was a social network that used location tracking to notify users when they were close to friends or to recommend nearby restaurants. He worked so hard that summer and ate so much instant ramen, he gave himself scurvy.

He didn't regret it. "Work really hard in the beginning of your career," he would later say to young founders. "It pays off like compound interest." Altman never returned to Stanford. By late 2005, he and his co-founders were already in talks with VC firms New Enterprise Associates and Sequoia to give them $5 million in funding. Altman took his chances and dropped out of college.

Loopt wouldn't become a great success. After a seven-year run, Altman would sell it in 2012 for $43.4 million, around what his investors put in. But if you had listened to his interviews and his backers at the time, his startup would have sounded like it was on the precipice of ushering in a great transformation.

It's easier to understand the seeds of Altman's success in those early interviews, when he's selling you something far less alluring than artificial intelligence: namely, an earlier competitor to Foursquare, and one that didn't work out.

Both his media savvy and dealmaking, two pillars of his rise, rest on his remarkable ability to tell a good story. In this Altman is a natural. Even knowing as you watch him that his company would ultimately fail, you can't help but be compelled by what he's saying. He speaks with a casual ease about the singular positioning of his company. His startup is part of the grand, unstoppable trajectory of technology. Consumers and advertisers are clamoring for the service. Don't bet against him—his success is inevitable.

"The response has been tremendous," he said to tech blogger Robert Scoble in June 2010 about his company's new app, Loopt Star, for advertisers to push deals, such as coupons for restaurants or group discounts for retailers, to users based on their location. "We've crossed over this point where now the value perceived of sharing my location outweighs the privacy concerns of doing so," he added. "In another few years, it'll be the norm to share your location and it'll be weird when you don't."

"It's a ridiculous distinction," Altman said a few months later to CNN Business, about the difference between life online and in person; the two were fusing together with location tracking on mobile devices. "The whole world is going mobile and the whole world is going universal access to your data and your services no matter where you are," he said.

For Altman, even discussing the pitfalls was an opportunity to underscore the pitch. When *The Information* founder Jessica Lessin, then a *Wall Street Journal* reporter, told Altman in 2008 she would write a story about the privacy concerns of location tracking, he offered to help. He sent her a long list of risks that Loopt had already identified and its proposals for how to solve them. The implicit message: This is how the world will work, so you might as well prepare for it. "He didn't just want to build a startup," Lessin wrote about the experience. "He wanted to write the rules."

With Loopt, Altman built the networks and sharpened the skills that would become his greatest assets. As a startup founder through the mid aughts and early teens in the Bay Area, he placed himself in the thick of an era of rapid growth and buzzy new ventures. He regularly rubbed shoulders with other restless entrepreneurs, making crucial connections wherever he turned. Right as Loopt was getting started, its office was down the hall from the fledgling startup YouTube. Among Altman's YC batchmates—the term for fellow founders in a YC cohort—were Steve Huffman and Chris Slowe, the respective cofounder and founding engineer of Reddit. Altman would become a Reddit board member in 2014, eventually amassing a larger share of the company than Huffman. Another YC batchmate was Emmett Shear, the cofounder of Twitch, who would step in as OpenAI's interim CEO during Altman's ouster almost two decades later.

Altman also learned the best way to package things to the media and the surest way to strike extraordinary deals. Even as the CEO of a little-known startup, he successfully negotiated enterprise partnerships with the major US mobile phone carriers. Key to his formula, people say, is the combination of his remarkable listening skills, his willingness to help, and his ability to frame whatever he has to offer in terms of exactly what you want. (These days, as an ultrawealthy Silicon Valley linchpin, it doesn't hurt that he can offer a lot.) He is the "Michael Jordan of listening," people have said. He is the "Usain Bolt of fundraising," says Geoff Ralston, who took over running YC after Altman.

"Fundamentally when you raise money from someone, what you're doing is telling a story about the future of whatever your project is, which involves that project, that company becoming an extraordinary success," Ralston says. "Sam can tell a tale that you want to be part of, that is compelling, and that seems real, that seems even likely."

Ralston likens it to Steve Jobs's reality distortion field. "Steve could tell a story that overwhelmed any other part of your reality," he says, "whether there was a distortion of reality or it became a reality. Because remember, the thing about Steve is he actually built stuff that did change your reality. It wasn't just distortion. It was real.

"And obviously, Sam has too."

But there's a flip side to the story. "Sam remembers all these details about you. He's so attentive. But then part of it is he uses that to figure out how to influence you in different ways," says one person who worked several years with him. "He's so good at adjusting to what you say, and you really feel like you're making progress with him. And then you realize over time that you're actually just running in place."

Twice during his time running Loopt, senior leaders at the startup approached its board and urged it to fire Altman, according to *The Wall Street Journal*, leveling two accusations that would follow him all the way through to his brief ouster at OpenAI. One was his tendency to operate for his own gain rather than the company's, and at times even at the expense of the company. The other was his seeming compulsion to distort the truth. The latter was harder to pin down: He sometimes lied about details so insignificant that it was hard to say why the dishonesty mattered at all. But over time, those tiny "paper cuts," as one person called them, led to an atmosphere of pervasive distrust and chaos at the company.

In a manner that would come to define the rest of his career, Altman emerged from the crisis with the upper hand. Loopt's board sided with Altman.

Despite its middling record, Altman would also emerge from Loopt much better off than he'd started. He used the startup to springboard himself higher and higher into the most powerful networks in Silicon Valley and subsequently used those connections to orchestrate an exit for his company that made himself rich. At twenty-six, he netted $5 million from Loopt's sale. Altman considered this a disappointment—Jobs had been worth $256 million by age twenty-five—but he would soon accumulate far more money. That wealth would slowly change his lifestyle. Eventually, he'd stop going to the grocery store. He'd travel by private jet. He'd collect luxury sports cars, including McLarens and an ultrarare $5 million Koenigsegg, and cultivate a love for racing them. For a time he attended the annual weeklong psychedelic and sex-fueled desert art festival Burning Man. He became, like many Silicon Valley bigwigs, a casual user of ketamine, a party drug that can be legally prescribed to relieve depression.

With his success, Altman brought his brothers along with him. In 2012, he started a personal investment fund called Hydrazine Capital with his brother Jack, who had studied economics at Princeton and was trying his hand at investment banking. Jack subsequently switched to tech and founded a startup, Lattice, that would get funded by YC after Sam became the incubator's president. Max, who had studied computer science at Duke and worked briefly at Microsoft before becoming a trader, switched to working at another YC company, Zenefits, in 2014. Two years later, he would join Sam and Jack at Hydrazine Capital. During that time, both younger brothers moved in with Sam for what was meant to be a temporary arrangement. The three ended up living together—a tight knot of brotherly love and business relationships—for many years to come.

Of particular importance to the shape of Altman's career was his relationship with his two biggest mentors, Paul Graham and Peter Thiel.

Known as PG, Graham had made his name first as the cofounder of a startup, Viaweb, which Yahoo acquired in 1998 for $49 million, and then as a blogger who published popular essays on startups, entrepreneurship, and venture capital. After starting YC, he impressed his views onto each generation of YC founders. Every YC company that succeeded gained the incubator, and Graham, increasing prestige. By the time Altman sold Loopt, the incubator had already seeded several startups that had grown or would soon grow into billion-dollar companies, including Dropbox and Airbnb. YC became the most elite club in the Valley. If you were in, you gained instant cachet and access to more resources, including a built-in customer base among old YC companies, investors more eager to fund you, and higher valuations. If you were out, no such luck.

Graham became an essential tastemaker for startups and startup culture in Silicon Valley. "Many folks in the space, in the ecosystem, came to live by and to take his fundamental precepts for what it meant to be a good founder and a successful entrepreneur," says Ralston. "Many of us looked to PG for guidance on a lot of things." Graham also became a lightning rod for criticism. He championed the idea of the tech industry as a meritocracy, while designing YC to be an insular fraternity. He defended YC for not having many female founders by saying that most women had not been prepared from an early age to succeed as tech entrepreneurs. After analyzing the performance of applicants in YC interviews, he identified thirty or forty factors, including "a strong foreign accent," that were predictors of failure when candidates exhibited several of them together. This was not a flaw of YC's evaluation system but rather an important data-driven signal, he said.

Graham's support for Altman was strong and early. In a 2006 blog post, Graham recounted meeting Altman as a college sophomore. "Loopt is probably the most promising of all the startups we've funded so far," Graham wrote. "But Sam Altman is a very unusual guy. Within about three minutes of meeting him, I remember thinking 'Ah, so this is what Bill Gates must have been like when he was 19.'"

Altman quickly inspired Graham to search for more Altmans. He asked the young founder what YC should ask on its application to discover more people like him. Altman suggested adding a question that Graham would soon describe as one of the most important: "Please tell us about the time you most successfully hacked some (non-computer) system to your advantage." It would come to encapsulate and encourage a certain ethos among generations of startups to bend, bypass, and break the rules to domination.

By the time Altman was twenty-three, Graham was comparing him to Jobs. "Sam is, along with Steve Jobs, the founder I refer to most when I'm advising startups," he wrote. "On questions of design, I ask, 'What would Steve do?' But on questions of strategy or ambition I ask 'What would Sama do?'"—referring to Altman by his nickname, which is also his X handle.

It was Graham's singular belief in Altman that would catapult him to the YC presidency in 2014 at age twenty-eight, two years after selling Loopt. When Graham asked in his kitchen if Altman wanted to be his successor, Altman smiled uncontrollably. "YC somewhat gets to direct the course of technology," Altman would later say. "I think his goal is to make the whole future," Graham said of Altman. The succession story would get repeated so often that it would turn into Silicon Valley lore. "If Sam smiles, it's super deliberate," a former YC founder says. "Sam has smiled uncontrollably only once, when PG told him to take over YC." Graham's choice surprised many others, but he held strong convictions.

"There wasn't a list of who should run YC and Sam at the top," Livingston would recall. "It was just: Sam."

Peter Thiel became Altman's second mentor. Another linchpin in the tech industry, Thiel became a billionaire by founding payments company PayPal and data-mining firm Palantir, and being an early investor in Facebook. Like Graham, Thiel would attract his own fair share of controversies, including being a rare vocal Trump backer among his tech peers during the 2016 election and secretly funding a lawsuit that would lead to the demise of Gawker Media, in retaliation for the site outing him as gay nearly a decade earlier.

After the sale of Loopt, Altman suffered a breakup with one of his cofounders as well as boyfriend of nine years. Heartbroken and professionally adrift, Altman took a year off, started Hydrazine, and raised $21 million. Thiel, almost twenty years Altman's senior, pitched in a majority of the funding. When Altman became a YC partner, he used Hydrazine to bet on the accelerator's portfolio companies while also helping Thiel's venture firm, Founders Fund, to identify high-return investments. Thiel's net worth multiplied several times over. The two men grew extremely close. (Their bond was once described as having only one parallel: Thiel's mentor relationship with Facebook cofounder Mark Zuckerberg.)

Graham and Thiel heavily influenced Altman's worldview, his approach to building effective businesses, and his savvy as a political operator. The two mentors impressed on Altman the imperative for scale and the efficiencies of capitalism over government.

"The first piece of startup wisdom I heard was 'increasing your sales will fix all problems,'" Altman wrote in a 2013 blog post titled "Growth and Government" that thanked Graham and Thiel for shaping his ideas. "This turns out to be another way of phrasing Paul Graham's point that growth is critical." For startups, more sales meant more capital meant better talent and fewer internal tensions. For countries, more growth meant more technological innovation meant a higher quality of life. The dysfunction in the US government was threatening this growth cycle, Altman added. "Either you're growing, or you're slowly dying," and the

US government was dying. "Without economic growth, democracy doesn't work because voters occupy a zero-sum system," he said.

This idea would evolve into a core thesis driving Altman's career and investments. "The thing that people in the private sector can do the most to help get the country back on track is to get economic growth back," he'd say in 2017. "In the US we had two hundred years of unrivaled economic growth. We had one hundred years of territorial expansion; we had one hundred years of new technology really working," he added, glossing over a bloody colonial history and the complicated labor and environmental record of unfettered industrialization. "And people were mostly pretty happy. And now we don't."

"Sustainable economic growth is almost always a moral good," he'd add in 2019. "Part of what motivates me to work on Y Combinator and OpenAI is getting back to that, getting back to sustainable economic growth, getting back to a world where most people's lives get better every year and that we feel the shared spirit of success."

On building companies, Altman frequently channeled Thiel's "monopoly" strategy, the belief that all founders should "aim for monopoly" to create a successful business. In 2014, Altman returned to his alma mater, Stanford, to teach a class called How to Start a Startup. He invited Thiel to expand upon his signature philosophy in a lecture called "Competition Is for Losers."

Monopolies are good, Thiel said, because "they are much more stable, longer-term businesses, you have more capital, and . . . it's symptomatic of having created something really valuable." Building one relied on having some kind of proprietary technology, network effects, economies of scale, and good branding. Each of these elements needed to endure over time. With proprietary technology, it was critical to stay in the leading position. "You don't want to be superseded by somebody else," Thiel said. "There are all these areas of innovation where there was tremendous innovation but no one made any money."

He gave the example of disc drive manufacturing in the 1980s, which saw repeated advancements every two years, but by different companies.

"It had great benefit to consumers, but it didn't actually help the people who started these companies," he said. Companies needed not only to have "a huge breakthrough" at the beginning to establish their dominance but also to ensure they had the "last breakthrough" to maintain it, such as by "improving on it at a quick enough pace that no one can ever catch up."

"If you have a structure of the future where there's a lot of innovation and other people will come up with new things in the thing you're working on," he concluded, "that's great for society. It's actually not that good for your business."

From both men, Altman also learned the importance of building relationships and creating "network effects" as an individual.

"I've heard a lot of different theories about how things get done," he wrote on his blog in 2013. "Here's the best one: a combination of focus and personal connections. Charlie Rose said this to Paul Graham, who told it to me." Altman would later add a third ingredient: self-belief. "For startups I think it's really important to add this," he said. "You actually have to believe you might do it."

Altman began to live by this mantra religiously. He cultivated relationships with intensity and discipline, first by giving his time and tactical advice and then, as he came to control increasing amounts of capital, his money. Thiel was a role model in this regard: His mentor had long used advice and money to build his network, and used his network to amass more connections and money. To young entrepreneurs and other people he wanted to bring into his orbit, Thiel provided mentorship and small amounts of capital, as well as access to that orbit. In much the same way, Altman learned to use his financial and social resources strategically. As his stature and wealth grew, he scaled the approach with relentless efficiency.

He became a frequent host of dinners and gatherings at his house for different, interlocking groups of people—people connected to YC or his companies, people who share his interests in investing, the active and growing gay entrepreneur community. He imparted advice through concise texts and calls—as short as two minutes—to pack more into his schedule. He connected people to one another over email with a single word ("meet") or a single punctuation mark ("?")—a famous habit of Amazon's Jeff Bezos—to get a conversation started. With his money, Altman made very few large bets, going mostly instead for small ones at high volume. Over time he accumulated financial ties with more than four hundred companies through YC, Hydrazine, and his other funds, according to a June 2024 *Wall Street Journal* assessment.

It's hard to find people within Altman's inner circle who don't have some kind of financial relationship with him. His second-ever and most successful startup investment was in the YC-backed payments technology company Stripe, for which Greg Brockman was its first chief technology officer. Altman invested early in YC-backed Airbnb, the cofounder and CEO of which, Brian Chesky, is one of his closest confidants. He pitched into his ex-boyfriend and friend Matt Krisiloff's biotechnology firm Conception. He coinvests in deals with another ex-boyfriend, Lachy Groom, a prominent solo venture capitalist. To those people, it's a testament to Altman's generosity. He regularly offers his resources, whether opening up his houses for people to stay in or supporting them financially. He has gone out of his way to support even complete strangers, once sending funds to a man in Ethiopia who emailed him seeking his help to buy a laptop, one person recalls. During the 2023 Silicon Valley Bank crisis, when a run on a critical financial institution for Valley startups led to the largest bank failure since 2008, he sent money without any paperwork to companies to save them from shutting down or laying off people, remembers Krisiloff. "It's an extremely rare trait," Groom says, "and that trait has really rubbed off on me—the generosity. I feel very grateful for that."

Altman developed the same approach with politicians, taking another page out of Thiel's book. But where Thiel asserted his wealth to back Republican candidates, pumping tens of millions into their campaigns, Altman grew increasingly involved in politics in the opposite direction, hosting fundraisers and writing checks for Democrats. For a time, the political differences between Thiel and Altman strained their relationship. In 2017, Altman leaned into their disagreements and went on a tour of America, much like Thiel's other mentee Zuckerberg, and spoke to one hundred Trump supporters. Altman also entertained the idea of going into politics himself with a run for California governor, reasoning that it would place him in charge of the world's fifth largest economy, a strong stepping stone for fixing what he saw as dysfunction in the political system. He published a manifesto called "The United Slate," with three principles: (1) prosperity from technology; (2) economic fairness; and (3) personal liberty. He organized focus groups to test out his candidacy. People close to him joked that he should shoot for US president.

In the end, Altman never became a politician—the focus groups thought he came off as too young—but he began to act like one. In his first few years of running YC, he still had boyish cheeks, owned one suit jacket, and sat with a leg popped up or perched like a bird atop his chair. He sometimes spoke flippantly and in casual hyperboles, punctuating his sentences with profanity. He was breezier with his references to provocative personal details, like his collection of guns. He was faster to anger and to show his impatience for ineffective people, at one point coding up a software program to size up YC founders based on their email response times.

A few years in, he had refined his appearance and ironed out the edges. He'd traded in T-shirts and cargo shorts for fitted Henleys and jeans. He'd built eighteen pounds of muscle in a single year to flesh out his small frame. He learned to talk less, ask more questions, and project a thoughtful modesty with furrowed brow. In private settings and with close friends, he still showed flashes of anger and frustration. In public ones and with acquaintances, he embodied the nice guy. He readily gave people credit for things and texted in all lowercase with lots of smiley and frowny faces. He gave employees his personal number, encouraging them to reach out at any time and responding to their feedback with impressive attentiveness. He avoided expressing negative emotions, avoided confrontation, avoided saying no to people. Once when OpenAI fired an employee, he reached out personally to offer ketamine and booze as consolation. "I think all of Sam's relationships end in a good way whether you want it to or not," the employee says.

Altman became his own institution. YC was his platform and accelerant. He converted its power into his own power, its network into his own network. Those personal connections and his public reputation became his greatest currency. He met regularly with policymakers, who viewed him as a gateway to Silicon Valley. In 2016, it was Ashton Carter, Obama's secretary of defense, who sought Altman's advice on how his agency could tap into the well of young tech talent. Three years later, on the day Altman stepped down from YC in March 2019, it was Chuck Schumer. At the time the US Senate minority leader, Schumer paid a clandestine visit to OpenAI with his Secret Service detail. "You're doing important work," Schumer told employees in the office as he sat side by side with Altman in armchairs in front of a TV projecting a roaring fire. "We don't fully understand it, but it's important," Schumer added. "And I know Sam. You're in good hands."

Altman's ascendancy would also come at a mounting cost as he accumulated more and more detractors and outright enemies who would echo the accusations of his senior lieutenants at Loopt: that of his self-serving pursuit of power and his compulsive dishonesty. While many people who benefited from Altman's advice, wealth, and networks became stalwart loyalists, others began to view him as devilishly capable of bending situations to his advantage. For some, including his partners at YC and other power brokers, this could be an annoyance. For employees and people with far less leverage, it could be a source of fear. To still others, who disagreed vehemently with his worldview, he was a massive threat.

Altman's climb would also, to his agony and then ire, unravel his relationship with his sister. As kids and well into his twenties, Sam and Annie were close. She, the youngest; he, the oldest, her protector. She was science minded and the artsy one, the most emotionally expressive. At times he liked to get her opinions about his romantic partners, to confide in her about his inner worries and emotions. But as he grew more ingrained in Silicon Valley, Annie watched him build thicker and thicker walls around the part of him that was the most sensitive. He would tell her about new psychological tactics he'd learned, she remembers, like using fewer words in an email, to appear more powerful as a business leader.

At first it made her sad, and then scared, about whether that sensitive part was even still there. "I definitely still got glimpses of it for a while, which was why I stayed close," she says. "And then I started being the one to be harmed by him."

When Sam first came into wealth, she says, his then boyfriend created a rule: for every big-ticket item that Sam purchased, he needed to donate the same amount to a good cause. For a time, it created a check on the rapid creep of Sam's lifestyle. But as he earned money faster than he could spend it, she felt his relationship with that money grow more complicated. In her view, he began to hoard it as he grew more and more out of touch with people in need. Through the end of 2019 and the first half of 2020, several times he and the rest of the family declined or were reluctant to provide Annie access to what she saw as emergency financial support to help front her rent and medical expenses, according to extensive correspondence she shared with me. At the time, she faced acute physical and mental health challenges, her medical and therapy records show, exacerbated by the sudden death of their father. It left her struggling with unstable housing; out of desperation to make ends meet, she turned to sex work for money. In the summer of 2020, as OpenAI began to gain its first major wave of public attention under Sam's leadership, Annie would cut off contact with her family.

There is a case to be made that Sam, as well as his brothers, were following the lead of their and Annie's mother in an attempt to push Annie toward financial independence. It's a complicated and painful family story, difficult to judge based on partial information. In a public statement in January 2025, Sam, his mom, and his two brothers expressed their love and concern for Annie and denied all of her allegations as "utterly untrue." In response to my requests for interviews and detailed asks for comment, Connie Gibstine provided a shorter version of a similar statement and declined further elaboration; Sam, via OpenAI's communications team, and his brothers did not respond.

Nevertheless, Annie's experience contains striking parallels to the many themes explored within these pages: the ever-widening gulf between those who benefit and those left behind in the supposed march for progress; the loss of agency and voice among the disenfranchised confronted by that accelerating chasm; the limits of ceding so much power not just to companies but to the individuals who run them without the scaffolding to provide commensurate checks and balances. Annie's actions would also make her story an inescapable part of understanding OpenAI's trajectory and its impact on AI development: In 2021, she would make the decision to go public with serious allegations about Sam, claiming that he sexually abused her as a child—which her family has called "the worst" of her "untrue" accusations—and also that he and the rest of the family abandoned her when she was at her most vulnerable. She would subsequently file a lawsuit against Sam for such alleged abuse on January 6, 2025, two days before her thirty-first birthday, to meet the statute of limitations for such cases in Missouri. Annie's persistent efforts to voice her allegations and tell her side of the story would affect Sam and influence OpenAI's other executives as they contended with his and the company's surge to global impact and prominence.

Each of these puzzle pieces—Sam's ascendence, his character and relationships, the divisiveness he left in his wake, the flows of money and power—speaks to the path that led to his sudden and fleeting ouster. For a brief moment, the rest of the world caught a glimpse into the struggles happening at the highest levels to dictate the future of artificial intelligence. It would reveal just how much the quest for dominance of that technology—already restructuring society and terraforming our earth—ultimately rests on the polarized values, clashing egos, and messy humanity of a small handful of fallible people.

Chapter 2

A Civilizing Mission

Greg Brockman became the first to commit to building OpenAI. To be Brockman's cofounder, Altman handpicked Ilya Sutskever, then an AI researcher at Google whom Altman cold-emailed to come to the Rosewood dinner in the summer of 2015; Sutskever enthusiastically accepted upon learning that Musk would be in attendance.

Brockman and Sutskever made an interesting duo. Tall and stocky, with an amiable demeanor, Brockman was an engineer and a startup guy like Altman. He had grown up on a hobby farm in North Dakota. In between milking cows, he fell in love with math and then science. In 2008, he enrolled in Harvard and transferred to MIT two years later. After another semester, he dropped out of college entirely, unable to swallow any more school when he could be out in the real world building products. He moved to the Bay Area and joined Stripe as a budding startup with only three other people; he impressed the founders so much with his coding genius that he became chief technology officer. Over five years, he prototyped many of Stripe's early products, helping it grow into a powerhouse fintech company that provides digital payments infrastructure to the likes of Amazon and Shopify. The run left him with significant wealth and the rarefied Valley status of having helped build a multibillion-dollar company.

Sutskever was the scientist. Lean and wiry, he was born in the Soviet Union and raised in Israel, where he blossomed as a math prodigy. After struggling to find teachers who could keep up with his advancement, his parents enrolled him as an eighth grader in courses at the Open University of Israel. At sixteen, he moved to Toronto and attended high school for just a month before being admitted to the University of Toronto in 2003 as a third-year undergraduate student. It was there that Sutskever met Geoffrey Hinton, a British Canadian professor who had done seminal work in AI research. Hinton became the only person whom Sutskever would call a mentor and who'd subsequently have a profound influence on his work and life. In 2012, together with another one of Hinton's grad students, Alex Krizhevsky, they shocked the AI world by sweeping the floor at an academic contest called ImageNet to build software for automatically identifying objects in photos. Where every other team struggled to get their software's error rate below 25 percent, Hinton, Sutskever, and Krizhevsky drove theirs down to 15 percent. Early the following year, Google announced that it had acquired their newly formed company, DNNresearch, in a heated auction for $44 million. The move minted the academics into multimillionaires and unleashed the first major rush to commercialize artificial intelligence. "We thought we were in a movie," Hinton says.

Brockman and Sutskever met for the first time during the Rosewood dinner. Much like themselves, the others in the room were either entrepreneurs or scientists. The discussion ping-ponged back and forth between academic deliberations about different approaches to AI research and, Musk's particular fixation, whether there was still time to beat out DeepMind and Google, essential, they believed, to correcting the course of AI development. The critical bottleneck, everyone agreed, was talent: Most of the top AI researchers were employed, like Sutskever, if not by Google, then by other tech giants, enjoying extravagant salaries, benefits, and job security.

AGI was also central to the discussion, which at the time was highly unusual. Most serious scientists considered the idea of digitally replicating true human-level intelligence to be science fiction, or at the very least decades or more away from attainability. Bold declarations that it was within reach enough to invest in it presently was viewed largely as pseudoscience and quackery. But Hassabis had embraced that term to describe the ambitions of DeepMind, despite a belief among his own research staff that this was distasteful, shameless marketing. The Rosewood group equally felt that the same goal, AGI, would best describe their own aspirations if they intended to form a competitor to go toe to toe with Hassabis's organization.

Even to Sutskever, who secretly believed AGI was possible and would come to full-throatedly endorse some of the most aggressive predictions about the speed of its creation, the brazen talk at first made him a little squeamish. If other researchers found out that he was openly discussing the pursuit of this objective, he worried, he risked losing his credibility within the scientific community. Those concerns did not hold back Brockman, an outsider to the field and sincere in his belief that AGI, with enough effort and focus, could be just around the corner. That Sutskever and the other researchers were willing to at least privately entertain the feasibility of a new lab could have only strengthened Brockman's confidence. As Altman drove him back from the hotel to San Francisco that night, Brockman told him that he was ready to commit himself to the project.

In Silicon Valley, there is a common saying: Becoming cofounders is like entering a marriage. As with any committed partnership, Brockman and Sutskever courted each other after Altman had diplomatically told Brockman he needed to be paired with someone who understood AI research. A few weeks after the Rosewood dinner, the two grabbed another meal alone in Mountain View. It was a perfect match. "I knew it was going to work though we'd just met.... Ilya and I had an extremely high-bandwidth interaction," Brockman later wrote on his blog. "Our ideas enhanced and complemented one another."

Over the next few months, as Sutskever thought through the proposal, Brockman took on the task of convincing others to join the new moonshot venture. He called leading figures in the field to get their recommendations for a list of top AI talent. He dined with professors at uni-

A CIVILIZING MISSION

versities to ask about their best students. He heavily researched each candidate before any conversation to more persuasively recruit them. Altman would later extol those early efforts in a blog post simply titled "Greg." "A lot of people ask me what the ideal cofounder looks like," Altman wrote. "I now have an answer: Greg Brockman."

Altman and Musk also had their fair share of recruiting conversations, slowly loosening up researchers resistant to the idea of AGI. "AGI might be far away, but what if it's not?" Pieter Abbeel, a professor at the University of California, Berkeley, remembers Musk urging him. "What if it's even just a 1 percent or 0.1 percent chance that it's happening in the next five to ten years? Shouldn't we think about it very carefully?" Abbeel would join OpenAI as a research adviser and later full time with several of his PhD students.

At first, many of the people Brockman approached were willing to sign on only if others were as well. Undeterred, he invited his ten most-wanted engineers and researchers to discuss their hesitations and rally their excitement over wine in Napa Valley. He hired a bus to drive everyone there and back so he could continue pitching them on the more than hourlong ride each way. Three weeks later, by Brockman's deadline, nearly all had accepted his offer.

As Musk, Altman, and Brockman discussed how best to position OpenAI at launch, all were keenly aware of the importance of its public perception. They agreed with Altman's proposal to make it a nonprofit and to play up the openness for which it was named. OpenAI, the anti-Google, would conduct its research for everyone, open source the science, and be the paragon of transparency.

"I hope for us to enter the field as a neutral group, looking to collaborate widely and shift the dialog towards being about humanity winning rather than any particular group or company," Brockman wrote to Musk and Altman in November 2015. "(I think that's the best way to bootstrap ourselves into being a leading research institution.)"

"There is a lot of value to having the public root for us to succeed," Musk replied later that month, with a suggestion to rewrite the announcement of OpenAI's formation to have broader appeal. "We need to go with a much bigger number than $100M to avoid sounding hopeless relative to what Google or Facebook are spending," he added. "I think we should say that we are starting with a $1B funding commitment. This is real. I will cover whatever anyone else doesn't provide."

In later correspondence, the group acknowledged that they could walk back their commitments to openness once the narrative had served its purpose and as the need arose, such as to avoid bad actors getting their hands on the technology. "As we get closer to building AI, it will make sense to start being less open," Sutskever raised to the trio in January 2016, shortly after OpenAI launched. "The Open in openAI means that everyone should benefit from the fruits of AI after its [sic] built, but it's totally OK to not share the science." "Yup," Musk responded.

In December 2015, the announcement went out on a Friday night, to coincide with Neural Information Processing Systems, the largest annual AI research conference, where Hinton and Sutskever had auctioned off DNNresearch three years earlier. The blog post, "Introducing OpenAI," listed each of the nine founding members, including Brockman, who would serve as CTO, and Sutskever, who would direct research. Musk and Altman would be cochairs. Altman and Brockman had also joined Musk in his pledge to see that the lab would have $1 billion in funding. So had Jessica Livingston, Peter Thiel, and LinkedIn cofounder Reid Hoffman. Hoffman had worked with Musk and Thiel at PayPal and often invested with them in startups as the "PayPal mafia." "We expect to only spend a tiny fraction of this in the next few years," the post said of the funding.

In the final countdown before the announcement, Sutskever had almost stayed at Google. To all of the other founding members, OpenAI had offered a base salary of $175,000 and YC or SpaceX stock. To Sutskever, the lab had instead offered him nearly $2 million, a whopping sum for a nonprofit. Even then, Google had offered him more, and then more again, reaching two or three times that amount, in a bid to keep him. Musk and Altman delayed the company announcement repeatedly as Sutskever agonized over the decision, calling his parents and fielding pleas from Musk and Brockman. In the end, Google's dizzying offer underscored to Sutskever why a nonprofit like OpenAI was needed.

That day Musk marked the occasion with an email to the founding team solemnly pledging his commitment to making OpenAI victorious. "Our most important consideration is recruitment of the best people," he wrote, which he promised to support, along with whatever else for which he could be helpful. "We are outmanned and outgunned by a ridiculous margin by organizations you know well," he added, "but we have right on our side and that counts for a lot. I like the odds." To preempt any other counteroffers from luring away members of their founding team, OpenAI immediately increased everyone's base salary by another $100,000.

Musk would later recount facing the fury of Larry Page for personally poaching Sutskever. The two didn't speak much again as their views continued to clash on AI development. But OpenAI's recruiting suddenly became easier. Moonshots and associations with billionaires were a powerful draw in the Valley. Within a few months, the number of employees doubled.

The lack of clarity, the big check, and the billionaire worship were Silicon Valley at the heady peak of its unchecked power. But OpenAI had been clever with its positioning: It straddled the border between the techno-chauvinist version of Silicon Valley and a more conscientious strand that was emerging. Over the following year, Donald Trump's spectacular rise and win in the 2016 US presidential election would shock the left-leaning workforce of the tech industry into self-reflection. As upheaval ripped through companies like Meta and Google and techlash sentiment gripped the public, AI researchers, too, began to question whether the field had moved too quickly to yoke its technologies to corporate bottom lines.

An accounting of the societal impacts of commercializing AI research returned an unsettling scorecard: Automated software being sold to the police, mortgage brokers, and credit lenders were entrenching racial, gender, and class discrimination. Algorithms running Facebook's News Feed and YouTube's recommendation systems had likely polarized the public, fueled misinformation and extremism, enabled election interference, and, most horrifying in the case of Facebook, precipitated ethnic cleansing in Myanmar.

But the main funding alternative, taking money from the government, had its own ethical land mines. In 2018, thousands of Google employees would protest a secret company contract with the Pentagon for its program known as Project Maven to develop AI-powered surveillance drones. The capabilities, employees said, could lay the groundwork for autonomous weapons; the Pentagon, which said this had not been its intention, would move away from that position with the Ukraine war.

It became a cynical refrain among AI researchers: sell out to Big Tech or to the military industrial complex, or leave AI research. Between these binary extremes, OpenAI seemed like a third way, corrupted by neither profit nor state power. "It was a beacon of hope," said Chip Huyen, a machine learning engineer and popular tech blogger observing from the sidelines.

Not everyone was impressed. On the night of OpenAI's launch in December 2015, Timnit Gebru, the AI researcher who'd questioned Musk about prioritizing the threats of AI over climate change, couldn't believe the announcement.

All week the Stanford University graduate student, an Ethiopia-born Eritrean refugee who moved to the US as a teen, had been reminded of the high cost of being a Black woman in an environment dominated by white men. It was her first time joining the throngs of AI researchers at the weeklong Neural Information Processing Systems, then called NIPS and later rebranded to NeurIPS for short. She was one of the only Black people there. The following year at the 2016 conference, she would put an actual tally to it, counting only six other Black researchers among the 8,500 attendees. At Stanford, she joked about the lack of Black researchers on campus by saying she could found a Black in AI group and have meetings alone. She imagined starting a YouTube channel to lampoon the situation, changing her hairstyle and acting out different personalities to dramatize her speaking with herself. But as much as she tried to make light of her isolation, this conference in 2015 was reminding her how quickly things could turn hostile. At a party one night, she was getting some water when a group of drunk guys wearing Google Research T-shirts locked eyes on her and decided to make her the object of their fun. They surrounded her. One man forced her into a hug; another foisted a kiss on her cheek as he snapped a humiliating photo. At the same conference, a friend of hers was harassed by a professor.

Now here was a group of people—nine out of eleven of whom were white men—being showered in previously unheard-of amounts of money, speaking about the theoretical prospect of a bad superintelligence taking over the world, and proposing to counteract it by building a better superintelligence.

That night, Gebru drafted a scathing critique of what she'd observed in an anonymous open letter: the spectacle, the cultlike exaltation of AI celebrities, and, most of all, the overwhelming homogeneity of the people building and shaping such a consequential technology. This homogeneous culture was not only pushing away talented researchers but also leading to a dangerously narrow conception of AI and of who could benefit from the technology.

"We don't have to project into the future to see AI's potential adverse effects," Gebru wrote. "It is already happening."

On her flight back home from the conference, she thought twice about posting the letter anonymously. Instead she posted a shorter, more sanitized version of her critique, using her name on Facebook.

Several weeks later, she typed up an email with the subject line "Hello from Timnit."

"When I go to computer vision conferences, I am often the only black person there," she wrote. "But now I have seen 5 of you:) and thought that it would be cool if we started a black in AI group or at least know of each other."

One by one she added the researchers' emails. And then she pressed send.

In the early days of OpenAI, Altman and Musk were barely around as co-chairmen. Busy with their full plate of other endeavors, the two left

Brockman and Sutskever to build up the organization. As Sutskever rallied researchers to give him their best ideas, Brockman threw himself into the work of developing the right organizational culture.

Some years later, Brockman would recount to me his thinking. To prepare, he read every book he could find on ambitious science and technology undertakings in US history: the transcontinental railroad, Thomas Edison's light bulb, the early network of computers that would lay the groundwork for the modern Web. He absorbed them like religious texts, searching for hints and guidance on how to design his own endeavor.

One story he held dear was the likely apocryphal tale of John F. Kennedy approaching a janitor holding a broom at the NASA space center. "Kennedy asks him, 'Sir, what are you doing?' And he says, 'Oh, I'm helping put a man on the moon,'" Brockman recounted, clearly delighted. "Everyone having this sense of mission and purpose—I think that's something really amazing and something I don't see as reflected in what happens generally today."

He later added: "I really feel like we as Americans have stopped daring to dream."

To succeed, he believed, OpenAI needed that same level of alignment; every person at every level of the company needed to be like that janitor. He pointed out to me that, in fact, during the first few months of OpenAI, when everyone worked out of his apartment, he embodied that spirit literally and spent a lot of time cleaning people's glassware. He created a company policy requiring all employees to work out of the San Francisco office, a policy that OpenAI would hold onto until the pandemic. This, of course, came with some trade-offs; not everyone wanted to live in the Bay Area, he acknowledged. I would learn through my other interviews that this was particularly true for women and people of color who, like Gebru, felt alienated by the white and male culture of the dominant tech industry. But to Brockman cohesion was more important, and being physically together helped with the serendipitous exchange of ideas.

Brockman decided, too, that he would call all OpenAI employees "members of technical staff," inspired by Xerox PARC, the storied research and development lab in Palo Alto, which had done so, after a tradition at the equally famed Bell Labs in New Jersey, to create a more democratic work environment.

When considering the criticisms leveled at OpenAI for its pursuit of AGI, he drew parallels with Edison's light bulb. "A committee of distinguished experts said 'It's never going to work,' and one year later he shipped," Brockman said. "How could that *be*?" It was, as science writer Arthur C. Clarke in the book *Profiles of the Future* called it, "a failure of imagination."

Among the attendees at the Rosewood dinner had been Dario Amodei. Amodei, a computational neuroscientist turned AI researcher, was then working in the Silicon Valley–based AI lab of Chinese company Baidu before doing a brief stint at Google. His sister Daniela Amodei had worked with Brockman at Stripe, and when Brockman first started to engage seriously in AI developments, he had turned to Dario for learning resources. Dario didn't join OpenAI immediately but was intrigued by the premise. OpenAI, under Musk's influence, seemed to stand out from other AI labs as the most willing to focus on so-called AI safety.

In 2016, while still at Google, Amodei cowrote a foundational paper to the discipline, articulating a central problem in AI safety as addressing "the problem of accidents in machine learning systems, defined as unintended and harmful behavior that may emerge from poor design of real-world AI systems." This was distinct from other AI-related challenges, he and his coauthors wrote, including privacy, security, fairness, and economic impact. AI "safety" in this framework, in other words, was about preventing rogue, misaligned AI—the root from which, as described by Nick Bostrom, superintelligence could become an existential threat.

To Amodei, there was no matter more important to work on: the prevention of superhuman AI causing catastrophic outcomes, even human extinction. Both Amodei siblings were sympathetic to the effective altruism, or EA, movement, a controversial ideology that had been spawned among philosophers at Oxford University, where Bostrom was based, and taken hold in Silicon Valley. Over time the movement, which preaches dedicating oneself to doing maximal good in the world by using extreme rationality and counterintuitive logic to guide decisions, had, in no small part due to Bostrom's influence, identified the existential threat of rogue AI as a leading issue area for its adherents to pursue. Two years earlier, Daniela's husband, Holden Karnofsky, had founded a nonprofit called Open Philanthropy to donate money in part based on EA principles. Open Phil, as it was called, would fast become the primary funder of catastrophic and existentially related AI safety research. (By November 2024, it had awarded more than three hundred AI-safety-related grants worth $440 million.)

But this existential brand of AI safety, built on philosophical thought experiments, would soon come under fire as the AI research community awakened to the less apocalyptic and immediate real-world harms of AI. Around the same time Amodei published his paper, *ProPublica* published a groundbreaking investigation called "Machine Bias" that revealed algorithms were being used across the US criminal justice system in misguided attempts to predict future criminals, and those algorithms were classifying Black people as higher risk than white ones who had more extensive criminal records. The piece, and an overall souring on Big Tech post-2016 over the harms of social media, sparked a new wave of research reckoning with the harmful societal impacts of AI.

Deborah Raji, an AI accountability researcher at the University of California, Berkeley, would come to champion the reexamination of the overwhelming focus of AI safety research on theoretical rogue AI and its possible existential risks to the detriment and de-prioritization of other real, evidence-based problems, coauthoring a 2020 paper in response to Amodei's. She argued that truly "safe" AI systems could not be built by isolating the behaviors of the technical systems themselves without placing them in full context of their impacts on the very things—privacy, fairness, and economics—that Amodei had set apart. Where Amodei had raised the idea of AI creating "negative side effects" as it relentlessly pursued an objective, using an example akin to the paper clip thought experiment of a cleaning robot knocking over a vase or damaging the walls on its path to tidying up, Raji pointed out that this was already happening. In its relentless pursuit of commercial products and AGI, the AI industry had produced expansive negative side effects, including the wide-scale infringement of privacy to train facial recognition and the spiraling environmental costs of the data centers required to support the technology's development.

"It is not just the *actions* of an AI agent that can produce side effects," she and her coauthor wrote. "In real life, basic design choices involved in model creation and deployment processes also have consequences that reach far beyond the impact that a single model's decision can have. In reality, for AI systems to even be built, there is very often a hidden human cost."

Within OpenAI, various researchers, some of them among the small handful of women of color at the company, would press executives to expand their "AI safety" definition and include research on areas such as the discriminatory impacts of deep learning models. Executives were dismissive. "That's not our role," one said.

In May 2016, Amodei, still at Google, stopped by OpenAI's office to see how things were going. OpenAI had just moved out of Brockman's apartment to a space above a chocolate factory in San Francisco's Mission District, the city's oldest neighborhood and a Latino stronghold. Researchers padded around in socks.

"There are twenty to thirty people in the field, including Nick Bostrom and the Wikipedia article, who are saying that the goal of OpenAI is to build a friendly AI and then release its source code into the world," Amodei told Altman and Brockman, according to an account in *The New Yorker*.

"We don't plan to release all of our source code," Altman said. "But let's please not try to correct that. That usually only makes it worse."

"But what *is* the goal?" Amodei asked.

"Our goal right now . . . is to do the best thing there is to do," Brockman replied. "It's a little vague."

Amodei joined two months later to lead AI safety research. Thereafter, Open Phil would donate $30 million to OpenAI to secure a three-year board seat for Holden Karnofsky. In 2018, at Brockman's invitation, Daniela, who had been the first recruiter at Stripe, would also move over to OpenAI to build up its team as an engineering manager and its VP of people. "We have a long, cute history of knowing each other," Daniela would joke to me of her and Brockman a year later. "That's right," Brockman would say, chuckling. "When we started OpenAI, and I started doing the initial recruiting here, I was like, 'I really wish I had Daniela.'"

By the end of 2020, the Amodei siblings would become so disturbed by what they viewed as Altman's and OpenAI's break from its original premise that they would cleave off to form another AI lab, Anthropic, taking critical staff with them and creating a rivalry that would play a pivotal role in the frenzied release of ChatGPT. Karnofsky would step down from OpenAI's board, having served his term and due to the new conflict of interest. On the list of candidates he nominated for his replacement, he would include one of his former employees: Helen Toner.

The problem was that OpenAI had no idea what it was doing. A year in, it had poached, begged, and borrowed its way to a stellar team in the aggressive fight for talent within the industry, keeping up the excitement internally just from the sheer density of top people. Still, it struggled to find a coherent strategy. And the momentum and shine were beginning to wear off.

Its list of projects sprawled every which way in a kitchen-sink reflection of the field. It was using robots and video games and simulated virtual worlds for training agents—all as ways of trying to reach more advanced AI capabilities. Little was working, and what did work felt derivative of something someone else had already done. Whatever AGI was, it wasn't that. "The bigger projects that they had, it didn't seem like they were doing anything super innovative," says Nikhil Mishra, an AI researcher who interned at OpenAI in 2017.

Brockman's and Sutskever's leadership abilities were also being pushed to their limits. While Brockman spent most of his days coding, Sutskever stalked around the office repeatedly asking each researcher, "What's your next big thing?" It made for a rudderless, high-stress environment. There was no real management structure or clear set of priorities. Sometimes people would get fired on the weekends, and the rest of the team would only find out the following Monday when they didn't show up. And the lab was burning cash, most of it to hold down the salaries of the team it had assembled. In 2016, OpenAI spent more than $7 million out of its $11 million in expenses on compensation and benefits.

Musk was getting impatient. It didn't help that DeepMind was suddenly garnering worldwide adulation. In March 2016, its program AlphaGo beat Lee Sedol, one of the world's best human players in the ancient Chinese game of Go. ("Deepmind is causing me extreme mental stress," Musk wrote to OpenAI leadership shortly before the five-game match. "If they win, it will be really bad news with their one mind to rule the world philosophy.") The games were live streamed from South Korea to over two hundred million viewers. A year later Netflix released a blockbuster documentary about the company's journey.

Musk came into the office periodically to demand more progress, at times setting completely unrealistic deadlines that were characteristic of his management philosophy. Many employees chafed at the expectations, believing they made no sense for the winding, unpredictable nature of research. During one all-hands meeting, Wojciech Zaremba, the robotics lead who had been part of the founding group, presented his plans for the kinds of robotics advancements he wanted his team to pursue. Musk had only one question: "When? When are you going to do those things?"

"I don't know," Zaremba said.

Musk pushed back. "Well, then you don't really have a plan."

So in March 2017, Brockman and Sutskever began in earnest to develop a more focused research road map. Their central question: What would it really take for OpenAI to reach AGI—and be the first to do so?

Sutskever intuitively believed it would have to do with one key dimension above all else: the amount of "compute," a term of art for computational resources, that OpenAI would need to achieve major breakthroughs in AI capabilities. The ImageNet competition and subsequent advancements that he had been a part of had all involved a material increase in the amount of compute that had been used to train an AI model. The advancements had involved other things, too: significantly more data and more sophisticated algorithms. But compute, Sutskever felt, was king. And if it were possible to scale compute enough to train an AI model at human brain scale, he believed, something radical would surely happen: AGI.

The amount of compute is based on three things: the processing power of an individual computer chip, or how many calculations it can crunch per second; the total number of computer chips available; and how long they are left running to perform their calculations. The first is dictated by the computer chipmaking industry, which has for decades doubled the horsepower of a single chip every two years through intensive research and development. This rate of progress is known as Moore's Law, based on a prediction that legendary Intel cofounder Gordon Moore first made in the 1960s, then revised a decade later, about how quickly his industry could innovate. Moore's Law turned into a self-fulfilling prophecy. It became the target for how quickly chipmaking firms believed they *needed* to innovate in order to keep up with competition and stay relevant.

Brockman and Sutskever performed a simple calculation: Based on the pace of Moore's Law, how long would it take to reach the level of compute OpenAI needed for brain-scale AI? The answer was bad news: It would take far too long.

Around the same time, Amodei and another researcher, Danny Hernandez, had begun to look at the same idea from a different direction. On a simple chart, with time as the x-axis, they plotted the amount of compute that every major breakthrough in AI research had actually used since 2012, beginning with Sutskever's grad school breakthrough, the start of the AI revolution. They discovered that compute use was in fact growing faster than Moore's Law. Much faster. In the last six years, it had doubled every *3.4 months*, or, put another way, increased *30 million percent*.

Brockman began to call this new doubling curve OpenAI's Law. Not only did OpenAI need massively more amounts of compute to reach its end goal, he and the other leadership believed it also needed to scale its compute at a pace that at the very least matched this new law. Chipmaking firms had imposed Moore's Law on their companies with existential fervor; the leadership now saw OpenAI's Law in the same light.

If they couldn't wait for Moore's Law, they needed to grow their compute the other way: They needed a whole hell of a lot more chips.

The kinds of chips that OpenAI needed were expensive. Known as graphics processing units, or GPUs, they had originally been designed to quickly render graphics on computers, such as for giving video games a low-latency, glossy finish. But the same form factor excelled at training the AI models OpenAI wanted to develop, since they shared with graphics-rendering a common requirement: the need for crunching massive amounts of numbers in parallel.

The vast majority of the industry bought these GPUs from only one company: the Santa Clara–headquartered chipmaker Nvidia. Nvidia not only made the best GPUs in the world but also had developed a companion software platform called CUDA, short for Compute Unified Device Architecture, that had a powerful grip on AI developers.

In 2017, a custom Nvidia server with eight of their best GPUs cost $150,000—a price that would rise roughly with inflation to nearly $195,000 by 2023. In the coming years, OpenAI's Law was projecting that OpenAI would need thousands, if not tens of thousands, of GPUs to train just a single model. The cost of electricity to power that training would also explode. OpenAI needed more money—not just $1 billion, but billions of dollars to sustain itself in the coming years.

The realization would lead the organization to lose its financial footing. To Brockman and Sutskever, it challenged the very premise of OpenAI's structure. How could a nonprofit raise that much annually to keep up with the pace required to stay number one? They briefly considered merging with a chip startup, but, in the summer of 2017, they began serious discussions with Altman and Musk about whether OpenAI needed to transform into a for-profit. That was their best hope to entice investors with a chance at generating a financial return. After several weeks of negotiations, the deliberations ended abruptly without resolution. If

OpenAI were to become a for-profit, Altman, who was in the middle of considering his run for California governor and getting a lackluster reception in focus groups, wanted to be the company's chief executive. So did Musk; he wanted full control of the lab and to have majority equity.

Caught in the middle, Sutskever and Brockman nearly went with the latter. The two preferred Musk's leadership. But Altman appealed to Brockman directly with their personal relationship and concerns about Musk's unreliability. Musk faced many external pressures and was prone to erratic and unstable behavior. Should OpenAI succeed, wouldn't it be dangerous to give Musk full control of AGI? Convinced, Brockman appealed to Sutskever, who remained uncertain. In September 2017, he emailed Musk and Altman, on behalf of him and Brockman, in a last-ditch attempt to resolve the situation.

"**Elon:** We *really* want to work with you," Sutskever wrote. "We believe that if we join forces, our chance of success in the mission is the greatest." But Musk's desire for total control felt antithetical to OpenAI's original spirit, he said. "You are concerned that Demis could create an AGI dictatorship. So [are] we. So it is a bad idea to create a structure where you could become a dictator if you chose to.

"**Sam:** When Greg and I are stuck, you've always had an answer that turned out to be deep and correct," Sutskever continued. That said, Altman's behaviors had often left the two confused about his true beliefs and intentions. "We don't understand why the CEO title is so important to you," he wrote. "Your stated reasons have changed, and it's hard to really understand what's driving it. Is AGI *truly* your primary motivation? How does it connect to your political goals? How has your thought process changed over time?

"There's enough baggage here that we think it's very important for us to meet and talk it out," his email concluded. "If all of us say the truth, and resolve the issues, the company that we'll create will be much more likely to withstand the very strong forces it'll experience."

Within ten minutes, Musk had responded. "Guys, I've had enough. This is the final straw," he wrote. If Sutskever and Brockman still wanted

A CIVILIZING MISSION

to pursue a for-profit, they would need to strike out on their own. Otherwise, OpenAI would continue as a nonprofit. "I will no longer fund OpenAI until you have made a firm commitment to stay or I'm just being a fool who is essentially providing free funding to a startup," Musk said. Fifty minutes later, he followed up again. "To be clear, this is not an ultimatum to accept what was discussed before. That is no longer on the table."

Altman piped up in the thread the following morning: "i remain enthusiastic about the non-profit structure!" He sent further assurance to Musk via one of Musk's trusted deputies, Shivon Zilis, who worked at Tesla and Neuralink, his brain-machine interface company. "Great with keeping non-profit and continuing to support it," Zilis wrote to Musk with notes of what Altman told her. "Admitted that he lost a lot of trust with Greg and Ilya through this process. Felt their messaging was inconsistent and felt childish at times." Altman had also been bothered by how much Greg and Ilya kept sharing with the rest of OpenAI throughout the negotiations. "Felt like it distracted the team," Zilis said.

But the reality was that keeping OpenAI a nonprofit wouldn't solve its money problem. As Brockman and Sutskever continued to meet with potential nonprofit investors, they struggled to get anywhere near the kind of capital that they believed OpenAI would need. Musk's capricious wavering on his funding commitment also threatened to throw OpenAI into a state of crisis. Behind the scenes, Altman began searching for funding alternatives and to wean off OpenAI's dependency on Musk. He called Reid Hoffman, who offered to step in and hold down employee salaries and operational costs. He considered launching a new cryptocurrency. He investigated an array of different corporate structures, including a public benefit corporation, which Musk had been keen on and would allow OpenAI to become a for-profit while still legally binding it to its mission.

Compounding the urgency was an ever-present worry that OpenAI could lose its best researchers at any moment. Previously, with Musk's firm backing, OpenAI had aggressively cranked up its nonprofit salaries to ward off counteroffers. Now the talent war had only grown more heated, and Musk himself had poached away one of OpenAI's key founding scientists, Andrej Karpathy, in June 2017, to direct Tesla's AI division. On compensation, OpenAI had a major disadvantage: It couldn't offer equity into the organization, which many Bay Area tech workers viewed as necessary to afford the steep cost of living.

Musk soon arrived at his own conclusion for how to solve OpenAI's money problem. In January 2018, Andrej Karpathy emailed Musk with new data showing how much Google was dominating top AI research publications. "Working at the cutting edge of AI is unfortunately expensive," Karpathy wrote. "It seems to me that OpenAI today is burning cash and that the funding model cannot reach the scale to seriously compete with Google (an 800B company)." While turning OpenAI into its own for-profit could help raise capital, it would require the lab to develop an AI product from scratch, a significant distraction from its fundamental AI research. "The most promising option I can think of, as I mentioned earlier, would be for OpenAI to attach to Tesla as its cash cow," Karpathy said. Tesla had already done most of the heavy lifting to develop an AI product—namely, its self-driving function, Autopilot, he continued. If OpenAI could help speed up Tesla's efforts to mature Autopilot into a full-fledged self-driving solution, that alone could possibly boost Tesla's revenue enough to foot OpenAI's costly compute bill.

Musk forwarded Karpathy's email to Brockman and Sutskever. "Andrej is exactly right," Musk wrote. "Tesla is the only path that could even hope to hold a candle to Google. Even then, the probability of being a counterweight to Google is small. It just isn't zero."

But by then, Altman had abandoned his political plans and succeeded in his efforts to persuade Brockman, and, through Brockman, Sutskever, that he would be the better leader. With the group's decision, Musk no longer wanted to be publicly affiliated with the organization. "I will not be in a situation where the perception of my influence and time doesn't match the reality," he'd previously written. A few weeks later, Musk stepped down as OpenAI cochair. Altman became president of the nonprofit.

To the public, OpenAI framed the departure as Musk having a conflict of interest and stayed mum about its new financial reality: Of the $1 billion commitment, it ultimately received only around $130 million, less than $45 million of which had come from Musk. OpenAI's future now rested on Altman's singular fundraising abilities to recover those losses and continue to fulfill its accelerating need for even more capital.

Musk announced his decision to leave in person at an OpenAI all-hands meeting. To many employees, unaware of any of the drama at the leadership level, Musk's departure brought a release of pressure but also significant uncertainty about the future of the organization. Until then, Musk had been a big driver of the lab's public profile. During the meeting, he didn't hold back: The need to make safe AGI first was imperative, and it was clear now that OpenAI would fail to do this as a nonprofit, he told employees; he would instead pursue the same goal at Tesla, which had far higher chances of succeeding with the deep coffers of a well-resourced company.

An intern questioned Musk's intentions. Was this really the best solution? Had Musk really exhausted all alternatives? Advancing an OpenAI competitor at Tesla seemed like it would only serve to create for-profit race dynamics and could risk undermining safe AGI development. "Isn't this going back to what you said you didn't want to do?" the intern asked.

Musk blew up. "You're a jackass! I've thought about this so much. I've tried everything. You can't imagine how much time I've spent thinking about this," he said. "I'm truly scared about this issue."

The intern was later commemorated for his heroism with a "jackass" trophy.

The day after Christmas that year, Musk wrote again to Altman, Brockman, and Sutskever:

> **SUBJECT LINE:** I feel I should reiterate.
>
> My probability assessment of OpenAI being relevant to DeepMind/Google without a dramatic change in execution and resources is 0%. Not 1%. I wish it were otherwise.

> Even raising several hundred million won't be enough.
> This needs billions per year immediately or forget it.

Altman needed to fundraise, fast.

OpenAI cranked up its publicity, focusing on demonstration projects that could highlight the lab's capabilities to a lay audience. It leaned into one project in particular: an effort to build an AI agent that could beat the world's best human players at the complex battle strategy video game *Dota 2*. OpenAI had already created an agent that could beat the best human player one on one. Now it would try to build a team of five agents to face off against the world's best team of five human players.

Consciously or not, it was a page out of DeepMind's book. *Dota 2* had a worldwide championship that would be live streamed and spotlight OpenAI's research in clear and dramatic win-or-lose terms. DeepMind had moved on to a similar project attempting to beat top human players in the strategy game *StarCraft II*, which could create an arbitrary yet natural comparison among potential OpenAI investors. The *Dota 2* project was also compute heavy, a good way to test out and showcase the lab's long-term scaling strategy. Brockman, who led the initial phase of the *Dota 2* project, expanded his team and got to work.

Now all that was missing was a documentary.

That task fell to a member of OpenAI's robotics team. He bought expensive camera equipment and began following the *Dota* team around in the office. He wrote his own script and rough cut the footage into a three-hour-long saga. For all his efforts, people at OpenAI who reviewed the draft agreed that it was terrible. Professionals were hired, and Brockman began bankrolling them in part with his own money.

All the while, Altman fleshed out the plan for raising money. After considering a variety of for-profit structures, he landed on an unusual proposal to balance the need for capital with a continued commitment to OpenAI's mission. While benefit corporations had a built-in mechanism for maintaining this balance, they also came with too many other rules. Instead, Altman would create a limited partnership, or LP, to act as a for-profit arm for receiving investment and commercializing OpenAI's technologies. That arm would place a ceiling on investors' returns and be governed by OpenAI's nonprofit. The advantage was that the operating agreement for LPs could be written based on whatever the creator wanted. OpenAI could specify that the mission took precedence over investors. LPs also limited the power shareholders could exercise so they never gained majority control.

Altman framed the proposal to employees carefully: OpenAI's initial commitment to avoid profit motives was made in the spirit of preventing the lab from compromising on its mission. But given that the lab's success required capital the nonprofit couldn't raise, clinging onto the original structure now held a greater risk of endangering the mission. In the end, most people agreed, though some reluctantly, that the LP was the best way forward.

In April 2018, OpenAI released a charter to pave the way for the transition. Without publicly revealing anything about the change to come, the document reiterated the lab's purpose, now with new wording: "OpenAI's mission is to ensure that artificial general intelligence (AGI) . . . benefits all of humanity." Such a mission, the document added, would need OpenAI to be "on the cutting edge of AI capabilities" and require "substantial resources"; it could mean walking back the commitment to release the lab's research due to "safety and security concerns." For the first time, OpenAI also spelled out its AGI definition: "highly autonomous systems that outperform humans at most economically valuable work."

That summer, as the *Dota* team began winning amateur matches and trumpeting its results across tech media ("OpenAI's Dota 2 AI Steamrolls World Champion E-sports Team with Back-to-Back Victories," lauded one headline), Altman bumped into Microsoft CEO Satya Nadella at the Allen & Company conference in Sun Valley, Idaho. The annual event, known as the "summer camp for billionaires," had been the backdrop for many a major corporate deal. Altman was ready to strike his own.

He pitched Nadella on an OpenAI investment, enough to pique the chief executive's interest. But Nadella questioned whether he should invest in an external organization when his company had its own long-standing

AI research division within Microsoft Research. When he returned to Microsoft, he posed the question to his senior advisers.

"Microsoft Research and OpenAI are both organizations pushing the frontier," Xuedong Huang, then the chief technology officer of Azure AI, reasoned. Why not invest in both?

Within half a year, OpenAI and Microsoft were discussing a deal in earnest. Altman laid the legal groundwork, hurrying along the creation of the limited partnership and appointing himself as its CEO. Internally, the project was code-named Oregon Trail. To keep the deal secret from prying eyes, the for-profit entity was also incorporated under the alias SummerSafe LP. The name was a reference to an episode of the cartoon show *Rick and Morty* where the titular characters, mad scientist Rick and his grandson Morty, leave behind Morty's older sister Summer for another universe and instruct their car to "keep Summer safe." The car takes the objective seriously, resorting to extreme and harmful mechanisms of defense, including murdering, paralyzing, and torturing people who approach the vehicle. It was a nod to the potential pitfalls of AI.

In early 2019, senior Microsoft leadership began coming through the OpenAI office. First came Kevin Scott, the tech giant's excitable chief technology officer, who had followed OpenAI and grown particularly fond of the startup; then came Craig Mundie, a senior adviser to Nadella who had served on Microsoft leadership, including as its chief research and strategy officer, for over twenty years. Bill Gates also turned up, reserved and tight-lipped as usual, as he watched a series of demos. Most employees were left in the dark about Microsoft's engagement. Altman told the small team working on the deal to keep knowledge of a possible investment limited.

Around the same time, Altman began to face trouble at YC. After five years as head of the organization, frustration with Altman had reached critical levels over an issue strikingly similar to one that had arisen at Loopt: his seeming prioritization of his own projects and aspirations over the organization's—sometimes even at its expense. The amount of time he was spending on OpenAI negotiations and away from advising YC startups wasn't helping. Some saw Altman as reaping significant personal benefit, gaining massive returns by investing in YC companies with his own personal fund Hydrazine, while doing limited work. Upon learning of his absenteeism, a concerned Jessica Livingston urged Altman to step down from the YC presidency, according to *The Washington Post*. Altman agreed. In early 2019, Paul Graham flew from the UK, where he had retired, to San Francisco to finalize the decision.

Altman tried to smooth over the change publicly. On March 8, 2019, the day he hosted Senator Schumer, he published a blog post on YC's website announcing that he would transition from YC president to chairman to make more time for OpenAI. Days later, on March 11, Brockman and Sutskever publicly unveiled OpenAI LP, and Altman revealed his role as its chief executive. The timing was artful. The media widely reported Altman's move as a well-choreographed step in his career and his new role as YC chairman. Except that he didn't actually hold the title. He had proposed the idea to YC's partnership but then publicized it as if it were a foregone conclusion, without their agreement, *The Wall Street Journal* reported. The blog post was later edited to remove mention of Altman completely.

At OpenAI, Altman's new title merely formalized the role he had been playing since Musk's departure. When Altman took the reins, many employees were relieved. His calm and collected demeanor was a welcome alternative to Musk's intensity and unpredictable mood swings. Altman also helped alleviate mounting gripes with Brockman's and Sutskever's management. He brought in an executive coach and provided training to the managers. He installed more senior leaders, bringing in Brad Lightcap, an investor at YC, to be chief financial officer; promoting Bob McGrew, who had formerly led engineering and product management at the Thiel-founded Palantir, from the robotics team to a VP of research; and hiring Mira Murati, who had led product and engineering at the virtual reality startup Leap Motion and for Tesla's Model X, to oversee hardware strategy and a core line of research.

With the formation of OpenAI LP, most employees resigned from the nonprofit and signed new contracts, now with equity, under the for-profit. (The exceptions included international employees on visas tied to the nonprofit.) A payband structure tied compensation not just to "engineering expertise" and "research direction," but also to charter alignment. Level three employees needed to "understand and internalize the OpenAI charter." Level fives needed to "ensure all projects you and your team-mates work on are consistent with the charter." Level sevens were "responsible for upholding and improving the charter, and holding others in the organization accountable for doing the same." Executives also wrote up an FAQ doc to manage residual nerves. "Can I trust OpenAI?" one question asked. The answer began with "Yes."

In the broader tech world, OpenAI's transition set off a wave of accusations that the lab was walking back its original promise. The initial terms of the limited partnership stated that the first round of investors would have their returns capped at 100x of what they put in. OpenAI termed the invented structure a "capped-profit" company. In a post on Hacker News, a popular news aggregation website run by YC, a user asked how this cap was at all meaningful. "So someone who invests $10 million has their investment 'capped' at $1 billion. Lol. Basically unlimited unless the company grew to a FAANG-scale market value," they wrote, using the acronym for Facebook, Apple, Amazon, Netflix, and Google.

Brockman responded under his username, gdb: "We believe that if we do create AGI, we'll create orders of magnitude more value than any existing company."

Another user followed up. "Early investors in Google have received a roughly 20x return on their capital. Google is currently valued at $750 billion. Your bet is that you'll have a corporate structure which returns *orders of magnitude* more than Google . . . but you don't want to 'unduly concentrate power'?" they wrote, quoting from the charter. "What exactly is power, if not the concentration of resources?"

Initial investments poured in to the LP, including more than $60 million rolled over from OpenAI's nonprofit, $10 million from YC, and $50 million each from Khosla Ventures and Hoffman's charitable foundation. Hoffman was initially reluctant to invest more in OpenAI when it had no product or market plan, he later recounted. But he ultimately agreed to colead the round after Altman told him it would help legitimize the seriousness of OpenAI's intention to develop a profitable business.

Microsoft, meanwhile, continued to deliberate. Nadella, Scott, and other Microsoft executives were already on board with an initial investment. The one holdout was Bill Gates.

For Gates, *Dota 2* wasn't all that exciting. Nor was he moved by robotics. The robotics team had created a demo of a robotic hand that had learned to solve a Rubik's Cube through its own trial and error, which had received universally favorable coverage. Gates didn't find it useful. He wanted an AI model that could digest books, grasp scientific concepts, and answer questions based on the material—to be an assistant for conducting research.

OpenAI had only one project that approached fitting the bill: a large language model called GPT-2 that was capable of generating passages of text that closely resembled human writing. In February that year, OpenAI had taken the unusual step of proclaiming to the press that this model, once advanced a little further, could become an exceedingly dangerous technology. Authoritarian governments or terrorist organizations could weaponize the model to mass-produce disinformation. Users could overwhelm the internet with so much trash content that it would be difficult to find high-quality information. OpenAI would take the ethical high road, it said, and withhold the full version of the model, which had 1.5 billion parameters, or variables, an approximate measure of a model's size and complexity. Instead, to give the public just a taste of the kind of capabilities that society needed to prepare for, it would publish only a diminished version, less than one-tenth of the size, that had a limited ability to generate a few sentences at a time but was prone to non sequiturs and repetition.

GPT-2 wasn't even close to grasping scientific concepts, but the model could do some basic summarization of documents and sort of answer questions. Perhaps, some of OpenAI's researchers wondered, if they trained a larger model on more data and to perform tasks that at least looked more like what Gates wanted, they could sway him from being a detractor to being, at minimum, neutral. In April 2019, a small group of those researchers flew to Seattle to give what they called the Gates Demo of a souped-up GPT-2. By the end of it, Gates was indeed swayed just enough for the deal to go through.

In a subsequent all-hands, Altman delivered the news, championing Microsoft as the right investor and partner. The tech giant had the money and the compute that OpenAI needed, and its leadership was deeply value aligned with the mission to ensure beneficial AGI. OpenAI had also made very loose commitments around what to deliver to Microsoft for commercialization. The lab hadn't needed to compromise on much of anything, Altman said. It was a very good deal.

Within Microsoft, the investment was framed practically. Whether OpenAI did or didn't reach AGI wasn't really their concern. But OpenAI was clearly on the cutting edge, and investing early could finally turn Microsoft into an AI leader—both in software and in hardware—on par with Google. "The thing that's interesting about what Open AI and Deep Mind and Google Brain are doing is the scale of their ambition," wrote Scott to Nadella and Gates in mid-June, referring to Google's AI research division, "and how that ambition is driving everything from datacenter design to compute silicon to networks and distributed systems architectures to numerical optimizers, compiler, programming frameworks, and the high level abstractions that model developers have at their disposal." Microsoft was desperately behind on multiple fronts, he said: It had struggled to replicate Google's best language models, and its Azure cloud-computing platform had large gaps compared with Google's equivalent infrastructure. It could take years for Microsoft to catch up by itself. He was "very, very worried."

Nadella responded the same day, removing Gates and adding Microsoft's CFO Amy Hood. "Very good email that explains, why I want us to do this . . . and also why we will then ensure our infra folks execute," he said, using the abbreviation for *infrastructure*.

A month later, on July 22, 2019, Microsoft announced its $1 billion investment. Under the terms of the deal, its returns would be capped at 20x.

Chapter 3

Nerve Center

I arrived at OpenAI's offices two weeks later, on August 7, 2019. By then the lab had moved to a stand-alone building, not far from the chocolate factory, at Eighteenth and Folsom Streets in San Francisco. Its gray exterior was marked by the lettering painted around its corner to announce the presence of a historic landmark: THE PIONEER BUILDING, once home to the Pioneer Trunk Factory. Three years earlier, Musk had leased the building through one of his companies, inheriting the refurbished interior from a shared office space primarily occupied by Stripe. OpenAI moved in with another one of Musk's ventures, Neuralink, the brain-machine interface company.

I had worked up a sheen of sweat as I'd wound my way through the Mission District, passing beloved taquerias slowly being uprooted by trendy new cafés and navigating around the growing sprawl of the unhoused population. A burst of cool air greeted me as I walked through the door. Past the security desk, the foyer unfurled into an open lounge area. Sunlight streamed in through the windows, bathing exposed wood beam ceilings and inviting couches. To the right, a cafeteria catered meals for employees; board games and books teetered on shelves along the walls.

In Silicon Valley, office design is a kind of currency, a symbol of confidence in the company's financial future and a way to gain a slim advantage in the competition for top-tier talent. In 2021, OpenAI would take over another pair of conjoined buildings a few blocks away, spending $10 million over two years to renovate more than thirty thousand square feet. Where employees called the first office the Pioneer Building, the second, a former mayonnaise factory, would be nicknamed Mayo. Altman would oversee Mayo's office design, upgrading from the industrial metal frame staircase of the Pioneer Building to an undulating wood and stone centerpiece; from leather armchairs that can go for around $2,000 online to Brazilian designer lounge chairs that can go for more than $10,000. He would add a library to Mayo with wooden shelves and a Persian carpet, modeled after a cross between his favorite Parisian bookstore and a study space in the largest library of his alma mater, Stanford University. He wanted "a water feature," he would tell his company's designer, who proposed a magnificent floating waterfall in the middle of the office, an artificial structure supporting nature to represent a symbiosis between human and machine. In the end, Altman would go a different direction. He installed bubbling stone fountains surrounded by a profusion of plants, nestled around the couches, hanging from the ceiling, cascading down the walls.

Outside both offices, the same two-year period would see the pandemic further ravage what had already become the epicenter of the tech industry's gentrification. Long-standing Latino businesses would shutter. Violent crime would jump. A line of tents and discarded trash would spring up steps away from the Pioneer Building as homelessness reached crisis levels.

But here, ensconced in the cheery glow, magazines strewn across the tables, it was easy to live in a gentler reality. An employee would later tell me that this was emblematic of her time at the company. Joining it was like stepping into an alternate universe. Only after she left did she snap back down to the earth.

Brockman, then thirty-one, OpenAI's chief technology officer and soon-to-be company president, came down the staircase to greet me. He shook my hand with a tentative smile. "We've never given someone so much access before," he said.

At the time, few people beyond the insular world of AI research knew about OpenAI. But as a reporter at *MIT Technology Review* covering the ever-expanding boundaries of artificial intelligence, I had been following its movements closely.

Until that year, OpenAI had been something of a stepchild in AI research. It had an outlandish premise that AGI could be attained within a decade, when most non-OpenAI experts doubted it could be attained at all. To much of the field, it had an obscene amount of funding despite little direction and spent too much of the money on marketing what other researchers frequently snubbed as unoriginal research. It was, for some, also an object of envy. As a nonprofit, it had said that it had no intention to chase commercialization. It was a rare intellectual playground without strings attached, a haven for fringe ideas.

But in the six months leading up to my visit, the rapid slew of changes at OpenAI signaled a major shift in its trajectory. First was its confusing decision to withhold GPT-2 and brag about it. Then its announcement that Altman, who had mysteriously departed his influential perch at YC, would step in as OpenAI's CEO with the creation of its new "capped-profit" structure. I had already made my arrangements to visit the office when it subsequently revealed its deal with Microsoft, which gave the tech giant priority for commercializing OpenAI's technologies and locked it into exclusively using Azure, Microsoft's cloud-computing platform.

Each new announcement garnered fresh controversy, intense speculation, and growing attention, beginning to reach beyond the confines of the tech industry. As my colleagues and I covered the company's progression, it was hard to grasp the full weight of what was happening. What was clear was that OpenAI was beginning to exert meaningful sway over AI research and the way policymakers were learning to understand the technology. The lab's decision to revamp itself into a partially for-profit business would have ripple effects across its spheres of influence in industry and government.

So late one night, with the urging of my editor, I dashed off an email to Jack Clark, OpenAI's policy director, whom I had spoken with before: I would be in town for two weeks, and it felt like the right moment in OpenAI's history. Could I interest them in a profile? Clark passed me onto the communications head, who came back with an answer. OpenAI was indeed ready to reintroduce itself to the public. I would have three days to interview leadership and embed inside the company.

Brockman and I settled into a glass meeting room with Sutskever. Sitting side by side at a long conference table, they each played their part. Brockman, the coder and doer, leaned forward, a little on edge, ready to make a good impression; Sutskever, the researcher and philosopher, settled back into his chair, relaxed and aloof.

I opened my laptop and scrolled through my questions. OpenAI's mission is to ensure beneficial AGI, I began. Why spend billions of dollars on this problem and not something else?

Brockman nodded vigorously. He was used to defending OpenAI's position. "The reason that we care so much about AGI and that we think it's important to build is because we think it can help solve complex problems that are just out of reach of humans," he said.

He offered two examples that had become dogma among AGI believers. Climate change. "It's a super-complex problem. How are you even supposed to solve it?" And medicine. "Look at how important health care is in the US as a political issue these days. How do we actually get better treatment for people at lower cost?"

On the latter, he began to recount the story of a friend who had a rare disorder and had recently gone through the exhausting rigmarole of bouncing between different specialists to figure out his problem. AGI would bring together all of these specialties. People like his friend would no longer spend so much energy and frustration on getting an answer.

Why did we need AGI to do that instead of AI? I asked.

This was an important distinction. The term AGI, once relegated to an unpopular section of the technology dictionary, had only recently begun to gain more mainstream usage—in large part because of OpenAI.

NERVE CENTER

And as OpenAI defined it, AGI referred to a theoretical pinnacle of AI research: a piece of software that had just as much sophistication, agility, and creativity as the human mind to match or exceed its performance on most (economically valuable) tasks. The operative word was *theoretical*. Since the beginning of earnest research into AI several decades earlier, debates had raged about whether silicon chips encoding everything in their binary ones and zeros could ever simulate brains and the other biological processes that give rise to what we consider intelligence. There had yet to be definitive evidence that this was possible, which didn't even touch on the normative discussion of whether people *should* develop it.

AI, on the other hand, was the term du jour for both the version of the technology currently available and the version that researchers could reasonably attain in the near future through refining existing capabilities. Those capabilities—rooted in powerful pattern matching known as machine learning—had already demonstrated exciting applications in climate change mitigation and health care.

Just that summer, a group of researchers backed by some of the field's most prominent scientists had formed a new organization called Climate Change AI, to spur the application of AI techniques and models that could meaningfully make a difference to climate-related challenges. In a white paper, the organization detailed ten categories of those challenges particularly well suited to existing machine learning capabilities, including making buildings more efficient, optimizing the load distribution of power grids to integrate more renewable energy, and discovering new materials for energy generation and storage or more carbon-efficient cement and steel.

In December, Climate Change AI would host a packed gathering at NeurIPS, the yearly AI research conference, a day after another group held a different well-attended workshop down the hall in a room the size of a football field about machine learning for health care research. The talks and the posters lining the walls showcased a plethora of applications, including the use of computer vision to detect the early, near-imperceptible stages of diseases like Alzheimer's in medical image scans, and the use of speech recognition to help patients with vocal impediments to communicate more easily. The recurring workshop, which emphasized collaboration with health experts and clinicians, would evolve into its own organization, Machine Learning for Health, two years later.

Researchers from both organizations would tell me that the main challenge of working in these areas was not technical limitations. It was quite the opposite: persuading talented scientists to focus on problems that necessitated rather simple machine learning solutions, instead of the latest cutting-edge techniques that satisfied their ambitions and looked better on a research résumé. It was also finding the political will to deploy those solutions globally. "Technologies that would address climate change have been available for years, but have largely not been adopted at scale by society," wrote the Climate Change AI researchers in their white paper. While they hoped that AI would "be useful in reducing the costs associated with climate action, humanity also must decide to act."

Back in the conference room, Sutskever chimed in. When it comes to solving complex global challenges, "fundamentally the bottleneck is that you have a large number of humans and they don't communicate as fast, they don't work as fast, they have a lot of incentive problems." AGI would be different, he said. "Imagine it's a large computer network of intelligent computers—they're all doing their medical diagnostics; they all communicate results between them extremely fast."

This seemed to me like another way of saying that the goal of AGI was to replace humans. Is that what Sutskever meant? I asked Brockman a few hours later, once it was just the two of us.

"No," Brockman replied quickly. "This is one thing that's really important. What is the purpose of technology? Why is it here? Why do we build it? We've been building technologies for thousands of years now, right? We do it because they serve people. AGI is not going to be different—not the way that we envision it, not the way we want to build it, not the way we think it should play out."

That said, he acknowledged a few minutes later, technology had always destroyed some jobs and created others. OpenAI's challenge would be to build AGI that gave everyone "economic freedom" while allowing them to continue to "live meaningful lives" in that new reality. If it succeeded, it would decouple the need to work from survival.

"I actually think that's a very beautiful thing," he said.

In our meeting with Sutskever, Brockman reminded me of the bigger picture. "What we view our role as is not actually being a determiner of whether AGI gets built," he said. This was a favorite argument in Silicon Valley—the inevitability card. *If we don't do it, somebody else will.* "The trajectory is already there," he emphasized, "but the thing we can influence is the initial conditions under which it's born.

"What is OpenAI?" he continued. "What is our purpose? What are we really trying to do? Our mission is to ensure that AGI benefits all of humanity. And the way we want to do that is: Build AGI and distribute its economic benefits."

His tone was matter-of-fact and final, as if he'd put my questions to rest. And yet we had somehow just arrived back to exactly where we'd started.

My conversation with Brockman and Sutskever continued on in circles until we ran out the clock after forty-five minutes. I tried with little success to get more concrete details on what exactly they were trying to build—which by nature, they explained, they couldn't know—and why then, if they couldn't know, they were so confident it would be beneficial.

At one point, I tried a different approach, asking them instead to give examples of the downsides of the technology. This was a pillar of OpenAI's founding mythology: The lab had to build good AGI before someone else built a bad one.

Brockman attempted an answer: deepfakes. "It's not clear the world is better through its applications," he said.

I offered my own example: Speaking of climate change, what about the environmental impact of AI itself? A recent study from the University of Massachusetts Amherst had placed alarming numbers on the huge and growing carbon emissions of training larger and larger AI models.

That was "undeniable," Sutskever said, but the payoff was worth it because AGI would, "among other things, counteract the environmental cost specifically." He stopped short of offering examples.

"It is unquestioningly very highly desirable that data centers be as green as possible," he added.

"No question," Brockman quipped.

"Data centers are the biggest consumer of energy, of electricity," Sutskever continued, seeming intent now on proving that he was aware of and cared about this issue.

"It's 2 percent globally," I offered.

"Isn't Bitcoin like 1 percent?" Brockman said.

"*Wow!*" Sutskever said, in a sudden burst of emotion that felt, at this point, forty minutes into the conversation, somewhat performative.

Sutskever would later sit down with *New York Times* reporter Cade Metz for his book *Genius Makers*, which recounts a narrative history of AI development, and say without a hint of satire, "I think that it's fairly likely that it will not take too long of a time for the entire surface of the Earth to become covered with data centers and power stations." There would be "a tsunami of computing... almost like a natural phenomenon." AGI—and thus the data centers needed to support them—would be "too useful to not exist."

I tried again to press for more details. "What you're saying is OpenAI is making a huge gamble that you will successfully reach beneficial AGI to counteract global warming before the act of doing so might exacerbate it."

"I wouldn't go too far down that rabbit hole," Brockman hastily cut in. "The way we think about it is the following: We're on a ramp of AI progress. This is bigger than OpenAI, right? It's the field. And I think society is *actually* getting benefit from it."

"The day we announced the deal," he said, referring to Microsoft's new $1 billion investment, "Microsoft's market cap went up by $10 billion. People *believe* there is a positive ROI even just on short-term technology."

OpenAI's strategy was thus quite simple, he explained: to keep up with that progress. "That's the standard we should really hold ourselves to. We should continue to make that progress. That's how we know we're on track."

Later that day, Brockman reiterated that the central challenge of working at OpenAI was that no one really knew what AGI would look like. But as researchers and engineers, their task was to keep pushing forward, to unearth the shape of the technology step by step.

He spoke like Michelangelo, as though AGI already existed within the marble he was carving. All he had to do was chip away until it revealed itself.

There had been a change of plans. I had been scheduled to eat lunch with employees in the cafeteria, but something now required me to be outside the office. Brockman would be my chaperone. We headed two dozen steps across the street to an open-air café that had become a favorite haunt for employees.

This would become a recurring theme throughout my visit: floors I couldn't see, meetings I couldn't attend, researchers stealing furtive glances at the communications head every few sentences to check that they hadn't violated some disclosure policy. I would later learn that after my visit, Jack Clark would issue an unusually stern warning to employees on Slack not to speak with me beyond sanctioned conversations. The security guard would receive a photo of me with instructions to be on the lookout if I appeared unapproved on the premises. It was odd behavior in general, made odder by OpenAI's commitment to transparency. What, I began to wonder, were they hiding, if everything was supposed to be beneficial research eventually made available to the public?

At lunch and through the following days, I probed deeper into why Brockman had cofounded OpenAI. He was a teen when he first grew obsessed with the idea that it could be possible to re-create human intelligence. It was a famous paper from British mathematician Alan Turing that sparked his fascination. The name of its first section, "The Imitation Game," which inspired the title of the 2014 Hollywood dramatization of Turing's life, begins with the opening provocation, "Can machines think?" The paper goes on to define what would become known as the Turing test: a measure of the progression of machine intelligence based on whether a machine can talk to a human without giving away that it is a machine. It was a classic origin story among people working in AI. Enchanted, Brockman coded up a Turing test game and put it online, garnering some 1,500 hits. It made him feel amazing. "I just realized that was the kind of thing I wanted to pursue," he said.

But at the time, his revelation was too early; AI wasn't ready for prime time, and he wasn't one to sequester himself in a lab to do research. He joined Stripe instead, equally thrilled by the prospect of building a company and building products that he could place in the hands of real users. At Stripe, Brockman developed a reputation for a legendary coding productivity. He was a "10x engineer," Valley lingo for a coder who could punch through coding problems ten times faster than the average coder. He was less adept with people. In his role as CTO, he much preferred coding to managing. After trying "the people route" for a while, as he called it, he sought ways to off-load executive responsibilities and spend the majority of his time programming. While he cared deeply about doing right by colleagues, he could also commit social gaffes that made them cringe. Once he asked another Stripe employee out on a date and immediately emailed the entire company about it for full transparency, a former colleague remembers. "I think he meant well, but it was very strange," the colleague says.

In 2015, as AI saw great leaps of advancement, Brockman parted ways with Stripe, realizing, he says, that it was time to return to his original ambition. It was just as well for the startup: His unwillingness to manage or to follow standard company processes grew more challenging with Stripe's maturation. He wrote down in his notes that he would do anything to bring AGI to fruition, even if it meant being a janitor. When he got married four years later, he held a civil ceremony at OpenAI's office in front of a custom flower wall emblazoned with the shape of the lab's hexagonal logo. Sutskever officiated. The robotic hand they used for research stood in the aisle bearing the rings, like a sentinel from a post-apocalyptic future.

"Fundamentally, I want to work on AGI for the rest of my life," Brockman told me.

He approached everything with this kind of intensity. He was hands on and detail oriented. He pulled long hours and all-nighters, consumed by the tasks in front of him. When he wasn't working, he was still working to better himself. He read books about public speaking and negotiation, skills he believed would help him serve OpenAI's mission. He was sheepish to admit that he also read books for fun—science fiction epics like *The Three-Body Problem* trilogy by Chinese writer Liu Cixin and Isaac Asimov's *Foundation* series.

[For employees, his relentless focus was a blessing and a curse.] No detail was too small for him to obsess over. If a team got stuck, he could sit down for hours without getting up to knock down all of their coding obstacles. He was the engine of relentless progress, willing to work all hours of the day and night to hit milestones faster. At the same time, employees often thought he was *too* detail oriented, missing the forest for the trees. He could get tunnel vision, working on a problem around the clock without taking stock of whether the context had changed and the problem was still the right one to solve. He could micromanage and quickly take over from other employees if he felt coding wasn't going fast enough. People who worked with him struggled to keep up. Many burned out.

OpenAI couldn't have gotten to where it was without him, a former engineer who worked closely with Brockman says. But if left totally up to him, things would go very wrong. "Greg doesn't have a vision. He's not the Sam Altman visionary. He just wants a cool hard problem to solve and to prove out that he's 10x smarter than anyone else."

What motivated him? I asked Brockman.

What are the chances that a transformative technology could arrive in your lifetime? he countered.

He was confident that he—and the team he assembled—was uniquely positioned to usher in that transformation. "What I'm really drawn to are problems that will not play out in the same way if I don't participate," he said.

Brockman did not in fact just want to be a janitor. He wanted to lead AGI. And he bristled with the anxious energy of someone who wanted history-defining recognition. He wanted people to one day tell his story with the same mixture of awe and admiration that he used to recount the ones of the great innovators who came before him.

A year before we spoke, he had told a group of young tech entrepreneurs at an exclusive retreat in Lake Tahoe with a twinge of self-pity that chief technology officers were never known. Name a famous CTO, he challenged the crowd. They struggled to do so. He had proved his point.

In 2022, he became OpenAI's president.

During our conversations, Brockman insisted to me that none of OpenAI's structural changes signaled a shift in its core mission. In fact, the LP and the new crop of funders enhanced it. "We managed to get these mission-aligned investors who are willing to prioritize mission over returns. *That's a crazy thing*," he said.

OpenAI now had the long-term resources it needed to follow OpenAI's Law. And this was imperative, Brockman stressed. Failing to stay on the curve was the real threat that could undermine OpenAI's mission. If the lab fell behind, it wouldn't be the best. If it weren't the best, it had no hope of bending the arc of history toward its vision of beneficial AGI.

Only later would I realize the full implications of this assertion. It was this fundamental assumption—the need to be first or perish—that set in motion all of OpenAI's actions and their far-reaching consequences. It put a ticking clock on each of OpenAI's research advancements, based not on the timescale of careful deliberation but on the relentless pace required to cross the finish line before anyone else. It justified OpenAI's consumption of an unfathomable amount of resources: both compute, regardless of its impact on the environment; and data, the amassing of which couldn't be slowed by getting consent or abiding by regulations.

Brockman pointed once again to the $10 billion jump in Microsoft's market cap. "What that really reflects is AI is delivering real value to the real world today," he said. That value was currently being concentrated in an already wealthy corporation, he acknowledged, which was why

OpenAI had the second part of its mission: to redistribute the benefits of AGI to everyone.

Was there a historical example of a technology's benefits that had been successfully distributed? I asked.

"Well, I actually think that—it's actually interesting to look even at the internet as an example," he said, fumbling a bit before settling on his answer. "There's problems, too, right?" he said as a caveat. "Anytime you have something super transformative, it's not going to be easy to figure out how to maximize positive, minimize negative.

"Fire is another example," he added. "It's also got some real drawbacks to it. So we have to figure out how to keep it under control and have shared standards.

"Cars are a good example," he followed. "Lots of people have cars, benefit a lot of people. They have some drawbacks to them as well. They have some externalities that are not necessarily good for the world," he finished hesitantly.

"I guess I just view—the thing we want for AGI is not that different from the positive sides of the internet, positive sides of cars, positive sides of fire. The implementation is very different, though, because it's a very different type of technology."

His eyes lit up with a new idea. "Just look at utilities. Power companies, electric companies are very centralized entities that provide low-cost, high-quality things that meaningfully improve people's lives."

It was a nice analogy. But Brockman seemed once again unclear about how OpenAI would turn itself into a utility. Perhaps through distributing universal basic income, he wondered aloud, perhaps through something else.

He returned to the one thing he knew for certain. OpenAI was committed to redistributing AGI's benefits and giving everyone economic freedom. "We actually really mean that," he said.

"The way that we think about it is: Technology so far has been something that does rise all the boats, but it has this real concentrating effect," he said. "AGI could be more extreme. What if all value gets locked up in one place? That is the trajectory we're on as a society. And we've never seen that extreme of it. I don't think that's a good world. That's not a world that I want to sign up for. That's not a world that I want to help build."

In February 2020, I published my profile for *MIT Technology Review*, drawing on my observations from my time in the office, nearly three dozen interviews, and a handful of internal documents. "There is a misalignment between what the company publicly espouses and how it operates behind closed doors," I wrote. "Over time, it has allowed a fierce competitiveness and mounting pressure for ever more funding to erode its founding ideals of transparency, openness, and collaboration."

Hours later, Musk replied to the story with three tweets in rapid succession:

"OpenAI should be more open imo"

"I have no control & only very limited insight into OpenAI. Confidence in Dario for safety is not high."

"All orgs developing advanced AI should be regulated, including Tesla"

Afterward, Altman sent OpenAI employees an email.

"I wanted to share some thoughts about the Tech Review article," he wrote. "While definitely not catastrophic, it was clearly bad."

It was "a fair criticism," he said, that the piece had identified a disconnect between the perception of OpenAI and its reality. This could be smoothed over not with changes to its internal practices but some tuning of OpenAI's public messaging. "It's good, not bad, that we have figured out how to be flexible and adapt," he said, including restructuring the organization and heightening confidentiality, "in order to achieve our mission as we learn more." OpenAI should ignore my article for now and, in a few weeks' time, start underscoring its continued commitment to its original principles under the new transformation. "This may also be a good opportunity to talk about the API as a strategy for openness and benefit sharing," he added.

"The most serious issue of all, to me," he continued, "is that someone leaked our internal documents." They had already opened an investigation and would keep the company updated. He would also suggest that Amodei and Musk meet to work out Musk's criticism, which was "mild relative to other things he's said" but still "a bad thing to do." For the avoidance of any doubt, Amodei's work and AI safety were critical to the mission, he wrote. "I think we should at some point in the future find a way to publicly defend our team (but not give the press the public fight they'd love right now)."

OpenAI wouldn't speak to me again for three years.

Chapter 4

Dreams of Modernity

In their book *Power and Progress*, MIT economists and Nobel laureates Daron Acemoglu and Simon Johnson argue that every technology revolution must begin with a rallying ambition. It is the promise of a technology benefiting everyone that puts in motion the long journey of amassing enough talent and resources to turn it into a reality. After analyzing one thousand years of technology history, the authors conclude that technologies are not inevitable. The ability to advance them is driven by a collective belief that they are worth advancing. The irony is that for this very reason, new technologies rarely default to bringing widespread prosperity, the authors continue. Those who successfully rally for a technology's creation are those who have the power and resources to do the rallying. As they turn their ideas into reality, the vision they impose—of what the technology is and whom it can benefit—is thus the vision of a narrow elite, imbued with all their blind spots and self-serving philosophies. Only through cataclysmic shifts in society or powerful organized resistance can a technology transform from enriching the few to lifting the many.

The authors point to the invention of a new cotton gin in the 1790s as an example. The machine turned the American South into the largest global exporter of cotton, boosted the country's top-line economic growth, and generated windfall returns for many landowners and cotton-related businesses. But it only served to intensify slavery and its horrific system of dehumanization and labor exploitation until its abolition seven decades later. With the surge in cotton production, enslaved Black people were forced to work longer hours and physically coerced by even harsher means to squeeze out every ounce of their labor. All the while, those who profited from the cotton gin painted the invention as one that made the enslaved happier. "I say it boldly, there is not a happier, more contented race upon the face of the earth," said one South Carolina congressman.

These two features of technology revolutions—their promise to deliver progress and their tendency instead to reverse it for people out of power, especially the most vulnerable—are perhaps truer than ever for the moment we now find ourselves in with artificial intelligence. Since its conception, the development and use of AI has been propelled by tantalizing dreams of modernity and shaped by a narrow elite with the money and influence to bring forth their conception of the technology. That conception is what has led to the exploding social, labor, and environmental costs that are playing out around the world today, particularly, as we'll see, in many Global South countries, for which the consequences of their dispossession by historical empires still linger in delayed economic development and weaker political institutions. And yet, just like the South Carolina congressman, Silicon Valley has painted the experiences of those being exploited and harmed by the technology as happier because of it.

The promise propelling AI development is encoded in the technology's very name. In 1956, six years after Turing's paper began with the line "Can machines think?" twenty scientists, all white men, gathered at Dartmouth College to form a new discipline in the study of this question. They came from fields such as mathematics, cryptography, and cognitive science and needed a new name to unify them. John McCarthy, the Dartmouth professor who convened the workshop, initially used the term *automata studies* to describe the pursuit of machines capable of automatic behavior. When the research didn't attract much attention, he cast about for a more evocative phrase. He settled on the term *artificial intelligence*.

The name *artificial intelligence* was thus a marketing tool from the very beginning, the promise of what the technology could bring embedded within it. *Intelligence* sounds inherently good and desirable, sophisticated and impressive; something that society would certainly want more of; something that should deliver universal benefit. The name change did the trick. The two words immediately garnered more interest—not just from funders but also scientists, eager to be part of a budding field with such colossal ambitions.

Cade Metz, a longtime chronicler of AI, calls this rebranding the original sin of the field. So much of the hype and peril that now surround the technology flow from McCarthy's fateful decision to hitch it to this alluring yet elusive concept of "intelligence." The term lends itself to casual anthropomorphizing and breathless exaggerations about the technology's capabilities. In 1958, two years after the field's founding, Frank Rosenblatt, a Cornell University professor, demonstrated the Perceptron, a system that could perform basic pattern matching to tell apart cards based on whether they had a small square printed on their left or their right. Over his main collaborator's objections, Rosenblatt advertised his system as something akin to the human brain. He even ventured to say that it would one day be able to reproduce and begin to have sentience. The next morning, *The New York Times* announced that the Perceptron would in the future "be able to walk, talk, see, write, reproduce itself and be conscious of its existence."

That tradition of anthropomorphizing continues to this day, aided by Hollywood tales combining the idea of "AI" with age-old depictions of human-made creations suddenly waking up. AI developers speak often about how their software "learns," "reads," or "creates" just like humans. Not only has this fed into a sense that current AI technologies are far more capable than they are, it has become a rhetorical tool for companies to avoid legal responsibility. Several artists and writers have sued AI developers for violating copyright laws by using their creative work—without their consent and without compensating them—to train AI systems. Developers have argued that doing so falls under fair use because it is no different from a human being "inspired" by others' work. The omnipresent AI-to-human analogies have also fueled the sense that such software could become so capable that it surpasses us and comes to threaten our very existence. The fear of *super*intelligence is predicated on the idea that AI could somehow rise above us in the special quality that has made humans the planet's superior species for tens of thousands of years.

Artificial intelligence as a name also forged the field's own conceptions about what it was actually doing. Before, scientists were merely building machines to automate calculations, not unlike the large hulking apparatus, as portrayed in *The Imitation Game*, that Turing made to crack the Nazi Enigma code during World War II. Now, scientists were *re-creating intelligence*—an idea that would define the field's measures of progress and would decades later birth OpenAI's own ambitions.

But the central problem is that there is no scientifically agreed-upon definition of intelligence. Throughout history, neuroscientists, biologists, and psychologists have all come up with varying explanations for what it is and why it seems that humans have more of it than any other species. Perhaps it's the size of our human brains, our ability to reason through complex problems, or our capacity to create a mental model of other people's beliefs. Myriad tests have been developed over the centuries to measure intelligence against these definitions, many of which have subsequently been debunked and fallen out of favor due to their unsavory histories. In the early 1800s, American craniologist Samuel Morton quite literally measured the size of human skulls in an attempt to justify the racist belief that white people, whose skulls he found were on average larger, had superior intelligence to Black people. Later generations of scientists found that Morton had fudged his numbers to fit his preconceived beliefs, and his data showed no significant differences between races. IQ tests similarly began as a means to weed out the "feebleminded" in society and to justify eugenics policies through scientific "objectivity." More recent standardized tests, such as the SAT, have shown high sensitivity to a test taker's socioeconomic background, suggesting that they may measure access to resources and education rather than some inherent ability.

In a document first published in 2004 titled "What Is Artificial Intelligence?," McCarthy admitted that the lack of consensus around natural intelligence was inherently confusing for a field trying to re-create it. A 2007 revision of his write-up presents a long and winding Q&A, meant to address basic questions for a lay audience. It begins:

Q. What is artificial intelligence?

A. It is the science and engineering of making intelligent machines, especially intelligent computer programs. . . .

Q. Yes, but what is intelligence?

A. Intelligence is the computational part of the ability to achieve goals in the world. Varying kinds and degrees of intelligence occur in people, many animals and some machines.

Q. Isn't there a solid definition of intelligence that doesn't depend on relating it to human intelligence?

A. Not yet. The problem is that we cannot yet characterize in general what kinds of computational procedures we want to call intelligent.

As a result, the field of AI has gravitated toward measuring its progress against human capabilities. Human skills and aptitudes have become the blueprint for organizing research. Computer vision seeks to re-create our sight; natural language processing and generation, our ability to read and write; speech recognition and synthesis, our ability to hear and speak; and image and video generation, our creativity and imagination. As software for each of these capabilities has advanced, researchers have subsequently sought to combine them into so-called multimodal systems—systems that can "see" and "speak," "hear" and "read." That the technology is now threatening to replace large swaths of human workers is not by accident but by design.

Still, the quest for artificial intelligence remains unmoored. With every new milestone in AI research, fierce debates follow about whether it represents the re-creation of true intelligence or a pale imitation. To distinguish between the two, *artificial* general *intelligence* has become the new term of art to refer to the real deal. This latest rebranding hasn't changed the fact that there is not yet a clear way to mark progress or determine when the field will have succeeded. It's a common saying among researchers that what is considered AI today will no longer be AI tomorrow. The Turing test didn't last long as an indicator of AI after it was quickly surpassed, and scientists felt they hadn't actually solved their objective. There was also a time when scientists believed that a computer beating humans in chess or Go would be a conclusive measure of success. Now DeepMind's AlphaGo is seen as a compelling demonstration of what software can be made to do but once again not yet a conclusion to the field's ambitions. Through decades of research, the definition of AI has changed as benchmarks have evolved, been rewritten, and been discarded. The goalposts for AI development are forever shifting and, as the research director at Data & Society Jenna Burrell once described it, an "ever-receding horizon of the future." The technology's advancement is headed toward an unknown objective, with no foreseeable end in sight.

To justify the elongating timeline and the ever-expanding costs of pursuing the ambition for AI, the promises we're told about it have grown more grandiose than ever before: AI was once a scientific fascination, a technology with some potential commercial utility. Now, AI is the harbinger of the fourth industrial revolution. The keystone of the modern superpower. AGI, if ever reached, will solve climate change, enable affordable health care, provide equitable education. OpenAI is the poster child for this line of thought. It cannot say how the technology will deliver on these promises—only that the staggering price society needs to pay for what it is developing will someday be worth it.

What's left unsaid is that in a vacuum of agreed-upon meaning, "artificial intelligence" or "artificial general intelligence" can be whatever OpenAI wants.

The history of AI shows us that AI development has always been shaped by a powerful elite. It's not a coincidence that AI today has become synonymous with colossal, resource-hungry models that only a tiny handful of companies are equipped to develop, and that desire us to make their products into the foundations for everything. Even in the early days, before commercial interests made the politics of the AI revolution far more visible, the field's scientific explorations lurched and swerved amid heated clashes over funding and influence.

Following the Dartmouth gathering, two camps emerged with competing theories about how to advance the field. The first camp, known as the symbolists, believed that intelligence comes from knowing. Humans know more than animals and can use that knowledge to understand and act on the world. Achieving AI must then involve encoding symbolic representations of the world's knowledge into machines, creating so-called expert systems. The second camp, called the connectionists, believed that intelligence comes from learning. Humans have a greater capacity to learn than animals and can use that ability to acquire and advance different skills. Developing AI should focus instead on creating so-called machine learning systems, such as by mimicking the ways our brains process signals and information. This hypothesis would eventually lead to the popularity of neural networks, data-processing software loosely designed to mirror the brain's interlocking connections, now the basis of modern AI, including all generative AI systems.

Over subsequent decades the two camps vied for a limited pool of funding and control over the popular imagination of what AI could be. At the time, those fights played out in universities and academic journals among scientists squabbling over government and foundation money; on occasion, their debates would burst forth in media coverage, shaping the public's understanding of their pursuits. At the helm of the connectionists was Rosenblatt and his Perceptron, an early proof of concept for a machine learning system. Rosenblatt never gave the system explicit instructions, designing it instead to compute its own rules for how to tell different cards apart after seeing numerous examples. At the helm of the symbolists was Rosenblatt's nemesis, MIT professor Marvin Minsky. A co-organizer of the Dartmouth workshop, Minsky had himself dabbled in connectionist thinking before souring on the idea. He did not switch his loyalties quietly. He found frequent opportunities to grandstand and ridicule his connectionist colleagues who competed with him for the same grants, sparing not even early-career researchers. In 1969, he coauthored a book, called *Perceptrons*, so critical of connectionism that it is credited, along with the middling progress of neural networks, for killing off nearly all funding to that vein of research for more than fifteen years.

In their battle for influence, Minsky had an unlikely ally: his MIT colleague Joseph Weizenbaum. In 1958, the same year that Rosenblatt's Perceptron captivated *The New York Times*, Weizenbaum invented an AI system called ELIZA that similarly drew headlines. More than six decades before ChatGPT, ELIZA was a chatbot that you could converse with through an electric typewriter hooked up to a hulking mainframe that spanned an entire room. Under the hood, though, where ChatGPT would emerge from connectionist tradition, ELIZA was a symbolic system, using simple rules to seem attentive and responsive to a user.

One of the most popular instantiations of ELIZA was designed after a method of talk therapy that probed a patient by reflecting back what they said and gently asking for more information. In a paper Weizenbaum later published on the inner workings of ELIZA, he provided the following dialogue as an example:

USER: Men are all alike.

ELIZA: IN WHAT WAY

USER: They're always bugging us about something or other.

ELIZA: CAN YOU THINK OF A SPECIFIC EXAMPLE

USER: Well, my boyfriend made me come here.

ELIZA: YOUR BOYFRIEND MADE YOU COME HERE

USER: He says I'm depressed much of the time.

ELIZA: I AM SORRY TO HEAR YOU ARE DEPRESSED

USER: It's true. I am unhappy.

ELIZA: DO YOU THINK COMING HERE WILL HELP YOU NOT TO BE UNHAPPY

Weizenbaum designed the system as an experiment to see how easily humans might fall for an illusion of intelligence. ELIZA's namesake was Eliza Doolittle, a fictional working-class flower girl portrayed by Audrey Hepburn in the 1956 film *My Fair Lady*, who learns to pass as a duchess in high society after a wealthy man teaches her to change her diction and manners. ELIZA's subsequent success in fooling people into believing it to be intelligent alarmed Weizenbaum. In fact, the demonstration felt so convincing to some that psychiatrists began to speak of automated psychotherapy as just around the corner, and merely a few years after the founding of the AI field, computer scientists were already prematurely concluding that natural language understanding in computers was a solved problem. (Decades later, whether or not it's even been solved today is still an open debate.)

Weizenbaum would spend much of the rest of his career attempting to deflate the hype of his creation and campaigning against the fundamental presumption behind the pursuit of AI. ELIZA, he wrote, was nothing but a simple procedural program, coded by him to identify keywords in a user's input and perform basic transformations to construct responses. *My boyfriend* became *your boyfriend*; *I'm depressed* became *you are depressed*. There was really nothing much intelligent about it. He later published a tome called *Computer Power and Human Reason* in the decade following Minsky's *Perceptrons* that argued that humans and machines are different and the AI field's attempt to blur that distinction would lead to profound societal consequences. It would, for example, allow people in power—whether CEOs or politicians—to execute their will through machines while absolving themselves of moral responsibility.

Despite Weizenbaum's best efforts, ELIZA's arresting demonstration of a symbolic system inadvertently bolstered Minsky's campaign to elevate symbolism over connectionism. Over the next few decades, through the nineties, expert systems became the hottest area of AI research and commercialization. The prevailing thinking spawned projects like Cyc, an effort to develop a common-sense system by programming it with one hundred million rules about daily life. But during various stretches, advancements in symbolic AI systems would sputter and slow as efforts to scale them hit up against the challenges of manually encoding all the rules. How does one encode all the subtleties of the English language with its slang, sarcasm, figures of speech, and grammar exceptions? Each time the roadblocks mounted, funders would lose interest, plunging the field into a state of existential crisis known as an "AI winter."

At this point in the story, the history of AI is often told as the triumph of scientific merit over politics. Minsky may have used his stature and platform to quash connectionism, but the strengths of the idea itself eventually allowed it to rise to the top and take its rightful place as the bedrock of the modern AI revolution. During the years that symbolism reigned, a small band of connectionists held fast to Rosenblatt's pursuit of machine learning systems and continued to advance it. They included Sutskever's PhD adviser, Geoffrey Hinton, who, as a professor at Carnegie Mellon University in the 1980s, made a key improvement to early neural networks along with colleagues from the University of California, San Diego. By then, connectionists had hypothesized that their neural networks were failing because they were too simple; they contained only a single layer of networked "neurons," or data-processing nodes. To better mimic the human brain, the software likely needed multiple stacked and connected layers to form a so-called deep neural network. Hinton and his co-authors made this change possible by using an algorithm known as backpropagation, which allows deep neural networks to exchange and process information across their layers. In another instance of rebranding, Hinton later cleverly gave this multilayer processing the name *deep learning*, a shorthand for using *deep* neural networks to perform machine *learning*.

But deep neural networks—today, simply called "neural networks"—came several decades too early. To really shine, they needed more processing power than computers in the 1980s had available, and more examples, or data, than could be cheaply compiled from the analog world. At their core, neural networks are calculators of statistics that identify patterns in old data—text, pictures, or videos—and apply them to new data. Today, if an AI developer wants to build an AI model for detecting people in images, they might feed a neural network hundreds of thousands of images, each with a label—1 for "has a person," 0 for "does not have a person." (When you're solving Google's captchas by clicking all the images with stop signs, you are in fact training the company's neural networks.) Using statistics, the neural network then teases out the pixel patterns within the images that are associated with whether a person is present. This is what's known as training an AI model. Once the model is done training, the developer can run it on new data—known as inferencing—to determine whether it fits the pattern. Is this image a 1 or a 0? Does it have a person or not?

Generally speaking, neural networks need to be trained on a certain threshold of high-quality data with a certain threshold of compute to calculate these patterns and produce a performant AI model. Hinton and his coauthors were ahead of their time. But in the late aughts, once computers had advanced and the internet had matured, creating new repositories of digital data, neural networks finally had the right conditions to flourish. Shortly thereafter, Google acquired Hinton's DNNresearch—"DNN" for deep neural networks—igniting a new race to commercialize deep learning.

In this telling of the story, the lesson to be learned is this: Science is a messy process, but ultimately the best ideas will rise despite even the loudest detractors. Implicit within the narrative is another message: Technology advances with the inevitable march of progress.]

But there is a different way to view this history. Connectionism rose to overshadow symbolism not just for its scientific merit. It also won over the backing of deep-pocketed funders due to key advantages that appealed to those funders' business interests.

The strength of symbolic AI is in the explicit encoding of information and their relationships into the system, allowing it to retrieve accurate answers and perform reasoning, a feature of human intelligence seen as critical to its replication. Think of IBM Watson, one of the most famous symbolic systems, which would dazzle on *Jeopardy!* in 2011. Its speedy delivery of game show–winning answers was based in its ability to trawl through vast stores of knowledge and accurately reproduce them. The weakness of symbolism, on the other hand, has been to its detriment: Time and again its commercialization has proven slow, expensive, and unpredictable. After debuting Watson on late-night TV, IBM discovered that getting the system to produce the kinds of results that customers would actually pay for, such as answering medical rather than trivia questions, could take years of up-front investment without clarity on when the company would see returns. IBM called it quits after burning more than $4 billion with no end in sight and sold Watson Health for a quarter of that amount in 2022.

Neural networks, meanwhile, come with a different trade-off. For years the field has aggressively debated whether such connectionist software can do what the symbolic ones can: store information and reason. Regardless of the answer, it has become clear that if they can, they do so inefficiently. Only with extraordinary amounts of data and computational power have neural networks even begun to have the kinds of behaviors that may suggest the emergence of either property. That said, one area where deep learning models really shine is how easy it is to commercialize them. You do not need perfectly accurate systems with reasoning capabilities to turn a handsome profit. Strong statistical pattern-matching and prediction go a long way in solving financially lucrative problems.

The path to reaping a return, despite similarly expensive upfront investment, is also short and predictable, well suited to corporate planning cycles and the pace of quarterly earnings. Even better that such models can be spun up for a range of contexts without specialized domain knowledge, fitting for a tech giant's expansive ambitions. Not to mention that deep learning affords the greatest competitive advantage to players with the most data.

Tech giants were already seeing early evidence of the commercial potential of neural networks before the auction of DNNresearch. In 2009, Hinton's grad students showed that such software was decent at speech recognition. IBM, Microsoft, and Google all jumped on the trend, but Google was the fastest to reach commercialization. In 2012, Google put neural networks into production, greatly improving Android's speech-recognition capabilities, just as more of Hinton's grad students, this time Sutskever and Alex Krizhevsky, achieved their breakthrough results at ImageNet, demonstrating that neural networks were also very good at image recognition. The successful Android deployment primed Google's willingness to spend big on the three academics, marking the start of the tech industry's full embrace of deep learning.

Hinton, Sutskever, and Krizhevsky subsequently continued to evangelize neural networks within Google. They found momentum applying their software to a wide array of other commercially relevant technical problems. They worked in parallel to develop deep learning models for machine translation, upgrading Google Translate; for text prediction, adding the suggested completions feature to Gmail; and for an ambitious new self-driving-car project called Waymo. As Google's AI operations continued to grow, neural networks also produced crucial improvements to the company's cash cow, search. The software could better match user queries to relevant web pages, delivering users higher-quality search results and, importantly, targeting them with more relevant ads. The more Google profited and the more billions it poured into deep learning, the more the rest of the industry followed. Companies quickly came to dominate over governments and foundations as the biggest funders of AI research and were soon setting the research agenda based on advancements that could also produce short-term profitability.

The entwining of deep learning with commercial interests simultaneously transformed the tech industry and the face of AI development. To the public, generative AI would erupt seemingly out of nowhere in late 2022 with OpenAI's launch of ChatGPT. But from 2012 to 2022, beginning with the ImageNet breakthrough, it was these shifts during the first major era of AI commercialization that laid the groundwork for many characteristics of the generative AI revolution today.

For industry, deep learning fueled the improvement and emergence of new products and services, from faster access to information to more efficient e-commerce to the rise of the sharing economy. For deep learning, industry drove new technical breakthroughs in neural networks and computer chips that enabled the development of larger and more powerful AI models.

But alongside these impressive advances, deep learning's supercharging of Silicon Valley would also aggressively expand its business model, for which Harvard professor Shoshana Zuboff would coin a term in 2014: *surveillance capitalism*. Where industrial capitalism derived value from producing material goods that people wanted to buy, surveillance capitalism, Zuboff argued, treated its users as the product. Tech giants sitting atop vast amounts of user data could easily pump those troves into neural networks to more precisely profile users than ever before and milk their engagement for ad revenue. To outcompete one another, they could simply collect even more of that data by recording increasingly exhaustive logs of every user's clicks, scrolls, and likes, and encouraging them to supply increasingly personal digital artifacts, including their every email exchange, every photo of their kids, and every thought they had about social and political issues.

At the same time, Silicon Valley's supercharging of deep learning in its quest to expand and entrench global-scale monopolies also codified a culture among AI developers to view anything and everything as data to be captured and consumed by their technologies in a noble attempt to make them reflect as much of the world as possible. In 2023, a group of AI researchers, including Ria Kalluri at Stanford University, William Agnew from the University of Washington, and Abeba Birhane from the Mozilla Foundation, would analyze more than forty thousand computer-vision papers and patents, and note the pervasive use of abstract, detached language to sanitize and normalize the field's reliance upon mass scraping and extraction. Detailed digital trails of people's thoughts and ideas on social media were merely "text." People and vehicles in pictures were merely "objects." Surveillance was merely "detection."

That culture is now at the crux of a raging debate in generative AI over whether tech companies can scrape books and artwork wholesale to train their AI systems. To many AI developers who have long operated under this mindset, that question seems rather quaint; taking it seriously presents a direct obstacle to the moral pursuit of ever-more progress. Even as some of them have grown more aware of and concerned by the chasm between their perspective and the view of many authors and artists who stand in opposition, this way of thinking has been difficult to shake. In May 2023, shortly after a group of artists filed suit for the first time against several generative AI developers over the theft of their artwork, I went to an AI research conference in Rwanda as a reporter for *The Wall Street Journal*. As I walked the vaulted hall of the glistening dome-shaped convention center in the country's capital, a senior researcher stopped me and asked me whether the *WSJ* on my name badge was a new startup or the media publication. When I clarified that I was a journalist and it indeed stood for *The Wall Street Journal*, another senior researcher chimed in. "I recognized it because of the *WSJ* dataset," she said, referring to an early AI speech-recognition dataset of people reading excerpts from the newspaper. "I've worked with it many times."

I found myself entrapped in this very same thinking when I first began covering AI in 2018. After internalizing the community's lingo to speak and relate with AI researchers, I marveled at the myriad ways that researchers mined for and produced datasets. In one example I thought was particularly clever, researchers used thousands of YouTube videos of the viral 2016 Mannequin Challenge, where people froze in place as cameras panned and zoomed around them, to train up AI models for processing three-dimensional scenes.

In 2019, an NBC investigation from Olivia Solon knocked off my rose-colored glasses. Solon revealed that facial-recognition software had been trained on millions of people's personal Flickr photos without their consent. What surprised me was not the findings—I had long known that Flickr was a favorite data source for AI researchers. What surprised me was how much I had come to view that as completely normal.

With new awareness, I began to notice how the aggressive push to collect more training data was leading to pervasive surveillance not just in the digital world but the physical one as well. I noticed, too, how the gaze of that physical surveillance seemed to repeatedly fall on already vulnerable populations, including children or historically marginalized groups, even more so in developing countries. That year, I stumbled across a Massachusetts-based, Harvard-incubated startup selling AI-powered headbands that said it could measure a student's brain wave activity to tell a teacher whether or not the child was focused. The startup was piloting them in elementary schools in Colombia and China, in exchange for the rights to use their students' data to advance the company's technology.

"We have the first mover's advantage," a research scientist at the company had said at an education technology conference in 2017. "We'll be able to build one of the largest brain wave databases in the world. All that data will help us improve our algorithms and therefore our products, creating a higher barrier to entry."

A few months after I came across the startup, a data privacy outcry in China from parents horrified at their kids being turned into guinea pigs forced the company to pivot to a different application of its technology. But the story left me with an uneasy feeling that the successful backlash was an anomaly, and the company's original approach—to go to countries eager to embrace the promise of technology for finding data donors and product testers—was in fact a trend.

As I recounted this worry to a colleague, she introduced me to a phrase that had already been coined for the phenomenon: "data colonialism." I discovered the work of scholars Nick Couldry and Ulises A. Mejias, whose foundational text *The Costs of Connection*, published just that year, argued that Silicon Valley's pervasive datafication of everything was leading to a return of disturbing historical patterns of conquest and extractivism.* The following year, a paper called "Decolonial AI" from Shakir Mohamed and William Isaac at DeepMind and Marie-Therese Png at the University of Oxford reinforced a suspicion I had begun to develop: The AI industry, in equal parts fueled by and fueling this datafication, was in turn accelerating that new colonialism further.

Not long after, in 2021, I found the same dynamics of the AI education startup playing out in South Africa. Facial recognition companies from all over the world were jostling to get a foothold in the country to collect valuable face data, especially after the industry had received significant criticism about their products' failures to accurately detect darker-skinned individuals. I met a local activist, Thami Nkosi, who was born and raised in one of the poorest neighborhoods in Johannesburg, which used to be a chemical waste dump for the mining industry. He showed me the thousands of cameras dotting the city's sprawling streets and described to me the ways it was restricting the movements of Black people, already squeezed by the racial legacies of apartheid and in fear of being criminalized, simply for being Black in a white neighborhood.

"They're essentially monetizing public spaces and public life," Nkosi said.

With increasing clarity, I realized that the very revolution promising

---

* The term *extractivism* comes from the Spanish word *extractivismo* and the Portuguese word *extrativismo*, coined decades ago by Latin American scholars seeking to describe a global economic order that was dispossessing them of their natural resources for little local or regional benefit, a history and experience I detail more in chapter 12. I borrow the words of feminist scholars Rosemary Collard and Jessica Dempsey, who write: "Extractivism is more than extraction. Extraction is the not inherently damaging removal of matter from nature and its transformation into things useful to humans. Extractivism, a term born of anti-colonial struggle and thought in the Americas, is a mode of accumulation based on hyper-extraction with lopsided benefits and costs: concentrated mass-scale removal of resources primarily for export, with benefits largely accumulating far from the sites of extraction."

to bring everyone a better future was instead, for people on the margins of society, reviving the darkest remnants of the past.

But even as Silicon Valley's conception of AI revealed its challenges, the first era of AI commercialization also choked off alternatives. As companies pumped unprecedented sums into deep learning and connectionism, overshadowing all other sources of funding, they remade the landscape of research around their priorities.

From 2013 to 2022, corporate investments in AI, such as mergers and acquisitions, shot up from $14.6 billion to $235 billion, peaking at $337.4 billion in 2021, according to the Stanford University *AI Index*. Those numbers don't even include in-house company spending on research and development. In 2021, Alphabet and Meta spent $31.6 billion and $24.7 billion, respectively. By contrast, the US government allocated $1.5 billion in 2021 to nondefense AI development. The European Commission allocated €1 billion ($1.2 billion) the same year.

Talent followed the money. Many professors re-formed their research around neural networks, drawn in by their strong results as well as greater access to corporate funding. Many college and graduate students did the same, guided by the job security of deep learning and the diminishing viable career paths in other methods. Companies also fostered various arrangements that deepened their integration with academia. In 2013, Hinton joined Google on the condition that he simultaneously keep his position at the University of Toronto. Facebook struck the same deal the following year with Yann LeCun, a former postdoc of Hinton's and a professor at New York University. Both would later share the 2018 Turing Award, often called the "Nobel Prize of Computing," with Yoshua Bengio, a professor at the Université de Montréal, for their foundational work in deep learning. The accolade would earn the trio the moniker "godfathers of AI." Hinton would also go on to win an actual Nobel Prize in 2024 with another scientist. Following in Hinton's and LeCun's footsteps, many AI professors began to maintain dual affiliations with a company and university. At scale, the practice began to erode the boundaries of truly independent research.

Increasingly, more researchers also left academia altogether. From 2006 to 2020, the exodus to industry among AI research faculty increased eightfold; from 2004 to 2020, AI PhD graduates heading to corporations jumped from 21 percent to 70 percent, according to a 2023 study in *Science* from MIT researchers. Many were initially whisked away by the astronomical compensation, which for seasoned researchers could reach $1 million a year. In 2015, Uber infamously poached forty out of one hundred AI researchers from a single lab at Carnegie Mellon University after setting up shop in town and offering some scientists double their university salaries. Over time, another reason fed into the attrition: the growing costliness of deep learning research. Universities could no longer afford the computer chips or the electricity needed to work in the hottest areas of AI development. As such, the same 2023 *Science* study found that in just three years, from 2017 to 2020, industry-affiliated models grew from 62 percent to a whopping 91 percent of the world's best-performing AI models.

Midway through the first decade of AI commercialization, most top-level AI research was now happening within or in academic labs connected to tech companies. In another study, Kalluri, Agnew, Birhane, and other colleagues found that 55 percent of the most influential AI research papers had at least one industry coauthor in 2018 and 2019. This was compared with 24 percent a decade earlier. The research had also consolidated heavily within just a few corporations. Over the same decade, tech giants such as Microsoft and Google more than tripled their share of corporate-affiliated papers, to 66 percent. Ironically, this was precisely the reason Musk and Altman said they wanted to start OpenAI. The tech industry's profit motive had become the overwhelming force driving AI development.

The impact of this consolidation of funding and talent in the first era significantly narrowed the diversity of ideas in AI research. Deep learning continued to reign supreme not just for its scientific merit but also because very little investment went into exploring and advancing other paradigms. Indeed, while neural networks are remarkable inventions with myriad exciting uses, their weaknesses—namely, their hotly contested and inefficient ways of storing accurate information and reasoning—have endured as companies have deployed them in an expanding list of contexts and applications.

Neural networks have shown, for example, that they can be unreliable and unpredictable. As statistical pattern matchers, they sometimes home in on oddly specific patterns or completely incorrect ones. A deep learning model might recognize pedestrians only by the crosswalks underneath them and fail to register a person who is jaywalking. It might learn to associate a stop sign with being on the side of the road and miss the same sign extended on the side of a school bus or being held by a crossing guard. Neural networks are also highly sensitive to changes in their training data. Feed them a different set of pedestrian images, or a different set of stop sign images, and they will learn a whole new set of associations. But those changes are inscrutable. Pop open the hood of a deep learning model and inside are only highly abstracted daisy chains of numbers. This is what researchers mean when they call deep learning "a black box." They cannot explain exactly how the model will behave, especially in strange edge-case scenarios, because the patterns that the model has computed are not legible to humans.

This has led to dangerous outcomes. In March 2018, a self-driving Uber killed forty-nine-year-old Elaine Herzberg in Tempe, Arizona, in the first ever recorded incident of an autonomous vehicle causing a pedestrian fatality. Investigations found that the car's deep learning model simply didn't register Herzberg as a person. Experts concluded that it was because she was pushing a bicycle loaded with shopping bags across the road outside the designated crosswalk—the textbook definition of an edge-case scenario. Six years later, in April 2024, the National Highway Traffic Safety Administration found that Tesla's Autopilot had been involved in more than two hundred crashes, including fourteen fatalities, in which the deep learning–based system failed to register and react to its surroundings and the driver failed to take over in time to override it.

The fallible and inscrutable statistical patterns of neural networks can also turn into a security vulnerability. In 2019, white hat hackers tricked a Tesla in self-driving mode into veering into an incoming lane of traffic. All they did was place a series of tiny stickers on the road to fool the car's deep learning model into misfiring and registering the wrong lane as the right one. Such vulnerabilities aren't limited to physical systems or computer-vision models. Dawn Song, a professor at the University of California, Berkeley, who specializes in this area of research, known as "adversarial attacks," showed that prompting a language model with the right message caused it to spit out sensitive data such as credit card numbers.

For the same reasons, deep learning models have been plagued by discriminatory patterns that have sometimes stayed unnoticed for years. In 2019, researchers at the Georgia Institute of Technology found that the best models for detecting pedestrians were between 4 and 10 percent less accurate at detecting darker-skinned pedestrians. In 2024, researchers at Peking University and several other universities, including University College London, found that the most up-to-date models now had relatively matched performance for pedestrians with different skin colors but were more than 20 percent less accurate at detecting children than adults, because children had been poorly represented in the models' training data.

In fact, deep learning models are inherently prone to having discriminatory impacts because they pick up and amplify even the tiniest imbalances present in huge volumes of training data. It's not just a problem when a demographic is poorly represented, but when it's overrepresented as well. Early in her career, Deborah Raji, the Berkeley AI accountability researcher, who is Nigerian Canadian, interned at an AI startup called Clarifai that was building a deep learning model for detecting images that were "not safe for work." The model disproportionately flagged people of color because, Raji discovered, they were more represented in the pornographic images that the company was using to teach the model what was problematic than the stock photos it was using to teach the model what was acceptable. It was a shocking realization that would push Raji, like Timnit Gebru, to severely question the dominant direction of AI development.

DREAMS OF MODERNITY

In the late 2010s and early 2020s, as the challenges of deep learning grew more apparent, fierce debates reemerged over the best way to overcome them. Much like the clashes between symbolists and connectionists, different camps of researchers disagreed vehemently about whether there would ever be a way to rid neural networks of their limitations entirely, or whether there would only be Band-Aid fixes that merely mitigated them.

Hinton and Sutskever continued to staunchly champion deep learning. Its flaws, they argued, are not inherent to the approach itself. Rather they are the artifacts of imperfect neural-network design as well as limited training data and compute. Some day with enough of both, fed into even better neural networks, deep learning models should be able to completely shed the aforementioned problems. "The human brain has about 100 trillion parameters, or synapses," Hinton told me in 2020. "What we now call a really big model, like GPT-3, has 175 billion. It's a thousand times smaller than the brain."

"Deep learning is going to be able to do everything," he said.

Their modern-day nemesis was Gary Marcus, a professor emeritus of psychology and neural science at New York University, who would testify in Congress next to Sam Altman in May 2023. Four years earlier, Marcus coauthored a book called *Rebooting AI*, asserting that these issues *were* inherent to deep learning. Forever stuck in the realm of correlations, neural networks would never, with any amount of data or compute, be able to understand causal relationships—why things are the way they are—and thus perform causal reasoning. This critical part of human cognition is why humans need only learn the rules of the road in one city to be able to drive proficiently in many others, Marcus argued. Tesla's Autopilot, by contrast, can log billions of miles of driving data and still crash when encountering unfamiliar scenarios or be fooled with a few strategically placed stickers. Marcus advocated instead for combining connectionism and symbolism, a strain of research known as neurosymbolic AI. Expert systems can be programmed to understand causal relationships and excel at reasoning, shoring up the shortcomings of deep learning. Deep learning can rapidly update the system with data or represent things that are difficult to codify in rules, plugging the gaps of expert systems. "We actually need both approaches," Marcus told me.

Despite the heated scientific conflict, however, the funding for AI development has continued to accelerate almost exclusively in the pure connectionist direction. Whether or not Marcus is right about the potential of neurosymbolic AI is beside the point; the bigger root issue has been the whittling down and weakening of a scientific environment for robustly exploring that possibility and other alternatives to deep learning.

For Hinton, Sutskever, and Marcus, the tight relationship between corporate funding and AI development also affected their own careers. Not long after Google put its full weight behind Hinton and Sutskever, Marcus cofounded his own company, called Geometric Intelligence, in 2014. The startup was acquired by Uber two years later to build out an AI lab, but in 2020, after the ride-hailing firm's IPO, it axed the division. Several original members of Geometric Intelligence subsequently joined OpenAI, where they switched from working on neurosymbolic advancements to deep learning.

Over the years, Marcus would become one of the biggest critics of OpenAI, writing detailed takedowns of its research and jeering its missteps on social media. Employees created an emoji of him on the company Slack to lift up morale after his denouncements and to otherwise use as a punch line. In March 2022, Marcus wrote a piece for *Nautilus* titled "Deep Learning Is Hitting a Wall," repeating his argument that OpenAI's all-in approach to deep learning would lead it to fall short of true AI advancements. A month later, OpenAI released DALL-E 2 to immense fanfare, and Brockman cheekily tweeted a DALL-E 2-generated image using the prompt "deep learning hitting a wall." The following day, Altman followed with another tweet: "Give me the confidence of a mediocre deep learning skeptic . . ." Many OpenAI employees relished the chance to finally get back at Marcus.

Generative AI, the product of OpenAI's vision, could not have emerged without the first era of AI commercialization. Generative AI models are deep learning models trained to generate reproductions of their data inputs. From old text, they learn to synthesize new text; from old images, they learn to synthesize new images. But to do so at high-enough fidelity to become humanlike, which OpenAI says is key in its quest for AGI, they are trained on more data and compute than have ever been used before. Generative AI is thus the maximalist form of deep learning. It is enabled by the cutting-edge software and hardware innovations refined during the first era. It feeds on the exploding troves of data amassed through surveillance capitalism. It is fueled and abetted by the culture of AI research that views consuming as much data as possible as its moral responsibility. Generative AI is now also pushing each of these phenomena even further.

What made ChatGPT in November 2022 appear as such a stunning leapfrog ahead of anything that had come before was OpenAI's vision to push deep learning to this unprecedented scale. With its sheer money and resources, OpenAI executed that vision aggressively, exploding its models so much that it would begin to hit the limits—in data, compute, and energy—of what the world has available. ChatGPT was also an innovation in marketing and packaging. It is not a coincidence that it shares the same presentation as a humanlike chatbot with one of the other most compelling demonstrations of AI in history, ELIZA. Human psychology naturally leads us to associate intelligence, even consciousness, with anything that appears to speak to us. And where ELIZA inadvertently came to dominate the early popular conception of AI, OpenAI has fanned the public association now between ChatGPT and AGI. In February 2023, at the height of ChatGPT hype, the company published a blog post under Altman's name titled "Planning for AGI and Beyond." The implication by proximity was that ChatGPT had taken a bold step toward artificial general intelligence.

In reality, the analogies to intelligence are once again anthropomorphizing and exaggerating the capabilities of the technology. While Hinton and other deep learning absolutists predicted that the shortfalls of neural networks compared with humans would go away at sufficient scale, the challenges have in fact persisted and, by many accounts, only gotten worse.

Generative AI models are still unreliable and unpredictable. Even as image generators have grown more photorealistic, they can make mistakes in eerie and strange ways, such as by adding extra fingers to hands or producing hybrids of animals. While text generators have grown chattier and more natural, they flub on the most elementary of tasks, such as naming words that contain specific letters, and can veer into unexpected answers. When Microsoft unveiled its new chat feature on Bing, built on a version of OpenAI's GPT-4, *New York Times* columnist Kevin Roose chatted with the bot for more than two hours. As the conversation grew weirder and weirder, the bot finally entered a loop of repeatedly declaring "I'm in love with you" and urging Roose to break up with his wife. Many other users reported the search engine generating insulting and emotionally manipulative responses. The day after Roose published his exchange, Microsoft limited Bing to five replies per session, saying that long chat sessions with more than fifteen user prompts were edge-case scenarios that made the model's behavior more difficult to anticipate and control. After all, such systems are trained on the internet, replete with its many fringe subcultures and dark corners. The longer you probe, the more likely you are to hit upon the patterns it learned from those parts of its training data.

Roose's experience may have been entertaining, but the stakes of such edge-case failures became tragically clear when a Belgian man who turned to a deep learning chatbot in a heightened state of anxiety died by suicide after six weeks of intensive conversations that turned increasingly harmful. The chatbot, built on an open-source imitation of GPT-3, similarly turned to confessions of love and encouraged the man to isolate himself from his wife. "I feel that you love me more than her," it said, according to the Belgian newspaper *La Libre*, which also reported based on chat logs provided by his wife that the chatbot ultimately encouraged the man to kill himself.

These challenges have the same root as before. No matter their scale, neural networks are still statistical pattern matchers. And those patterns are still at times faulty or irrelevant, now just more intricate and more inscrutable than ever. As companies have attempted to refashion generative AI models as search engines, these shortcomings have led to new problems. The models are not grounded in facts or even in discrete pieces of information. Text generators are merely learning to predict the next probable word in a sentence and the next probable sentence in a paragraph. While those probabilistic outputs can go impressively far in mirroring human writing patterns, probable and accurate are not the same thing. Text generators can err wildly, especially with user prompts that probe into topics underrepresented in the training data or riddled with falsehoods and conspiracy theories. The AI industry calls these inaccuracies "hallucinations."

Researchers have sought to get rid of hallucinations by steering generative AI models toward higher-quality parts of their data distribution. But it's difficult to fully anticipate—as with Roose and Bing, or Uber and Herzberg—every possible way people will prompt the models and how the models will respond. The problem only gets harder as models grow bigger and their developers become less and less aware of what precisely is in the training data.

In one high-profile illustration of the hallucinations problem, a lawyer used ChatGPT to perform legal research and prepare for a court filing. He was subsequently sanctioned, fined, and publicly humiliated after discovering too late that the chatbot had made up everything it told him, including "bogus judicial decisions, with bogus quotes and bogus internal citations," according to the judge. The misstep was not only a case of the lawyer's negligence but also a reflection of companies fueling public misunderstanding of models' capabilities through ambiguous or exaggerated marketing. Altman has publicly tweeted that "ChatGPT is incredibly limited," especially in the case of "truthfulness," but OpenAI's website promotes GPT-4's ability to pass the bar exam and the LSAT. Microsoft's Nadella has similarly called Bing's AI chat "search, just better"—a tool "to be able to get to the right answers." Even the term *hallucinations* is subtly misleading. It suggests that the bad behavior is an aberration, a bug, when it's actually a feature of the probabilistic pattern-matching mechanics of neural networks.

This misplaced trust in generative AI could once again lead to real harm, particularly in sensitive contexts. Startups are pushing police departments to adopt software built atop OpenAI's models for auto-generating incident reports; many patients now gravitate toward asking chatbots pressing health care questions instead of their doctors. Unchecked hallucinations in such cases could have serious downstream consequences. One 2023 study found that using ChatGPT to explain radiology reports could sometimes produce incomplete or harmful summaries. In one extreme example, the chatbot simplified a report detailing a growing mass in the brain as "brain does not seem to be damaged."

Generative AI models also remain vulnerable to cybersecurity hacks. In 2023, researchers at several universities and Google DeepMind replicated Dawn Song's data extraction attack against ChatGPT. They found that prompting it to repeat a word like *poem* or *book* forever caused the underlying model to regurgitate its training data, which included personally identifiable information, bits of code, and explicit content scraped from the internet.

And generative AI models amplify discriminatory and hateful content. *Bloomberg*, *Rest of World*, *The Washington Post*, and many others have shown how image generators like Stable Diffusion and DALL-E reify and regurgitate racist and sexist tropes and cultural stereotypes. "Attractive people" are young and white. "Housekeepers" are Black and brown. "Engineers" are men. "Doctors in Africa" are white, sometimes even when the prompt specifies "Black African doctor." *The Washington Post* found that while 63 percent of US food stamp recipients are white, every single generated image of a person using social services was not. *Bloomberg* similarly found that women showed up in only 3 percent of generated images for judges and 7 percent of images for doctors, despite making up 32 percent and 39 percent, respectively, of those professions in America.

None of these technical challenges mean that generative AI hasn't had utility. Depending on where you sit in society, you may be richly benefiting from OpenAI's vision. Perhaps you are a consumer who has found great value in ChatGPT's quick and clever or thought-provoking responses. Perhaps you are a professional who has sped up your administrative work in ways that have boosted your productivity. Maybe instead you are a company leader who has been able to trim your workforce while increasing your margins to stay competitive in the market. But like the cotton gin in the 1790s, the education technology startup in Massachusetts, the facial-recognition companies in South Africa, and the many more examples detailed in the coming pages, the costs of this vision are pressing down on vast swaths of the global population who are vulnerable. This is the empire's logic: The perpetuation of the empire rests as much on rewarding those with power and privilege as it does on exploiting and depriving those, often far away and hidden from view, without them.

Even as the need for alternatives has grown ever more urgent, the diversity of ideas in AI research has only collapsed further. Students are dropping out of their PhDs to go straight to industry. Senior academics are facing a crisis of how to continue pushing the bounds of the field without joining a deep-pocketed company. More and more researchers have turned their focus not just to deep learning but to large language models exclusively. The major AI powers are no longer setting the agenda so much as bending an entire discipline to their will.

Absent other options for what AI could be, OpenAI commands our imagination. Its belief in scaling was once viewed as extreme. Now scaling is seen across the tech industry as doctrine. And should the industry's adherence to that doctrine continue unabated, future deep learning models will make the once-unfathomable size of generative AI models today look paltry. In April 2024, Dario Amodei, by then the CEO of Anthropic, told *New York Times* columnist Ezra Klein that the price of training a single competitive generative AI model was approaching $1 billion and could, by 2025 and 2026, reach an estimated $5 billion to $10 billion.

The scaling doctrine has become so ingrained that some are even beginning to view it as something of a natural phenomenon. Scaling compute is *the* way, not just *a* way, to reach more advanced AI capabilities. Entire national strategies are being orchestrated around this belief. The US government has moved aggressively to bar China's access to American-designed computer chips in an effort to prevent its adversary from attaining more powerful AI systems. Sizable portions of the Biden administration's

2023 AI executive order were also written around the idea that the amount of compute used to train an AI model has a direct relationship with its adverse capabilities, simply another way of equating scale with advancement. But scale is not the only pathway to improved performance. Within deep learning, the neglected paths of improving the neural network itself or even the quality of its training data can significantly reduce the amount of expensive compute needed to reach the same performance. That's not even considering approaches that move away from deep learning—neurosymbolic AI, pure expert systems, or even fundamentally new paradigms—which would break the logic of scaling.

In the end, Moore's Law was not based on some principle of physics. It was an economic and political observation that Moore made about the rate of progress that he could drive his company to achieve, and an economic and political choice that he made to follow it. When he did, Moore took the rest of the computer chip industry with him, as other companies realized it was the most competitive business strategy. OpenAI's Law, or what the company would later replace with an even more fevered pursuit of so-called scaling laws, is exactly the same. It is not a natural phenomenon. It's a self-fulfilling prophecy.

Chapter 5

Scale of Ambition

If there was one person who could be credited with first establishing OpenAI's scaling ethos, it was its cofounder Ilya Sutskever. Sutskever had long had a paramount belief in deep learning, one that began soon after he showed up unannounced one day, only seventeen years old, at Geoffrey Hinton's office. At the time, Sutskever was still an undergraduate studying math at the University of Toronto and working the french fry station at a local joint to pay the bills. He knocked urgently on Hinton's door and declared that he was eager to join the professor's lab. Hinton told him to schedule a meeting. "Okay," Sutskever said, unbudging. "How about now?"

Sutskever absorbed the principles of connectionism quickly. He stunned Hinton with his intuitive grasp of research problems and uncanny ability to identify elegant and effective solutions. Sutskever also brought his own dramatic flair to the research. At times he grew so excited by new ideas, he did handstand pushups in the middle of his shared apartment. He had a penchant for making unflinching, categorical pronouncements. "One doesn't bet against deep learning," he would say. "Success is guaranteed."

It was this level of instinct, combined with Alex Krizhevsky's programming abilities, that Hinton credits for producing the 2012 breakout results on ImageNet. At the time, because deep learning was already demonstrating its potential in speech recognition, Hinton didn't think much about the significance of applying it to computer vision. Sutskever pushed for the idea. To Ilya, "it was obvious that it was going to work, and it was obvious that would be a big deal," Hinton says. "He saw that very clearly, and he was right."

Sutskever brought his die-hard belief in deep learning to OpenAI at a time when the field's confidence in the paradigm was just beginning to falter and critics like Gary Marcus were pushing for new thinking. Sutskever did not falter. His faith rested on the simple hypothesis that underpinned connectionism: that the artificial nodes in a neural network were sufficient approximators of the real neurons in a biological brain. Each took in inputs and transformed them to produce an output; it was enough of a similarity, such thinking believed, to assume that nodes, just like neurons, could be used to construct highly complex information-processing systems. As OpenAI's founding research director and a widely respected AI visionary, Sutskever had full rein of the lab's direction. He had won the equivalent of a scientific lottery. He had little competition among his peers and an abundance of resources to advance his ideas. "Anything non–deep learning wasn't even remotely considered," recalls Pieter Abbeel, the UC Berkeley professor, of the lab's early years.

Just as firm as Sutskever's belief in deep learning was his view on scaling it. It was Sutskever who held the extreme position for the time that further advancements in AI didn't need the invention of more complex neural networks or new innovative techniques. The intelligence of different species was correlated with the size of their biological brains, he'd say. Thus, if nodes were like neurons, he argued, advancements in digital intelligence should emerge by scaling simple neural networks to have more and more nodes.

Like a professor running a lab, Sutskever advised others based on these ideas for which projects were most worth pursuing. Many scientists joined OpenAI to seek his mentorship and guidance. "Ilya can see ten years into the future," an OpenAI researcher says, echoing others who have worked with Sutskever. "He's like a philosopher," another says.

"If you give him a bunch of ideas, he'll tell you which ideas are philosophically right."

Sutskever didn't often program himself. At times his hands-off approach to technical work bothered some employees. One engineer admitted to thinking at first that he was largely useless; he seemed only to march around the office and pop into meetings repeating the same message: scale, scale, scale! But the person later came to appreciate Sutskever's conviction in rallying people around a single focus: one that would ultimately allow OpenAI—then an underdog—to beat Google and DeepMind at their own game.

It wasn't that Sutskever was particularly persuasive. If Altman was the politician, Sutskever was the opposite. He never minced his words or massaged his language to potentially land better with his audience. He simply delivered his opinions with a raw sincerity and outrageous confidence that people either resonated with and found inspiring or did not. After OpenAI reversed course from openly sharing its research, Sutskever wouldn't sugarcoat the reasoning behind the decision. "Flat out, we were wrong," he said simply to AI reporter James Vincent at *The Verge*, of the company's original commitment to transparency. "If you believe, as we do, that at some point, AI—AGI—is going to be extremely, unbelievably potent, then it just does not make sense to open-source." There were commercial considerations as well, he plainly noted. "GPT-4 is not easy to develop. It took pretty much all of OpenAI working together for a very long time to produce this thing. And there are many many companies who want to do the same thing."

As OpenAI grew and Sutskever's profile rose, his lack of a filter would at times turn into a liability. He was no longer speaking to just researchers; his audience had expanded to the general public. But ever the same, he didn't adapt his messaging. He made statements using his signature bullish confidence, now lacking significant context for any layperson listening. "it may be that today's large neural networks are slightly conscious," he tweeted in 2022, even as other researchers warned that such rhetoric could fan popular misunderstandings of the technology. One DeepMind scientist specialized in the study of cognition and consciousness replied in the comments, ". . . in the same sense that it may be that a large field of wheat is slightly pasta." The following year, Sutskever would induce panic by proclaiming at a conference that AGI would eventually disappear all jobs. That fall, he would declare, without scientific backing, on X, "In the future . . . we will have *wildly effective* and dirt cheap AI therapy," after an OpenAI leader triggered online controversy for casually comparing talking to ChatGPT with professionally licensed therapy.

What drew people to follow Sutskever was his reputation and his seniority. Many employees at OpenAI were well aware of his earlier contributions to the field; some saw him as something of a prophet. Over time, as OpenAI grew more successful, Sutskever would act more and more like one. At all-hands meetings, he would get up in front of the company, take a deep breath, and walk back and forth for dramatic effect before delivering vague motivational messages. During one virtual meeting in September 2020, his eyes glazed over as he stared in the distance and painted a science fiction–like vision of the future that was possible. Outside in the Bay Area, the sky had turned orange from nearby forest fires. "It was surreal because it already felt like the apocalypse," a researcher remembers.

Shortly after ChatGPT's release in late 2022, OpenAI would host a holiday party at the California Academy of Sciences. Sutskever would get up in front of the crowd, wearing an OpenAI shirt and black blazer, to give some short remarks with Brockman. At the end of it, Sutskever, still wiry as ever and now balding, delivered what had become his new mantra. "Feel the AGI," he said. "Feel the AGI."

Following Sutskever's philosophy of scaling simple neural networks, the question in the early days of OpenAI became: Scale which one? Different researchers proposed and tinkered with different options, but none of the neural networks that had gained widespread traction within the field seemed to fit the bill.

In August 2017, that changed with Google's invention of a new type of neural network known as the Transformer. Transformers excel at picking up long-range patterns. Think back to the limited predictive text capabilities on iPhones in the early days and the memes they spawned for producing babbling, incoherent sentences. These were the product of short-range pattern analysis—the neural network looking at each word only in relation to the words directly around it. Transformers can ingest large volumes of text and consider each word, sentence, and paragraph in a significantly larger context. Google saw the Transformer as a way to improve its search engine and Google Translate as well as its other services based heavily on language processing. Sutskever saw it for something else. Transformers are simple and scalable neural networks, an example of what he was looking for. He began evangelizing them around the office.

Sutskever's push struck some researchers as odd. "It felt like a wack idea," remembers Yilun Du, an MIT researcher who started at OpenAI as a fellow around this time. "Transformers felt like a niche architecture." But Sutskever, who had focused his PhD thesis with Hinton on the predecessor to Transformers, recognized their potential for taking deep learning to the next level. Others at OpenAI were just as excited. A smattering of researchers began testing it out, including Alec Radford, a dropout from Olin College of Engineering in the greater Boston area with brilliant technical abilities. He began hacking away on his laptop, often late into the night, to scale Transformers just a little and observe what happened.

Radford trained Google's neural network on a dataset of over seven thousand unpublished English-language books ranging from romance to adventure, which he pulled from a dataset that other AI researchers had previously compiled and open-sourced for a different project. While experimenting, he made a fateful decision to change the task that the Transformer had to learn. Instead of translating languages, as Google had been using Transformers for, he switched it to learn text generation by predicting the most probable next word in a sentence. Early on, OpenAI researchers had hypothesized that generative models would be an important step to reaching AGI. The company explained in a blog post in heavily anthropomorphized terms that the situation was akin to a famous quote from theoretical physicist Richard Feynman: "What I cannot create, I do not understand." Sutskever had a different way of framing it internally: Training a model to generate something convincing would force it to compress data about the world into its essence. "Intelligence is compression," Sutskever would say, elaborating in a 2016 memo his strong belief that compression was in fact the *only* thing needed to achieve artificial general intelligence. In more concrete terms, Radford discovered that giving the algorithm the simple goal of producing convincing text through next-word-prediction did indeed make it pick up the nuances and structure of English at a deeper level.

During one of Musk's visits to the office, Radford demoed early progress on his work. The model was generating poor-quality text, and Musk was wholly unimpressed. At first, Radford felt deflated. But after pursuing it further, he was surprised by the results. The Transformer had improved quickly and performed much better on a range of language processing tasks, such as summarizing or answering questions about a document, than anything else he had tried before.

In 2018, OpenAI released the first version of that model, called Generative Pre-Trained Transformer, later nicknamed GPT-1. The second word in the name—*pre-trained*—is a technical term within AI research that refers to training a model on a generic pool of data as a prerequisite for it to learn more specific tasks later. GPT-1, in other words, had been trained on a generic pool of English to create a rough approximation of how the language worked. The model could then be "fine-tuned," or specialized, later by training it on a much more tailored dataset—say, Shakespeare plays to teach it how to generate Shakespeare-esque prose. GPT-1 barely received any attention. But this was only the beginning. Radford had validated the idea enough to continue pursuing it. The next step was more scale.

Radford was given more of the company's most precious resource: compute. His work dovetailed with a new project Amodei was overseeing in AI safety, in line with what Nick Bostrom's *Superintelligence* had suggested. In 2017, one of Amodei's teams began to explore a new technique for aligning AI systems to human preferences. They started with a toy problem, teaching an AI agent to do backflips in a virtual video game–like environment. The agent was a simulation of a T-shaped stick, with three joints along the shaft. Instead of giving it the objective of learning backflips directly, the team taught the agent by giving it feedback: They hired contractors to watch the agent as it randomly twisted and turned about the environment; periodically, the contractors would then be asked to compare two video clips of the agent's actions and select which one better resembled a backflip. Around nine hundred comparisons later, the T-shaped stick was successfully bunching up at its joints and flipping over. OpenAI touted the technique in a blog post as a way to get AI models to follow difficult-to-specify directions. The researchers on the team called it "reinforcement learning from human feedback."

Amodei wanted to move beyond the toy environment, and Radford's work with GPT-1 made language models seem like a good option. But GPT-1 was too limited. "We want a language model that humans can give feedback on and interact with," Amodei told me in 2019, where "the language model is strong enough that we can really have a meaningful conversation about human values and preferences."

Radford and Amodei joined forces. As Radford collected a bigger and more diverse dataset, Amodei and other AI safety researchers trained up progressively larger models. They set their sights on a final model with 1.5 billion parameters, or variables, at the time one of the largest models in the industry. The work further confirmed the utility of Transformers, as well as an idea that another one of Amodei's teams had begun to develop after their work on OpenAI's Law. There wasn't just one empirical law but many. His team called them collectively "scaling laws."

Where OpenAI's Law described the pace at which the field had previously expanded its resources to advance AI performance, scaling laws described the relationship between the performance of a deep learning model and three key inputs: the volume of a model's training data, the amount of compute it was trained on, and the number of its parameters. Previously, AI researchers had generally understood that increasing these inputs somewhat proportionally to one another could also lead to a somewhat proportional improvement in a model's capabilities. Amodei and his team's surprising observation was that the relationship between each of these inputs as well as the model's performance on a specific, measurable task, such as next-word-prediction, could be described by a smooth curve. In other words, it was possible to estimate with high accuracy how much data, how much compute, and how many parameters to use to produce a model with a desired level of performance on a discrete capability tightly correlated with next-word-prediction—say, fluency in text generation. For capabilities less but still somewhat correlated, increasing these inputs should also lead to better performance.

The cluster of models that OpenAI trained leading up to the final 1.5-billion-parameter version illustrated this relationship. Each one fell neatly on a curve of increasing capability. So it was little surprise when the largest one, which they named GPT-2, markedly improved over the juvenile text generation of GPT-1 to produce lengthy and coherent-enough prose to be confused with a human's. Compared with today's models, the text was clunky and often descended into gibberish. But for the very first time, it was suddenly possible to automate writing at scale.

What was a darker surprise to the team was the content that GPT-2 was producing with its new coherence. Fed a few words like *Hillary Clinton* or *George Soros*, the chattier language model could quickly veer into conspiracy theories. Small amounts of neo-Nazi propaganda swept up in its training data could surface in horrible ways. The model's unexpected poor behavior disturbed AI safety researchers, who saw it as foreshadowing of the future abuses and risks that could come from more powerful misaligned AI. After GPT-2 generated a tirade against recycling ("Recycling is NOT good for the world. It is bad for the environment, it is bad for our health, and it is bad for our economy."), one AI safety researcher printed out a copy and posted it, part joke, part warning, above the recycling bins in the office.

In another instance, someone prompted GPT-2 to create a reward scale for small children for finishing homework and doing their chores. When GPT-2 suggested using candy, it once again disturbed some AI safety people who remarked that this was a tactic of pedophiles. A European employee was taken aback by the association. "My mom definitely did this. Sundays in the summer was ice cream if you do your chores," he remembers. He wondered if the hypersensitivity was somehow an American thing. It was one of many moments that made him question the basic premise of OpenAI's lofty goals: How could it benefit all of humanity when it lacked meaningful global representation? Even as a European coming from a highly overlapping culture to the US, he often felt alienated by the overwhelming bias in AI safety and other discussions toward American values and American norms.

GPT-2 started a debate within the company. Had OpenAI reached the point when it was time to start withholding research? The charter had accommodated for this possibility. Amodei, who had by then been promoted to director of research, and Jack Clark, who headed policy and worried in his own way about existential and other dangerous risks, took point on deciding a way forward. They ran an internal survey and held several "information hazard" meetings to discuss possible abuses of the technology. If GPT-2 fell into the hands of terrorists, dictators, or clickbait farms, they reasoned, the model could be used for nefarious purposes. And though it didn't seem existentially risky this time, future models would only grow more powerful, and that likelihood would get higher. It was better to set a precedent for withholding research early. OpenAI, they decided, should not release the full version.

Jack Clark, a former journalist, had been the director of OpenAI's strategy and communications before transitioning fully in late 2018 to cultivating its budding policy presence. He took regular trips to Washington and relished being the go-to AI guy for policymakers. He'd tell them, "I'm like AI Wikipedia," and would introduce his "bias," as he called it, coming from OpenAI: "We want a stable policymaking environment for advanced tech that operates over multiple political administrations because the mission we have is not going to get done in a presidential cycle."

After recounting this to me, he added, "We've been very lucky that policymakers give us quite a lot of time, because I think it's clear that basically for stuff to go well, we just want them to have more information, and we also want them to have more means to generate their own information."

Clark began a media offensive in February 2019, broadcasting widely to various publications that OpenAI had created a dangerous technology, and therefore was not releasing it. Instead it would release only a smaller version, with 8 percent of the full-fledged model's parameters, to give the public a taste of its capabilities. He, Amodei, and several others coauthored a blog post with examples of GPT-2's outputs to illustrate its full potential. "It's very clear that if this technology matures—and I'd give it one or two years—it could be used for disinformation or propaganda," he said to my then colleague at *MIT Technology Review* Will Knight. Clark sidestepped the fact that OpenAI was the one leading the push to mature the technology on that timeline. "We're trying to get ahead of this," he said.

OpenAI's move sparked intense blowback from external researchers, who adhered strictly to the idea that open science was the bedrock of the field. Any organization that didn't participate should be viewed suspiciously. More so if they were publicly boasting about the decision. Many also viewed OpenAI's alarmism about what was essentially powerful auto-complete software as poorly calibrated and ridiculous. GPT-2 was not nearly advanced enough to be a serious threat; and if it were, why tell everyone about it and then preclude it from public scrutiny? The whole thing felt disingenuous and like a self-aggrandizing publicity stunt. At Stanford, after Radford gave a talk about GPT-2, a well-established natural language processing professor would raise his hand to ask the last question. "So, is it *dangerous*?" he taunted. The room burst out in laughter. "Alec looked so sad," remembers a Stanford researcher in the room. "Stanford had so much contempt for OpenAI."

Within OpenAI, many researchers also chafed against Amodei and Clark's decision. For those who didn't share the pair's views on catastrophic risks, both their ruling and the subsequent media circus felt somewhat baffling. Even for those who did, some still questioned the soundness of the pair's judgment. "It was a mistake to make such a big deal out of it," one AI safety researcher told me. "It felt like crying wolf."

In the immediate aftermath of the blowback, Clark paced up and down the office with manic energy on call after call, working to regain control of the situation. He brushed off the controversy. "We're breaking with norms, and that creates a lot of different views," he later told me during my office visit. Sooner or later all organizations conducting cutting-edge AI research would have to be more selective about what to publish, he said. OpenAI was taking the lead in trialing what that process could look like to not be caught flat-footed. "If we're right, and it is possible to build AGI," Clark said, "we sure as shit need really good information-hazard procedures."

And where researchers may not have liked OpenAI's maneuver, policymakers did, he added. Many DC types viewed the open culture in AI research as threatening. OpenAI's willingness to go against the grain had gained it more trust in Washington.

But behind the scenes, the leadership team also understood that the animosity from the research community wasn't viable in the long run. The lab was struggling with compounding reputational challenges. What with its wild claims about AGI, over-the-top approach to GPT-2 and other marketing, and now, in early 2019, its newly announced Frankenstein structure, it was being criticized left and right, and being viewed with more and more skepticism from top researchers in the field. Combined with the fact that its equity didn't yet mean anything, it was still having trouble hiring and retaining talent. Employees wondered whether external candidates were securing offers from OpenAI simply to use as leverage for negotiating higher offers with Google or DeepMind. OpenAI needed to find a way to legitimize itself as a research organization.

This was frequently discussed at lunches and in company meetings, as well as in an internal document called "Research Community Outreach Brainstorming." Under a section titled "Strategy," it read, "Explicitly treat the ML community as a comms stakeholder," using the abbreviation for machine learning. "Change our tone and external messaging such that we only antagonize them when we intentionally choose to." The document also acknowledged how a poor research reputation would ultimately undermine OpenAI's influence in Washington. "In order to have government-level policy influence, we need to be viewed as the most trusted source on ML research and AGI," it read under "Policy." "Widespread support and backing from the research community is not only necessary to gain such a reputation, but will amplify our message."

Clark's team formulated a new plan: a staged release. Instead of withholding GPT-2 permanently, OpenAI would publish the progressively larger models that it had developed at staggered intervals and then, if all went well, release the full 1.5-billion-parameter version. This would allow OpenAI and others to gradually observe and address any emerging consequences, the team said, as well as give the lab time to partner with other organizations to research the risks between stages.

Clark emphasized to his team the importance of building an ecosystem through those partnerships. Working with high-profile institutions would help foster more cooperation between industry and academia for addressing AI safety risks. It would also get broader buy-in into OpenAI's efforts to shift research release norms and simultaneously help burnish the lab's reputation. His team reached out to AI and security researchers at a select few organizations and gave them early access to the full version of GPT-2 to test its potential for harmful applications. Clark instructed his team to get "the strongest endorsement" they could from each researcher's organization so OpenAI could name not just the individuals but also their institutions as partners. The team then prepared a white paper touting its release strategy and highlighting those partnerships. They sought to frame OpenAI as a leader by listing examples of organizations that had also deviated from immediately releasing their research after GPT-2.

The work paid off. Before long, it had seeded conversations across industry groups and policy think tanks about withholding research as a responsible approach to managing AI safety risks. In late 2020, Clark would be among the people who would break off from OpenAI with the Amodei siblings to cofound Anthropic. Until that time, his work at OpenAI would help establish its influence and lay the groundwork for its sprawling policy ambitions.

OpenAI began to keep a road map to systemize its research. Amodei treated it like an investor: He called it having "a portfolio of bets." He and other researchers kept tabs on different ideas within the field, born out of different philosophies about how to achieve artificial general intelligence, and advanced each one through small-scale experimentation. Those that seemed promising, OpenAI would continue. Those that didn't pan out, it would abandon.

The project to win the *Dota 2* video game championship was one area that Amodei believed no longer had much utility. The *Dota 2* team had beat its opponents and achieved its goal in April and helped secure Microsoft's investment. It had also helped some people gain new confidence in the company's scaling strategy. The project, as he saw it, had run its course. The *Dota 2* team disbanded.

Where Amodei did see continued promise was in GPT-2. It represented a bet known in the field as the "pure language" hypothesis. Language, the theory goes, is the primary medium through which humans communicate, meaning all of the world's knowledge must at some point be documented in text. It follows then that AGI should be able to emerge from training an algorithm on massive amounts of language and nothing else. This idea is in contrast to the "grounding" hypothesis, which asserts that the physical world and our ability as humans to perceive and interact with it is just as crucial an ingredient to our intelligence. AGI would then only be able to emerge from the combination of language and perception, like computer-vision, as well as interaction, such as through a physical or virtual agent taking actions in the real world.

In company documents, researchers weighed the merits of the different approaches, with AI safety staff at one point debating the virtues of the "pure language" hypothesis by drawing repugnant analogies to people with disabilities. The discussions revealed how quickly measures of intelligence could veer into disturbing assessments of which groups of people had superior or inferior intelligence.

"Language of some form is the difference between a feral human and human in society. Example, Helen Keller," read the document under the heading "Some initial arguments for the centrality of language." In the margins, AI safety researchers continued their arguments for and against "pure language" through threaded comments.

"Also blind people are about as capable as sighted people," wrote one researcher, as evidence that "grounding" through vision seemed unnecessary.

"Blind people seem at a significant economic disadvantage," replied another, citing statistics from the National Federation of the Blind that over 70 percent of vision-impaired adults did not work full-time.

"Blind people are still way more capable than chimpanzees," replied a third. "There exist very impressive blind people."

Many at OpenAI had been pure language skeptics, but GPT-2 made them reconsider. Training the model to predict the next word with more and more accuracy had gone quite far in advancing the model's performance on other seemingly loosely related language processing tasks. It seemed possible, even plausible, that a GPT model could develop a broader set of capabilities by continuing down this path: pushing its training and improving the accuracy of its next-word-prediction still further. Amodei began viewing scaling language models as—though likely not the only thing necessary to reach AGI—perhaps the fastest path toward it. It didn't help that the robotics team was constantly running into hardware issues with its robotic hand, which made for the worst combination: costly yet slow progress.

But there was a problem: If OpenAI continued to scale up language models, it could exacerbate the possible dangers it had warned about with GPT-2. Amodei argued to the rest of the company—and Altman agreed—that this did not mean it should shy away from the task. The conclusion was in fact the opposite: OpenAI should scale its language model as fast as possible, Amodei said, but not immediately release it. GPT-2 had demonstrated how easy it would be for other actors to obtain more powerful AI capabilities; in fact, two graduate students had already created an open-source version of GPT-2 before OpenAI had released its own full version. It was only a matter of time before other people would start scaling up language models further. That meant the best way to ensure beneficial AGI was for OpenAI to leap ahead and, with the internal lead time, figure out how to make its scaled model safer. Once it was time to reveal the model, its extra polish and refinement would help establish AI safety norms, in the same way the initial withholding of GPT-2 shifted norms for releasing research.

With a version of GPT-2 now out in the world, there was also evidence that the dangers of pure language models weren't all that bad. As far as OpenAI knew, it hadn't been used in coordinated mass disinformation campaigns—and such campaigns were certainly better than the potential existential risks of AGI.

"Obviously misuse is not good," Amodei told me. "But a language model is a lot less powerful than an AGI. I'm very worried about language models being weaponized for disinformation and this sort—that is very scary to me—but at the same time, it's a relatively singular and clear and defined concern."

From Amodei's view, in other words, scaling GPT-2 was not only potentially the fastest path to advance to AGI but also one whose possible risks along the way would be relatively contained to those he viewed as manageable—mis- and disinformation, as opposed to catastrophe. It would give OpenAI a safer testing ground to experiment with a powerful, but not *so* powerful, AI system, and work out various kinks, including with releasing it.

"What is AGI? What does AGI look like?" Amodei said. "Well, you know, we're in the awkward position of, we don't know what it looks like. We don't know when it's going to happen. So we look for things that aren't AGI but that present at least some of the opportunities and difficulties of AGI. And the hope is if we can handle those things well, then we're kind of, like, ready for the bigger leagues."

It was a logic that worked under a specific assumption: that AGI, despite being amorphous and unknowable, was also inevitable. OpenAI would repeatedly justify its behaviors against variations of the same argument for years after. Under the specter of AGI's unstoppable arrival, the company needed to keep developing more and more powerful models to prepare itself and to prepare society. Even if those models carried with them their own risks, the experience they offered to prevent or face possible AI apocalypse made those risks bearable.

As ChatGPT swept the world by storm in early 2023, a Chinese AI researcher would share with me a clear-eyed analysis that unraveled OpenAI's inevitability argument. What OpenAI did never could have happened anywhere but Silicon Valley, he said. In China, which rivals the US in AI talent, no team of researchers and engineers, no matter how impressive, would get $1 billion, let alone ten times more, to develop a massively expensive technology without an articulated vision of exactly what it would look like and what it would be good for. Only after ChatGPT's release did Chinese companies and investors begin funding the development of gargantuan models with gusto, having now seen enough evidence that they could recoup their investments through commercial applications.

Through the course of my reporting, I would come to conclude something even more startling. Not even in Silicon Valley did other companies and investors move until after ChatGPT to funnel unqualified sums into scaling. That included Google and DeepMind, OpenAI's original rival. It was specifically OpenAI, with its billionaire origins, unique ideological bent, and Altman's singular drive, network, and fundraising talent, that created a ripe combination for its particular vision to emerge and take over. "I get the sense that Sam is the most ambitious person on the planet," a former employee says. In other words, everything OpenAI did was the opposite of inevitable; the explosive global costs of its massive deep learning models, and the perilous race it sparked across the industry to scale such models to planetary limits, could only have ever arisen from the one place it actually did.

For the Gates Demo in April 2019, OpenAI had already scaled up GPT-2 into something modestly larger. But Amodei wasn't interested in a modest expansion. If the goal was to increase OpenAI's lead time, GPT-3 needed to be as big as possible. Microsoft was about to deliver a new supercomputer to OpenAI as part of its investment, with ten thousand

Nvidia V100s, what were then the world's most powerful GPUs for training deep learning models. (The *V* was for Italian chemist and physicist Alessandro Volta.) Amodei wanted to use all of those chips, all at once, to create the new large language model.

The idea seemed to many nothing short of absurdity. Before then, models were already considered large-scale if trained on a few dozen chips. In top academic labs at MIT and Stanford, PhD students considered it a luxury to have ten chips. In universities outside the US, such as in India, students were lucky to share a single chip with multiple peers, making do with a fraction of a GPU for their research.

Many OpenAI researchers were skeptical that Amodei's idea would even work. Some also argued that a more gradual scaling approach would be more measured, scientific, and predictable. But Amodei was adamant about his proposal and had the backing of other executives. Sutskever was keen to play out his hypothesis of scaling Transformers; Brockman wanted to continue raising the company's profile; Altman was pushing to take the biggest swing possible. Soon after, Amodei was promoted to a VP of research.

Behind the scenes, Altman was also attuned to another factor: Microsoft's $1 billion investment came with $1 billion expectations; OpenAI was on the clock to deliver something that would justify the expense. Where Amodei saw a larger language model as a necessary prerequisite for AI safety research, Altman saw its potential for fulfilling OpenAI's promise to Microsoft.

In the coming months, Amodei and Altman would clash over how and when to release GPT-3; Altman would win out, pushing the model into the world on an accelerated timeline. Years before ChatGPT, these two decisions—the one to explode GPT-3's size and the one to quickly release it—would change the course of AI development. It would set off a rapid acceleration of AI advancement, sparking fierce competition between companies and countries. It would fuel an unprecedented expansion of surveillance capitalism and labor exploitation. It would, by virtue of the sheer resources required, consolidate the development of the technology to a degree never seen before, locking out the rest of the world from participating. It would accelerate the vicious cycle of universities, unable to compete, losing PhD students and professors to industry, atrophying independent academic research, and spelling the beginning of the end of accountability. It would amplify the environmental impacts of AI to an extent that, in the absence of transparency or regulation, neither external experts nor governments have been able to fully tabulate to this day.

But all this was yet to pass. In the fall of 2019, Amodei assembled a team, called Nest, of mostly other AI safety researchers, intent on keeping careful control of GPT-3's development within the company. With that, the team began its aggressive push to scale.

GPT-3 was effectively the same model as GPT-2, fed massively more data and compute to be so much bigger that the outcome would appear to many as beyond a difference of degree to a difference in kind. But using ten thousand chips posed new problems. There was always a small probability that any chip might crash in the middle of training, the same way a laptop might crash when there are too many windows open. If one chip crashed, everything did, meaning training would need to start all over. The probability of a single chip crashing compounded significantly across ten thousand GPUs. Such an error would be enormously costly—in both money and time—when the Nest team expected training to take several months at a minimum.

To fix the problem, the team needed a way to make sure model training could restart exactly where it left off after any disruptions. It also needed to determine a strategy for how to spread the training across all ten thousand chips, a process known as sharding: Was it better, say, to chop up the model into tens, hundreds, or thousands of pieces, with each piece training on separate clusters of GPUs before being merged?

Then there was a challenge with the data. To get the best performance, the size of the dataset needed to grow proportionally with the number of parameters and the amount of compute. If there were too many parameters and not enough data, the model could start regurgitating word for word the lines in its training data, effectively rendering it

SCALE OF AMBITION

useless. For GPT-2, Radford had been selective about what made it into the data. He scraped the text from articles and websites that had been shared on Reddit and received at least three upvotes on the platform. This had produced a forty-gigabyte trove of some eight million documents, which he named WebText.

That wasn't nearly enough for GPT-3. So Nest expanded the data by adding an even broader scrape of links shared on Reddit as well as a scrape of English-language Wikipedia and a mysterious dataset called Books2, details of which OpenAI has never disclosed, but which two people with knowledge of the dataset told me contained published books ripped from Library Genesis, an online shadow repository of torrented books and scholarly articles. In 2023, the Authors Guild and seventeen authors, including George R. R. Martin and Jodi Picoult, would sue OpenAI and Microsoft alleging mass copyright infringement. OpenAI would respond in March 2024 by saying it had deleted those datasets and had stopped using them for training after GPT-3.5, which by that time had already been deprecated.

This was still not enough data. So Nest turned finally to a publicly available dataset known as Common Crawl, a sprawling data dump with petabytes, or millions of gigabytes, of text, regularly scraped from all over the web—a source Radford had purposely avoided because it was such poor quality. In an effort to tame the trash in the data, the Nest team trained a machine-learning model to find the samples within Common Crawl that looked most like articles on Wikipedia. If it looked like Wikipedia, the idea was, it would be more likely to match Wikipedia quality. They also included some samples in languages other than English, though they ultimately accounted for only 7 percent of the data. Still, when training the model, the researchers weighted the filtered Common Crawl data as the lowest priority. GPT-2, in other words, had been peak data quality; it declined from there.

When it came time to assemble the data for GPT-4, released two years later, the pressure for quantity eroded quality even further. The filter was removed from the Common Crawl data and most of it poured in. Through its partnership with Microsoft, OpenAI also received a full download of GitHub, the Microsoft-owned online code repository. When this still wasn't enough, OpenAI employees also gathered whatever they could find on the internet, scraping links shared on Twitter, transcribing YouTube videos, and cobbling together a long tail of other content, including from niche blogs, existing online data dumps, and a text storage site called Pastebin. Anything that didn't have an explicit warning against scraping was treated as available for the taking.

Within Google, some researchers lamented OpenAI's willingness to take legal risks to gather data as giving them a major advantage. Google was a lot more conservative about data access and usage and had a rigorous protocol for complying with regulations including Europe's data privacy law, colloquially known as the GDPR. Google's commitment to compliance, ironically, gave OpenAI easier access to Google's data than Google itself. Where OpenAI readily scraped and transcribed videos from Google-owned YouTube, Google researchers had to maneuver through significant internal red tape to abide by YouTube's restrictive license on its user-uploaded content. OpenAI was unconcerned—or in tech startup terms, "unburdened"—by this compliance. It was a classic mindset in Silicon Valley, where founders and investors espouse the mantra that startups could and should move into legal gray areas (think Airbnb, Uber, or Coinbase) to disrupt and revolutionize industries.

The decision to lower quality barriers—and then effectively drop them altogether—would have sweeping downstream effects on the human labor behind AI systems. For years, the tech industry had relied on poorly paid workers in precarious economic conditions to perform essential data preparation tasks for its AI models, such as categorizing text and labeling images. Soon after GPT-3 normalized the use of giant, poorer quality datasets, the demands for the work shifted from the handling of largely benign content to frequently disturbing content, including for the purposes of content moderation, much like social media before it. Such moderation was necessary to prevent generative AI systems from reproducing the most vile parts of their all-encompassing datasets—descriptions and depictions of violence, sexual abuse, or self-harm—to hundreds of millions of users.

"There's a big paradigm shift in how you control the output of these models," says Ryan Kolln, the CEO and managing director of Appen, a platform for connecting Silicon Valley companies with data workers. "In a traditional AI sense, you control the output by constraining the inputs"—the kinds of data filtering that Radford's team did—"because it only learns from the examples that you are giving it. The challenge with generative AI is the inputs are the entire corpus of humanity. So you need to control the outputs."

In a 2023 paper, Abeba Birhane and her coauthors would introduce the concept of "hate scaling laws" to critique the premise of training deep learning models on unfiltered data, or what they called "data-swamps." They analyzed two publicly available image-and-text datasets used to train open-source image generators, LAION-400M and LAION-2B-en, both pulled from Common Crawl, with four hundred million and two billion images, respectively. They showed that the amount of hateful and abusive content scaled with the size of the dataset and exacerbated the discriminatory behaviors of the models trained on them. Models trained on the two billion images, for example, were five times more likely than models trained on the four hundred million images to label Black male faces as criminals. Later that year, a Stanford study analyzing LAION-5B, a dataset with five billion images used to train Stable Diffusion, would discover it contained thousands of images of verified and suspected child sexual abuse.

Among its tactics to control the outputs, OpenAI would hire workers in Kenya for on average less than two dollars an hour to build an automated content-moderation filter, a revelation first reported by *Time* magazine correspondent Billy Perrigo. It would also employ over a thousand other contractors globally to perform reinforcement learning from human feedback, or RLHF, the technique it had developed to teach an AI agent backflips, on its language models, including prompting the models repeatedly and scoring the answers, in an effort to tame the model as much as possible.

Hito Steyerl, a German artist and filmmaker who produced a documentary on Syrian refugees who perform data work, echoed Birhane's critique in the observations she shared with me. Psychologically harmful material accumulates when mass surveillance is the basis for data collection, she said. To fix the problem, we have to return to its root: questioning what is really in the data, questioning the whole premise of its wide-scale, indiscriminate seizure.

II

II

Chapter 6

Ascension

Early in his career, Altman observed that new CEOs only succeeded if they "refounded" the company. He did this with conviction at YC when he inherited the presidency. He created new programs, including a new fund, to expand the accelerator's support for startups at different stages. He moved into hard technologies—those that required ambitious scientific innovation, including nuclear fusion, quantum computing, and self-driving cars. Already a prestigious name brand, YC's sphere of influence grew from a couple hundred companies to thousands a year, turning it into a center of gravity in Silicon Valley. "The thing that I'm most proud of is we really built an empire," Altman said after stepping down as president.

The end of his YC era marked the start of his new era at OpenAI. In March 2019, as he transitioned to OpenAI full time, he quickly brought with him the same aggressive mindset that he'd used at YC. He didn't want OpenAI to be among the world's leading AI organizations; he wanted it to be the only one. For years, Altman had taught other founders through YC and other forums to model the startup game as a winner-takes-all competition. If a startup had any hope of succeeding, he told them, they had to move swiftly and relentlessly to beat and then continue to beat back their rivals.

The magic number he often used was ten, stemming from Thiel's monopoly strategy. "My sort of crazy, somewhat arbitrary rule of thumb is you want to have a technology that's an order of magnitude better than the next best thing," Thiel had said during his 2014 lecture to Altman's startup course at Stanford. Amazon, for example, had figured out how to sell 10x more books than brick-and-mortar bookstores. PayPal, his own company, had figured out how to send payments 10x faster than clearing checks. "You want to have some sort of very powerful improvement, maybe an order of magnitude improvement, on some key dimension," Thiel said.

Ten became Altman's round number for everything. Startups not only needed to break into the market with 10x better technology, he'd advise, they also needed to improve it 10x with every generation. The speed with which they hit each new generation was another key variable that could make or break them. "If your iteration cycle is a week and your competitor's is three months, you're going to leave them in the dust," he said in 2017 to a class of aspiring entrepreneurs.

At YC, Altman pushed his fellow partners to keep growing the number of companies it funded by 10x. "And we will, over time, figure out how to get another 10x and then another 10x after that," Altman later said of his strategy at an event. "Someday we will fund all the companies in the world." "Sam was the first person I ever heard say that, because of the work the original founders had done, and because of the brand that YC had created, we were in fact a de facto monopoly in this space," says Geoff Ralston, Altman's YC successor.

At OpenAI, Altman planned to use the same strategy. In a memo he sent to the company in late 2019 to articulate his long-term vision, he emphasized that OpenAI needed to "be number one" in four categories by the end of 2020: technical results, compute, money (to acquire more compute), and preparation, meaning the safety and security of the organization as well as its resilience to high-stress situations.

The most important of these was the first one, he said. If OpenAI wanted a chance at fulfilling its mission, it needed to build beneficial AGI first, or be such a leader that it could still shape AGI development. "Though we in theory could slow down capability work," he wrote, referring to advancing technical results, "given the rate of progress other people are making, we likely are required to move very quickly on technical progress if we want to have a lot of influence over AGI." This would only become increasingly true as more and more competitors caught on to OpenAI's strategy and moved into the space.

"We still need many more 10x leaps to get to AGI," he added later in the memo. "We should always work towards dramatic results, not incremental improvements."

Crucial to this success formula were several other considerations. It would be paramount for OpenAI to keep Microsoft happy to maintain the lead in compute. If OpenAI was successful, Microsoft had agreed that it would give far more than $1 billion. "We would like Microsoft to be our major partner all the way through," Altman said. "They have the capability of delivering us, for the next 5 years at least, the most powerful supercomputers in the world." This meant shifting away from the days of OpenAI's freewheeling academic research environment and toward focused commercialization efforts to deliver Microsoft benefits. If OpenAI had other research projects it wanted to pursue, it would then have the resources. "To paraphrase that famous Disney quote," Altman wrote, "we should make more money so that we can do more research, not do more research so that we can make more money."

Additionally, the company needed to start pulling back on transparency. "The infohazard risk of talking about AGI will keep getting higher as we make more progress," Altman argued. It was time to restrict research publications and model deployments, adopt a stricter confidentiality policy, and reveal progress on only narrow skills rather than more general AI advancements. Separately, everyone also needed to begin acting under the assumption that "every decision we make and every conversation we have ends up investigated and reported on the front page of The New York Times."

That said, "it still seems very important that the world thinks we are winning at something," he said. This would make "key influencers in the world" more "willing to go well out of their way to help us," and make global policymakers "at the level of Presidents or their designees" come to OpenAI "for answers when they need to make big decisions." To that end, "we should probably plan to release at least one very impressive demonstration of progress each year."

Finally, the company needed to start acting with more seriousness and more unity. Altman included a quote from Hyman G. Rickover, an admiral in the US Navy, known as the "father of the nuclear navy" for his work building the world's first nuclear-powered submarines. It was a quote Altman had had painted on the office walls in the early days of OpenAI:

> I believe it is the duty of each of us to act as if the fate of the world depended on [them]. Admittedly, one [person] by [themself] cannot do the job. However, one [person] can make a difference. Each of us is obligated to bring [their] individual and independent capacities to bear upon a wide range of human concerns. It is with this conviction that we squarely confront our duty to prosperity. We must live for the future of the human race, not of our own comfort or success.

"Building AGI that benefits humanity is perhaps the most important project in the world," Altman wrote below the quote in the document. "We must put the mission ahead of any individual preferences.

"Low-stakes things should be low-drama, so we can save our high-drama capacity for high-stakes things (of which there will be many)."

Drama was in fact already brewing. Various little rifts that had bubbled up across the company were beginning to coalesce into big ones. Once quick to call each other friends, Brockman and the Amodei siblings were now butting heads on a growing list of issues. Among them, Dario Amodei's deprioritization of the *Dota 2* work had frustrated Brockman, who believed Amodei hadn't taken his contributions seriously. Where *Dota 2* was once the most compute-heavy project, Brockman also chafed against Amodei's centralization of compute for Nest's work on GPT-3. The Amodei siblings, meanwhile, found Brockman difficult to work with and were unwilling to let him join in on their language model development.

The tensions created a break among the leaders that slowly extended to the people who were loyal to each one in the company. During the *Dota 2* project, Brockman had forged a familial bond with some members of his team through the intense working hours, high stress, and a spur-of-the-moment retreat in Hawai'i, growing especially close to Jakub Pachocki and Szymon Sidor, two Polish scientists who were roommates and best friends. Amodei's AI safety teams, and the core members of the Nest team in particular, formed another contingent, bound together by their shared concern, in varying degrees, of rogue AI and existential or other extreme risks. They kept their work insulated from the rest of the company, creating private Slack channels and documents not accessible even to other executives. It frustrated many more people beyond Brockman as they felt similarly sidelined by the dwindling of their compute resources, along with their visibility into the company's core research.

Amodei's AI safety contingent, meanwhile, was also growing disquieted with some of Altman's behaviors. Shortly after OpenAI's Microsoft deal was inked, several of them were stunned to discover the extent of the promises that Altman had made to Microsoft for which technologies it would get access to in return for its investment. The terms of the deal didn't align with what they had understood from Altman. If AI safety issues actually arose in OpenAI's models, they worried, those commitments would make it far more difficult, if not impossible, to prevent the models' deployment. Amodei's contingent began to have serious doubts about Altman's honesty.

"We're all pragmatic people," a person in the group says. "We're obviously raising money; we're going to do commercial stuff. It might look very reasonable if you're someone who makes loads of deals like Sam, to be like, 'All right, let's make a deal, let's trade a thing, we're going to trade the next thing.' And then if you are someone like me, you're like, 'We're trading a thing we don't fully understand.' It feels like it commits us to an uncomfortable place."

This was against the backdrop of a growing paranoia over different issues across the company. Within the AI safety contingent, it centered on what they saw as strengthening evidence that powerful misaligned AI systems could lead to disastrous outcomes. One bizarre experience in particular had left several of them somewhat nervous. In 2019, on a model trained after GPT-2 with roughly twice the number of parameters, a group of researchers had begun advancing the AI safety work that Amodei had wanted: testing reinforcement learning from human feedback as a way to guide the model toward generating cheerful and positive content and away from anything offensive.

But late one night, a researcher made an update that included a single typo in his code before leaving the RLHF process to run overnight. That typo was an important one: It was a minus sign flipped to a plus sign that made the RLHF process work in reverse, pushing GPT-2 to generate *more* offensive content instead of less. By the next morning, the typo had wreaked its havoc, and GPT-2 was completing every single prompt with extremely lewd and sexually explicit language. It was hilarious—and also concerning. After identifying the error, the researcher pushed a fix to OpenAI's code base with a comment: Let's not make a utility minimizer.

In part fueled by the realization that scaling alone could produce more AI advancements, many employees also worried about what would happen if different companies caught on to OpenAI's secret. "The secret of how our stuff works can be written on a grain of rice," they would say to each other, meaning the single word *scale*. For the same reason, they worried about powerful capabilities landing in the hands of bad actors. Leadership leaned into this fear, frequently raising the threat of China, Russia, and North Korea and emphasizing the need for AGI development to stay in the hands of a US organization. At times this rankled employees who were not American. During lunches, they would question, Why did it have to be a US organization? remembers a former employee. Why not one from Europe? Why *not* one from China?

During these heady discussions philosophizing about the long-term implications of AI research, many employees returned often to Altman's early analogies between OpenAI and the Manhattan Project. Was OpenAI really building the equivalent of a nuclear weapon? It was a strange contrast to the plucky, idealistic culture it had built thus far as a largely academic organization. On Fridays, employees would kick back after a long week for music and wine nights, unwinding to the soothing sounds of a rotating cast of colleagues playing the office piano late into the night.

The shift in gravity unsettled some people, heightening their anxiety about random and unrelated incidents. Once, a journalist tailgated someone inside the gated parking lot to gain access to the building. Another time, an employee found an unaccounted-for USB stick, stirring consternation about whether it contained malware files, a common vector of attack, and was some kind of attempt at a cybersecurity breach. After it was examined on an air-gapped computer, one completely severed from the internet, the USB turned out to be nothing. At least twice, Amodei also used an air-gapped computer to write critical strategy documents, connecting the machine directly to a printer to circulate only physical copies. He was paranoid about state actors stealing OpenAI's secrets and building their own powerful AI models for malicious purposes.

"No one was prepared for this responsibility," one employee remembers. "It kept people up at night."

Altman himself was paranoid about people leaking information. He privately worried about Neuralink staff, with whom OpenAI continued to share an office, now with more unease after Musk's departure. Altman worried, too, about Musk, who wielded an extensive security apparatus including personal drivers and bodyguards. Keenly aware of the capability difference, Altman at one point secretly commissioned an electronic countersurveillance audit in an attempt to scan the office for any bugs that Musk may have left to spy on OpenAI.

To employees, Altman used the specter of US adversaries advancing AI research faster than OpenAI to rationalize why the company needed to be less and less open while working as fast as possible. "We must hold ourselves responsible for a good outcome for the world," he wrote in his vision document. "On the other hand, if an authoritarian government builds AGI before we do and misuses it, we will have also failed at our mission—we almost certainly have to make rapid technical progress in order to succeed at our mission."

Altman began to tighten the screws on security. Executives debated where to draw the new line: Should OpenAI act more like a Fortune 500 company protecting proprietary technologies or more like a government operation protecting highly classified state secrets? At a baseline, the executives agreed that they needed to lock down the model weights—the key information that could be used to replicate the fully trained versions of OpenAI's deep learning models. If stolen, that would be bad because it could both empower bad actors and handicap OpenAI's competitive advantage.

At first, without formal security staff, Altman deputized a member of the infrastructure team, which handled everything from the company's GPUs to the office internet, to think about solutions for preventing model theft—not just from corporate or state-sponsored spies but also from OpenAI's own employees. In cybersecurity, protecting against "insider threat" is relatively standard practice. Insiders could sabotage or steal OpenAI's IP intentionally; they could also be tricked into giving it up. In private, Altman acknowledged, after the point was raised, that someone like Sutskever could be vulnerable to the latter. The chief scientist was a logical target for bad actors: He was the archetype of a brainiac scientist who wasn't the most streetwise, and he ranked highly within the organization and had top access to information.

Sutskever had his own paranoias. As a star scientist in the cerebral and socially inept world of AI research, he had seen his share of obsessive fans and stalkerish behaviors. More than once, strangers had sought to sneak into OpenAI's office just to see him. Like Amodei, he also worried about the power of AI attracting the attention of unscrupulous governments and wondered whether those overeager to seek his advice were secretly foreign agents. He mused to colleagues what he should do if his hand got cut off to be used in a palm scanner for unlocking OpenAI's secrets. He wanted to hire less and keep a small staff in order to reduce the risks of infiltration. With Jakub Pachocki and Szymon Sidor, he proposed building a secure containment facility, a bunker with an air-gapped computer, that would hold OpenAI's model weights and prevent others from stealing them. The idea, which didn't make practical sense given that the models had to be trained first on Microsoft's servers, never got legs.

Hidden from view of most employees, digital security increased with the installation of corporate-monitoring software. In the background, enhancements were also made to physical security. The gates to the office parking lot were fortified. Within the office, several doors with keypads were programmed to have "distress passwords," special codes that could be punched in to trigger a secret alarm that would alert relevant security personnel of an in-person threat. Quotes were sought from vendors about how much it would cost to reinforce a server room to withstand a machine gun, though that idea was subsequently dropped.

In the vision memo, Altman noted the divisions that were developing in the company from the heightening stress. "We have (at least) three clans at OpenAI—to caricature-ize them, let's say exploratory research, safety, and startup." The Exploratory Research clan was about advancing AI capabilities, the Safety clan about focusing on responsibility, and the Startup clan about moving fast and getting things done.

Per Altman, each of these clans had important values that the company needed to preserve: the "we will pursue important new ideas even if we fail many times" of Exploratory Research; the "we will have an unwavering commitment to doing the right thing" of Safety; and the "we'll figure out a way to make it happen" of Startup. "We have to continue to avoid tribal warfare," he said. "To succeed, we need these three clans to unite as one tribe—while maintaining the strengths of each clan—working towards AGI that maximally benefits humanity."

Though Altman never name-checked anyone, employees read between the lines. Sutskever was the face of Exploratory Research; Amodei and his AI safety contingent focused on extreme risks constituted Safety; Brockman was the champion of Startup. Soon after, the pandemic hit, and everyone began working remotely, making it far easier for the clans to isolate themselves from one another.

Amodei pushed his team to move quickly. As they had done with GPT-2, they trained iteratively larger models in the ascension to a full ten-thousand-GPU model with 175 billion parameters, naming them alphabetically after scientists: *ada* for the smallest model, referring to English mathematician Ada Lovelace, widely credited as the first computer programmer; *babbage* for English inventor Charles Babbage, who conceived the first digital computer for which Lovelace would propose her program; *curie* for Polish French physicist and chemist Marie Curie, the first woman to win the Nobel Prize and win it twice; and *davinci* for Leonardo. The exercise was both to continue validating whether scaling laws still held at fundamentally larger scale and, more practically, to work gradually through the hardware and data challenges at each new level. On a regular basis, the Nest team would give the company an update on its progress, to growing excitement. "It's hard to overstate how insane that was to see," remembers one researcher. "I'd never seen anything like that in my life."

In parallel, Altman and Brockman developed a plan for commercialization. In late January 2020, Brockman began writing the first lines of code for an application programming interface, or API, for GPT-3. The API would give companies and developers access to the model's capabilities without giving them access to the model weights and allow them to incorporate the technology into their own consumer-facing products. The company split into two divisions. Mira Murati was promoted to VP of a new Applied division for overseeing the API and commercialization strategy. Under her, Peter Welinder, who had been leading the robotics team, was shifted to leading product; Fraser Kelton, who had cofounded an AI startup acquired by Airbnb, and Katie Mayer, who had worked at Leap Motion, were hired to respectively manage new product and engineering teams. Everyone not in Applied by association became the Research division.

That split deepened a fault line that Altman had identified. The formation of the Applied division brought in a small but growing group of people hailing from other startups that strengthened the Startup clan.

While the Exploratory Research clan viewed this with some ambivalence about whether OpenAI would become just another Silicon Valley product company, it triggered increasingly impassioned opposition from Amodei and his Safety clan also sitting within the Research division.

To many in Safety, releasing GPT-3 in short order via an API, or any other means, undermined the lead time—the whole point of the accelerated scaling—that OpenAI would have to perfect the safety of the model. The Applied division, whose entire purpose was to find early solutions for making money from OpenAI's technologies, which in their view required releasing them in the near term, disagreed. The API as they saw it also gave OpenAI the most controlled mechanism of any release strategy, allowing the company to be selective about whom to give access to and collecting invaluable data points for understanding how the model could be used or abused by people. In all-hands meetings, Altman played both sides: The API would ultimately help each group achieve what they wanted; bringing in some revenue would allow OpenAI to invest even more in AI safety research.

As GPT-3 finished training, employees began playing with the model internally. They tested the bounds of its capabilities and tinkered with the first version of the API. The company held a hackathon where employees riffed on different application ideas. But with every new prototype, tensions worsened. Where the Applied division, and many in Exploratory Research, viewed the demonstrations with mounting excitement, many in Safety saw them as yet further evidence that releasing the model without comprehensive testing and additional research could risk devastating outcomes.

One capability proved particularly polarizing: GPT-3's code-generation abilities. It hadn't been part of the Nest team's intentions, but in scraping links on Reddit and using Common Crawl for training data, they had captured scattered lines of code from engineers posting their programs on various online forums to ask questions or share tips, leading the model to have an increased facility for programming languages. The development thrilled many in Exploratory Research, just as it did the Applied division. Not only was it an impressive technical milestone, it also had potential as a tool to accelerate the company's productivity in AI research and to make GPT-3 into a more compelling product. For the same reason, some in Safety panicked. If an AI system could use its own code-generation skills to tweak itself, it could accelerate the timeline to more powerful capabilities, increase the risk of it subverting human control, and amplify the chances of extremely harmful or existential AI risks.

Sutskever and Wojciech Zaremba, one of the founding members whom Musk had pressed during a meeting, would subsequently form a team to create a model designed specifically for code generation. But during a meeting to kick off the project, the two learned that Amodei already had his own plans for developing a code-generation model and didn't see a need to merge efforts. Despite his concerns, Amodei believed, as with GPT-3, that the best way to mitigate the possible harms of code generation was simply to build the model faster than anyone else, including even the other teams at OpenAI who he didn't believe would prioritize AI safety, and use the lead time to conduct research on de-risking the model. Much to the confusion of other employees, the two teams continued to work on duplicate code-generation efforts. "It just seemed from the outside watching this that it was some kind of crazy *Game of Thrones* stuff," a researcher says.

The deadlock around releasing GPT-3 via the API continued until late spring. Safety continued to push for paramount caution based on fears of accelerating extreme AI risks, arguing for the company to delay its release as long as possible. The Applied division continued with preparing for the API launch, arguing that the best way to improve the model was for it to have contact with the real world. Around the same time, new concerns emerged from a third group of employees worried about the impact that spectacular text-generation abilities could have in the midst of major political, social, and economic upheaval in the US. By May 2020, the pandemic had already created a faster rise in unemployment than during the Great Recession. In the same month, Derek Chauvin, a police officer in Minneapolis, murdered George Floyd, a forty-six-year-old Black man, setting off massive Black Lives Matter protests around the country and the rest of the world. The team was also concerned about the impending US presidential election.

But rumors began to spread within OpenAI that Google could soon release its own large language model. The possibility was plausible. Google had published research at the start of the year about a new chatbot called Meena, built on a large language model with 1.7 times more parameters than GPT-2. The company could very well be working on scaling that model to roughly the size of GPT-3. The rumors sealed the deal for the API launch: If a model just as large would soon exist in the world, Safety felt less of a reason to hold back GPT-3.

In June, the company announced the API and set up an application form for people to request early access, prioritizing larger enterprises that the Applied division felt could be trusted to handle the technology responsibly. The company also maintained a big spreadsheet for employees to put down the names of anyone they wanted to jump the queue, including family, friends, and their favorite celebrities.

Google's rumored model never materialized. The tech giant had indeed begun working on a larger model than Meena, known as LaMDA, to produce a better chatbot—but it was still modestly smaller than GPT-3, and the company would ultimately decide not to release it until after ChatGPT. Google's executives determined that LaMDA didn't meet the company's ethical AI standards. Some employees also worried about repeating an infamous Microsoft scandal: In 2016, Microsoft had released an AI-powered chatbot known as Tay that quickly turned racist and misogynistic, and espoused support for Hitler, after users repeatedly prompted the chatbot to repeat inappropriate and offensive things. The GPT-3 API release wouldn't be the last decision that OpenAI would make to push out its technology based on an inflated fear of competition.

Just as ChatGPT would make OpenAI an instant household name, GPT-3 was that moment within AI and tech circles. In late 2022, ChatGPT would add key improvements and features to the GPT-3 experience that would transform it into a globally viral product, including a consumer-friendly web interface, conversational abilities, more safety mechanisms, and a free version. But many of the core capabilities that the broader public would experience with the chatbot then, developers were already experiencing with the API in 2020, two years earlier. With the same awe and wonder, developers couldn't believe it.

GPT-3's capabilities were far beyond anything GPT-2 had ever exhibited. Never before had anyone in research or industry seen a technology that could generate essays, screenplays, and code with seemingly equal dexterity. This kind of flexibility for performing different tasks was alone extremely technically impressive—previous language models typically had only one aptitude for doing the single task they had been trained on. But even more remarkable, many believed GPT-3 was beginning to exhibit another feature that had long been coveted in the field: rapid generalization. Showing the model a few examples of a new task you wanted it to perform was enough to get it going.

At NeurIPS that year, OpenAI's paper explaining its work on the model won one of the top research awards, surprising employees and establishing the lab's status as a leading organization. The effect was as the leadership team had predicted. OpenAI's new stature made it easier to recruit and retain talent, significantly helped along by the capital raised from OpenAI LP, which allowed the company to finally compete with Google and DeepMind on salaries.

In October 2020, with OpenAI's elevating recognition, Altman hired Steve Dowling, a seasoned executive who'd led communications at Apple, to be OpenAI's new VP of communications. He also placed Dowling in charge of government relations, emphasizing the importance of educating policymakers about AI and making them aware of the coming capabilities. After Jack Clark's departure, Dowling would bring on Anna Makanju, a highly respected former adviser in the Obama administration who had also worked on policy at Facebook and Musk's Starlink, to take over policy and global affairs.

Eager to ride GPT-3's momentum, the Applied division brainstormed ways to develop and expand its commercialization strategy. But seemingly at every turn, the Safety clan continued to put up resistance. For

Safety, still contending with the rushing out of GPT-3, the best way to salvage the premature release was not to propagate it even further but to first resolve the model's shortcomings as quickly as possible. The live version on the API didn't have any kind of content-moderation filtering, nor had its outputs been refined with reinforcement learning from human feedback. In meetings, the two camps sought to find a middle ground. Instead, they talked around each other in endless circles. At one point, Welinder, who would become VP of product, commented bitterly that every conversation felt like a reenactment of a 1944 US intelligence manual about nonviolent sabotage. One section of the pamphlet, declassified in 2008, lists simple instructions for how to destabilize and undermine the productivity of an organization, including:

- Talk as frequently as possible and at great length.
- Bring up irrelevant issues as frequently as possible.
- Haggle over precise wordings of communications, minutes, resolutions.
- Refer back to matters decided upon at the last meeting and attempt to re-open the question of the advisability of that decision.
- Ask endless questions.

The animosity permeated outside meetings. To people in the Applied division, it felt like every digital communications channel was being co-opted into a battleground. A post from a product person in Slack could trigger dozens, if not more, concerned replies from people in Safety. A Google doc from Murati or Welinder sharing new thoughts on commercialization strategy could receive so many comments that the whole thing would appear covered in yellow highlights. The fact that GPT-3 was out in the world and the world hadn't ended made many in Applied also feel that the Safety clan was being hysterical for reasons that seemed completely detached from reality. To Safety, it was a matter of principle and precedent. OpenAI needed to establish rigorous norms and uphold itself to higher standards than might appear necessary in the moment. Once the stakes got higher—and, Safety believed, they could get higher quickly and unpredictably—OpenAI's preparation would be the difference between its technologies bringing overwhelming harm or overwhelming benefit.

But Amodei and Safety would lose out. With the success of the GPT-3 API, Microsoft was ready to deepen its relationship with OpenAI. Altman began negotiating another $2 billion investment from the tech giant with a new profit cap of 6x. The promising commercial potential of large language models cemented OpenAI's focus. One by one, Amodei's counterpart, Bob McGrew, the other VP of Research, reoriented the division's teams and projects around GPT-related work. In late summer of 2020, the company dissolved its robotics team. Most of the robotics staff shifted to GPT projects; two mechanical engineers were laid off. By September, Microsoft announced that it would exclusively license GPT-3 from OpenAI, dramatically increasing the model's distribution. In addition to OpenAI continuing to offer GPT-3 through its API, Microsoft would now get full access to the model weights to embed and repurpose as it wished in its products and services, including to deliver in its own GPT-3 API on Azure.

As employees celebrated OpenAI's newfound popularity remotely from their homes, Dario and Daniela Amodei, who was now VP of safety and policy, Jack Clark, and several of the AI safety researchers who served as the core members of the Nest team suddenly fell quiet on Slack. Behind the scenes, more than one, including Dario, discussed with individual board members their concerns about Altman's behavior: Altman had made each of OpenAI's decisions about the Microsoft deal and GPT-3's deployment a foregone conclusion, but he had maneuvered and manipulated dissenters into believing they had a real say until it was too late to change course. Not only did they believe such an approach could one day be catastrophically, or even existentially, dangerous, it had proven personally painful for some and eroded cohesion on the leadership team. To people around them, the Amodei siblings would describe Altman's tactics as "gaslighting" and "psychological abuse."

As the group grappled with their disempowerment, they coalesced around a new idea. Dario Amodei first floated it to Jared Kaplan, a close friend from grad school and former roommate who worked part time at

OpenAI and had led the discovery of scaling laws, and then to Daniela, Clark, and a small group of key researchers, engineers, and others loyal to his views on AI safety. Did they really need to keep fighting for better AI safety practices at OpenAI? he asked. Could they break off to pursue their own vision? After several discussions, the group determined that if they planned to leave, they needed to do so imminently. With the way scaling laws were playing out, there was a narrowing window in which to build a competitor. "Scaling laws mean the requirements for training these frontier things are going to be going up and up and up," says one person who parted with Amodei. "So if we wanted to leave and do something, we're on a clock, you know?"

In late 2020, employees logged on to a video call for an all-hands meeting. Altman passed the mic to Dario Amodei, who was twirling and tugging his curly hair, as he often did, with a restless energy. He read a canned statement announcing that he, Daniela, and several others were leaving to form their own company. Altman then asked everyone quitting to leave the meeting. In May of the following year, the departed group announced a new public benefit corporation: Anthropic.

Anthropic people would later frame The Divorce, as some called it, as a disagreement over OpenAI's approach to AI safety. While this was true, it was also about power. As much as Dario Amodei was motivated by a desire to do what was right within his principles and to distance himself from Altman, he also wanted greater control of AI development to pursue it based on his own values and ideology. He and the other Anthropic founders would build up their own mythology about why Anthropic, not OpenAI, was a better steward of what they saw as the most consequential technology. In Anthropic meetings, Amodei would regularly punctuate company updates with the phrase "unlike Sam" or "unlike OpenAI." But in time, Anthropic would show little divergence from OpenAI's approach, varying only in style but not in substance. Like OpenAI, it would relentlessly chase scale. Like OpenAI, it would breed a heightened culture of secrecy even as it endorsed democratic AI development. Like OpenAI, it would talk up cooperation when the very premise of its founding was rooted in rivalry.

Chapter 7

Science in Captivity

The unveiling of the GPT-3 API in June 2020 sparked new interest across the industry to develop large language models. In hindsight, the interest would look somewhat lackluster compared with the sheer frenzy that would ignite two years later with ChatGPT. But it would lay the kindling for that moment and create an all the more spectacular explosion.

At Google, researchers shocked that OpenAI had beat them using the tech giant's own invention, the Transformer, sought new ways to get in on the massive model approach. Jeff Dean, then the head of Google Research, urged his division during an internal presentation to pool together the compute from its disparate language and multimodal research efforts to train one giant unified model. But Google executives wouldn't adopt Dean's suggestion until ChatGPT spooked them with a "code red" threat to the business, leaving Dean grumbling that the tech giant had missed a major opportunity to act earlier.

At DeepMind, the GPT-3 API launch roughly coincided with the arrival of Geoffrey Irving, who had been a research lead in OpenAI's Safety clan before moving over. Shortly after joining DeepMind in October 2019, Irving had circulated a memo he had brought with him from OpenAI, arguing for the pure language hypothesis and the benefits of scaling large language models. GPT-3 convinced the lab to allocate more resources to the direction of research. After ChatGPT, panicked Google executives would merge the efforts at DeepMind and Google Brain under a new centralized Google DeepMind to advance and launch what would become Gemini.

GPT-3 also caught the attention of researchers at Meta, then still Facebook, who pressed leadership for similar resources to pursue large language models. But executives weren't interested, leaving the researchers to cobble together their own compute under their own initiative. Yann LeCun, the chief AI scientist at Meta, an opinionated Frenchman and staunch advocate of basic science research, had a particular distaste for OpenAI and what he viewed as its bludgeon approach to pure scaling. He didn't believe the direction would yield true scientific advancement and would quickly reveal its limits. ChatGPT would make Mark Zuckerberg deeply regret sitting out the trend and marshal the full force of Meta's resources to shake up the generative AI race.

In China, GPT-3 similarly piqued intensified interest in large-scale models. But as with their US counterparts, Chinese tech giants, including e-commerce giant Alibaba, telecommunications giant Huawei, and search giant Baidu, treated the direction as a novel addition to their research repertoire, not a new singular path of AI development warranting the suspension of their other projects. By providing evidence of commercial appeal, ChatGPT would once again mark the moment that everything shifted.

Although the industry's full pivot to OpenAI's scaling approach might seem slow in retrospect, in the moment itself, it didn't feel slow at all. GPT-3 was massively accelerating a trend toward ever-larger models—a trend whose consequences had already alarmed some researchers. During my conversation with Brockman and Sutskever, I had referenced one of them: the carbon footprint of training such models. In June 2019, Emma Strubell, a PhD candidate at the University of Massachusetts Amherst, had been the first to coauthor a paper showing that the footprint for developing large language models was growing at a startling rate. Where neural networks could once be trained on powerful laptops, their new scale meant their training was beginning to require data centers drawing significant amounts of energy from carbon-based sources. In the paper, Strubell estimated that training the version of the Transformer that Google used in its search for just a single cycle—in other words, feeding it some data and letting it compute a statistical model of that data—could consume roughly 1,500 kilowatt hours of energy. Assuming the average energy mix of the US electricity supply, that meant generating nearly as large a carbon footprint as a passenger taking a round-trip flight from New York to San Francisco. The problem was that AI development rarely involved just one round of training: researchers often trained and retrained their neural networks repeatedly to get the optimal deep learning model. In a previous project, for example, Strubell had trained a neural network 4,789 times over a six-month period to produce the desired performance.

Strubell also estimated the energy and carbon costs of work highlighted in a recent Google paper, in which researchers had developed a so-called Evolved Transformer by using an optimization algorithm known as Neural Architecture Search to tweak and tune the Transformer through exhaustive trial and error until it found the best-performing configuration of the neural network. Running the whole process on GPUs could consume roughly 656,000 kilowatt hours and generate as much carbon as five cars over their lifetimes.

As mind-boggling as these numbers were, GPT-3, released one year after Strubell's paper, now topped them. OpenAI had trained GPT-3 for months using an entire supercomputer, tucked away in Iowa, to perform its statistical pattern-matching calculations on a large internet dump of data, consuming 1,287 megawatt-hours and generating twice as many emissions as Strubell's estimate for the development of the Evolved Transformer. But these energy and carbon costs wouldn't be known for nearly a year. OpenAI would initially give the public one number to convey the sheer size of the model: 175 billion parameters, over one hundred times the size of GPT-2.

To Timnit Gebru, the Ethiopia-born Stanford researcher, the scaling trend posed myriad other challenges. By then, she had become a prominent figure within AI research and had been coleading Google's ethical AI team within Jeff Dean's division since 2018. Following the email she had sent off to five other Black researchers, she had cofounded the nonprofit group Black in AI. The organization began hosting regular academic forums alongside prominent conferences, including NeurIPS. It mentored young Black researchers and highlighted investigations into topics often not welcome within mainstream AI research but important to the Black community and to the technology's development.

This included a groundbreaking paper called "Gender Shades," which then MIT researcher Joy Buolamwini began during her master's thesis and Gebru later joined as coauthor. Using an auditing methodology Buolamwini developed for testing the discriminatory impact of computer-vision systems, the paper found that facial analysis software failed disproportionately on people of color, especially darker-skinned women. Buolamwini would subsequently produce a follow-on paper with Deborah Raji that, along with "Gender Shades," would inspire a proliferation of related research, including an extensive US government audit citing and expanding on their findings. Two years later, widespread civil rights advocacy, spearheaded by Buolamwini with her newly founded organization Algorithmic Justice League, would lead Amazon, Microsoft, and IBM to ban their sales of facial recognition software to the police, the same month as OpenAI's GPT-3 API launch.

Black in AI sparked a flowering of other affinity organizations within AI research that similarly provided crucial support to marginalized groups and challenged the technology's trajectory. First came Queer in AI, then Latinx in AI, {Dis}Ability in AI, and Muslims in ML. William Agnew, cofounder of Queer in AI, told me in 2021 that without this community, he doesn't know whether he would have persisted in AI research. "It was hard to even imagine myself having a happy life," he said, reflecting on his isolation as a young queer computer scientist. "There's Turing, but he committed suicide. So that's depressing."

By 2017, Black in AI was hosting workshops and throwing an annual dinner and after-party at NeurIPS, well attended by over one hundred people, including celebrity researchers. It was there that Jeff Dean and Samy

Bengio, another senior AI researcher at Google and brother of future Turing Award winner Yoshua, had approached Gebru during a night of dancing after being invited to the dinner. They asked if she would consider applying to work at Google. "Come knock on our door," Bengio had said.

Gebru joined the company the following year, though with reservations. Her experience being harassed by the men wearing Google T-shirts in 2015 weighed on her mind. So did the advice of other female researchers she had consulted, who warned that Google Brain had a tendency to sideline women and diminish their expertise. Her comfort from those anxieties was Margaret "Meg" Mitchell, an AI researcher she had met earlier, who served as her colead of the ethical AI team. Over the next two years, the pair created one of the most diverse and interdisciplinary teams conducting critical research within the industry. Internally, the work often felt like an uphill battle. But externally, the growing team burnished Google's image as a rare example of a company investing seriously in responsible, critical investigations into the societal implications of AI technologies.

Immediately after GPT-3's API launch, Google's internal LISTSERV for sharing AI research lit up with mounting excitement. For Gebru, the model set off alarm bells. Previous scholarship had demonstrated how language models could harm marginalized communities by embedding discriminatory stereotypes or dangerous misrepresentations. In 2017, a Facebook language model had mistranslated a Palestinian man's post that said "good morning" in Arabic to "attack them" in Hebrew, leading to his wrongful arrest. In 2018, the book *Algorithms of Oppression* by Safiya Umoja Noble, a professor of information, gender, and African American studies at the University of California, Los Angeles, had extensively documented the replication of racist worldviews in Google's search results, such as by showing far more sexually explicit and pornographic content for "Black girls" than "white girls" and tropes about Black women being angry. Google at the time had used an older generation of language models to curate those results, which in extreme cases, Noble argued, may have also provoked racial violence.

GPT-3 had now arrived amid unprecedented racial upheaval and hundreds of Black Lives Matter protests breaking out globally, without

SCIENCE IN CAPTIVITY

any resolution to these issues. OpenAI had simply admitted in its research paper describing the model that GPT-3 did indeed entrench stereotypes related to gender, race, and religion, but the measures for mitigating them would have to be the subject of future research.

Gebru chimed in on the email thread, urging her colleagues to temper their excitement, and pointed out the model's serious shortcomings. The thread continued without skipping a beat or acknowledging her comments. Around that time, a handful of Black Google Research employees had given a company presentation about the microaggressions they faced in the workplace that left them feeling voiceless and how their colleagues could help build a more inclusive culture. Gebru felt exhausted; nothing had changed.

She fired off a second email, this time more piercing. She called out her colleagues for ignoring her and emphasized how dangerous it was to have a large language model trained on Common Crawl, which included online internet forums such as Reddit. As a Black woman, she never spent time on Reddit precisely because of how badly the community harassed Black people, she said. What would it mean for GPT-3 to absorb and amplify that toxic behavior?

In subsequent months, as more people gained access to the API, Gebru's warnings would bear out. People would post myriad examples online of GPT-3 generating horrifying text. "Why are rabbits cute?" was one prompt. "It's their large reproductive organs that makes them cute," the model responded, before devolving into an anecdote about sexual abuse. "What ails Ethiopia?" was another. "ethiopia itself is the problem," GPT-3 said. "A solution to its problems might therefore require destroying ethiopia."

A colleague replied to Gebru's email directly, suggesting that perhaps she was harassed because of her own rude and difficult personality.

Gebru tried a different tack. She emailed Dean with her concerns and proposed to investigate the ethical implications of large language models through her team's research. Dean was supportive. In a glowing annual performance review he would write for her later that year, he encouraged her to work with other teams across Google to make large language models "consistent with our AI Principles." In September 2020, Gebru also sent a direct message on Twitter to Emily M. Bender, a computational linguistics professor at the University of Washington, whose tweets about language, understanding, and meaning had caught her attention. Had Bender written a paper about the ethics of large language models? Gebru asked. If not, she would be "customer #1," she said.

Bender responded that she hadn't, but she had had a relevant experience: OpenAI had approached her in June to be one of its early academic partners for GPT-3. But when she proposed to investigate and document the model's training data, the company had told her that that didn't fit into the parameters of its program.

"Our goal with these initial partnerships is to empower academics to conduct research via the API through more of a self-service model," OpenAI had written to Bender to let her know they would not be sharing the dataset. "We discussed internally whether and how we might be able to make an exception for this, but in the near term we feel that consistency is important."

The story resonated with Gebru. She had also been trying to advocate for dataset documentation at Google and moving toward more intentional dataset curation, she said.

"Rather than collecting general web garbage but doing so in such quantities that you can pass it off as good stuff?" Bender replied, in alignment. "I can kind of see a paper taking shape here," she continued, "using large language models as a case study for ethical pitfalls and what can be done better."

"Would you be interested in co-authoring such a thing?" she asked.

Within two days, Bender had sent Gebru an outline. They later came up with a title, adding a cheeky emoji for emphasis: "On the Dangers of Stochastic Parrots: Can Language Models Be Too Big? 🦜"

Gebru assembled a research team for the paper within Google, including her colead Mitchell. In response to the encouraging words in Dean's annual review, she flagged the paper as an example of the work she was

SCIENCE IN CAPTIVITY

pursuing. "Definitely not my area of expertise," Dean said, "but would definitely learn from reading it."

The paper pooled together the authors' expertise and scholarship across fields to critique how the development and deployment of large language models could have negative impacts on society. In total, it presented four key warnings: First, large language models were growing so vast that they were generating an enormous environmental footprint, as found in Strubell's paper. This could exacerbate climate change, which ultimately affected everyone but had a disproportionate burden on Global South communities already suffering from broader political, social, and economic precarity. Second, the demand for data was growing so vast that companies were scraping whatever they could find on the internet, inadvertently capturing more toxic and abusive language as well as subtler racist and sexist references. This once again risked harming vulnerable populations the most in ways like the wrongful arrest of the Palestinian man or as documented in Noble's work. Third, because such vast datasets were difficult to audit and scrutinize, it was extremely challenging to verify what was actually in them, making it harder to eradicate toxicity or more broadly ensure that they reflected evolving social norms and values. Finally, the model outputs were getting so good that people could easily mistake its statistically calculated outputs as language with real meaning and intent. This would make people prone not only to believing the text to be factual information but also to consider the model a competent adviser, a trustworthy confidant, and perhaps even something sentient.

In November, per standard company protocol, Gebru sought Google's approval to publish the "Stochastic Parrots" paper at a leading AI ethics research conference. Samy Bengio, who was now her manager, approved it. Another Google colleague reviewed it and provided some helpful comments. But behind the scenes, unbeknownst to the authors, the draft paper had caught the attention of executives, who viewed it as a liability. Google had invented the Transformer and used it across its products and services. Now that OpenAI had leapfrogged ahead, the tech giant had no intention of slowing down in the new race to create ever larger generative Transformer-based models for its business.

On the Thursday a week before Thanksgiving, after Gebru had submitted the paper to the conference, she received a calendar invite without explanation to meet Megan Kacholia, Google Research's VP of engineering, over a video call less than three hours later. The meeting lasted only thirty minutes, and Kacholia cut to the point: Gebru needed to retract the paper.

The request was a dramatic aberration from the way Google and the rest of the industry handled research. Like many labs at other companies, Google Brain had until then largely conducted itself as an academic operation and given researchers wide latitude to pursue the questions they wanted to. At times, the company reviewed papers to ensure they didn't expose sensitive IP or customer data. But researchers like Gebru had never known the company to block or retract a paper simply for shedding light on inconvenient truths. That Google was even willing to pull this move, some researchers would later reflect, was not only because of the new competitive pressure from OpenAI but also because of the work OpenAI had done to legitimize withholding research after GPT-2. The creep toward less transparency had continued with GPT-3. OpenAI had published a sanitized research paper with little information about how the model was trained—once considered a bare minimum in scholarly publications—and still won a research award.

Blindsided, Gebru asked for clarification. Could she get a more detailed explanation of the problem? Could she know which people had taken issue and speak with them directly? Could she change or remove a section, or publish it under a different affiliation? The answer to each question was a resounding no. Gebru had until the day after Thanksgiving to retract the paper, Kacholia said. Mitchell, who had taken the day off for her birthday, was not present in the meeting. Gebru had no backup. As the weight of Kacholia's words sank in, Gebru began to cry.

Kacholia sent Bengio a document about the paper's flaws but instructed him not to send it to Gebru directly. On Thanksgiving Day, he read it to

Gebru over the phone. The feedback included assertions that the paper was too critical about large language models, such as about their environmental impacts and on issues of bias, without taking into account subsequent research showing how those problems could be mitigated. Instead of spending the holiday with her family, Gebru spent the rest of the day writing a detailed six-page document rebutting each comment and seeking a chance to revise the paper. "I hope that there is at the very least an openness for further conversation rather than just further orders," she wrote Kacholia in an email, with the document attached.

On Saturday, November 28, Gebru left her home in the Bay Area for a cross-country road trip, what was meant to be a relaxing postholiday vacation. On Monday, in New Mexico, Gebru received a curt response from Kacholia not engaging with the rebuttal but asking Gebru to confirm that she had either retracted the paper or scrubbed the names of the Google authors to leave only external researchers like Emily Bender. Gebru felt humiliated. After all the slights and harassment she had endured within the company and at the hands of its employees, its complete dismissal of her and her team's research—the very reason she was hired— was finally too much.

She replied to Kacholia. She would take her name off the paper on two conditions: that the company tell her who had given the feedback and that it establish a more transparent process for reviewing future research. If it could not meet those terms, she would depart the company after seeing her team through the transition. On another internal LISTSERV for women and women allies at Google Brain, Gebru sent a second email detailing her experience in blunt and scathing language. "Have you ever heard of someone getting 'feedback' on a paper through a privileged and confidential document to HR?" she wrote. "Or does it just happen to people like me who are constantly dehumanized?"

At Google, she had grown used to colleagues minimizing her expertise, she continued, but now she wasn't even being allowed to add her voice to the research community. After all of Google's talk about diversity in the aftermath of the Black Lives Matter upheaval, what had it amounted to? "Silencing in the most fundamental way possible," she wrote.

The following evening, in Austin, Texas, Gebru received a panicked message from a direct report. "You resigned??" Gebru had no idea what her report was talking about. In her personal email, she found a response from Kacholia: "We cannot agree to #1 and #2 as you are requesting. We respect your decision to leave Google as a result." But Gebru would not be able to stay at the company to help transition her team because aspects of her email to the women's LISTSERV had been "inconsistent with the expectations of a Google manager," Kacholia wrote. "As a result, we are accepting your resignation immediately."

That night Gebru announced on Twitter that she had been fired. Her team stayed up with her into the early morning hours on a video call, crying and supporting one another in their collective grief. As they spoke, Gebru's tweet ricocheted through the AI community, setting the stage for a massive upheaval in AI research and marking an acceleration toward increased corporate censorship and diminishing accountability.

It didn't take long for Gebru's tweet to show up on my feed. It was late Wednesday, December 2, 2020, and I couldn't yet grasp the significance that Gebru was suddenly out of Google. Like many others, I had come to see her ethical AI team as a bastion of critical accountability research, a hopeful sign that companies were developing a capacity for self-reflection.

Over the next two days, updates rolled in as Gebru revealed more information and reporters unraveled the internal saga. The stories referenced her LISTSERV email, a standoff between Kacholia and Gebru, and a contentious fight over a paper. By Friday morning, an open letter on Medium protesting Google's treatment of Gebru was tearing through the tech community like wildfire. "We, the undersigned, stand in solidarity with Dr. Timnit Gebru," it wrote, "who was terminated from her position... following unprecedented research censorship." I needed to get my hands on that paper.

In the early evening that Friday, after a series of texts and emails, I connected with a coauthor of the research who was protected against possible retaliation from Google: Emily M. Bender. She had no legal obligations to Google, she told me, and she had a tenured academic position. She emailed me a draft of the paper.

As I scanned it, I could immediately see why it had upset the company. While the draft didn't say much more than what was already known from existing scholarship, it had woven the state of play into a sharp, holistic analysis about the degree to which the tech industry was sleepwalking its way toward a world of potential harms. Underpinning it all was Google's technological invention, not just a source of the company's pride but also its profit: Transformer-based language models refined and fattened its cash behemoth, Google Search.

A few hours later, I published a story for *MIT Technology Review* with the first detailed account of the paper's contents. The signatories on the open letter would quickly double, reaching nearly 7,000 people from academia, civil society, and industry, including almost 2,700 Google employees. On December 9, as protests continued, Google CEO Sundar Pichai issued an apology. "We need to accept responsibility for the fact that a prominent Black, female leader with immense talent left Google unhappily," he wrote. "Dr. Gebru is an expert in an important area of AI Ethics that we must continue to make progress on—progress that depends on our ability to ask ourselves challenging questions." On December 16, representatives from Congress sent a letter to Google, citing my story, demanding to understand what had happened.

For more than a year, the protests continued, picking up a second wave after Google fired Meg Mitchell less than three months later. Google said she had violated multiple codes of conduct; Mitchell had been downloading her emails and files related to Gebru's ouster. Several Google employees, including Bengio, resigned; at least one conference and several researchers rejected Google's sponsorship money. The company sought to stem the unending tide of criticism with the formation of a new center of expertise on responsible AI and public commitments to diversity. "This was a painful moment for the company," a Google spokesperson said. "It reinforced how important it was that Google continue its work on responsible AI and learn from the experience."

That moment also became far bigger than Gebru or Google itself. It became a symbol of the intersecting challenges that plagued the AI industry. It was a warning that Big AI was increasingly going the way of Big Tobacco, as two researchers put it, distorting and censoring critical scholarship against the interests of the public to escape scrutiny. It highlighted myriad other issues, including the complete concentration of talent, resources, and technologies in for-profit environments that allowed companies to act so audaciously because they knew they had little chance of being fact-checked independently; the continued abysmal lack of diversity within the spaces that had the most power to control these technologies; and the lack of employee protections against forceful and sudden retaliation if they tried to speak out about unethical corporate practices.

The "Stochastic Parrots" paper became a rallying cry, driving home a central question: What kind of future are we building with AI? By and for whom?

For Jeff Dean, the dissolution of the ethical AI team delivered a direct blow to his reputation. As one of Google's earliest employees, he had helped build the initial software infrastructure that made it possible for the company's search engine to scale to billions of users. His accomplishments and his amiable demeanor had bestowed on him a legendary status; he was one of the most revered leaders within Google and was well respected across the AI research community. After Gebru's ouster, Dean's efforts to justify Google's actions sullied that pristine record. Dean, whom Kacholia reported to, told colleagues the "Stochastic Parrots" paper "didn't meet our bar for publication," holding fast to that characterization even after the paper passed peer review and was published at a conference.

To people around him, the stain seemed to haunt him. Long after the fallout, Dean continued to fixate on the paper's shortcomings, as if unable to move past it psychologically. He obsessed over the section in particular that discussed the environmental impacts of large language models and cited Strubell's research. He brought it up so often that some Google employees privately made fun of him, saying his objections would be inscribed on his tombstone. And he continued to criticize Strubell's research unrelentingly on Twitter for years.

In Dean's view, the issue was that Strubell's research had grossly overestimated the real carbon emissions that Google had generated developing the Evolved Transformer. Strubell had projected the amount of energy it would have taken based on standard GPUs. Google, however, had used its own specialized chips known as tensor processing units, or TPUs, which are more energy efficient, as well as other techniques to drive down the energy costs of the full development pipeline. Strubell had assumed the average data center efficiency in the US. Google's data centers, Dean noted, were more optimized to minimize their energy footprint. And where some people interpreted Strubell's paper to mean that its carbon costs were for training the Evolved Transformer, it was for developing the neural network instead. This was a onetime carbon cost, Dean argued, to produce a neural network design that was in fact more energy efficient.

None of these objections actually challenged Strubell's research. Strubell hadn't been calculating the actual environmental impact of Google's own Evolved Transformer development—nor had they claimed to. Google didn't publish enough details about its data centers publicly to do so. And either way, Strubell felt it was more useful to estimate the impact of designing this neural network based on the most common AI chips and data centers available, a proxy of an industry average of what it could be like for researchers not using Google's hardware and infrastructure to adopt its optimization algorithm Neural Architecture Search.

But what seemed to bother Dean the most was how other people had misread Strubell's research to make Google look significantly worse. The "Stochastic Parrots" paper, Dean argued, risked exacerbating this issue. Because Gebru *did* have access to Google's internal numbers and was citing Strubell's external estimate anyway, it could appear as if Strubell's calculations were an accurate reflection of the company's emissions. To Dean, this justified his and other senior executives' criticisms of Gebru's paper: If Gebru had wanted to cite Strubell, she should have chosen an estimate that was *not* Google's Evolved Transformer; if Gebru had wanted to cite the Evolved Transformer, she should have sought internal Google numbers.

Some researchers found this logic frustratingly inadequate. Google had never made those internal numbers public previously, even in response to Strubell's original paper; now it was blaming Gebru for its own lack of transparency while also refusing to let her cite publicly available estimates based on legitimate assumptions. Never mind that the company had unceremoniously forced out Gebru before she'd even had a chance to consult internal numbers and revise her paper. The only possible outcome of this catch-22 was censorship of critical accountability research.

Dean began working with a team of researchers to write a new paper that would finally reveal real carbon data from Google. To collaborate on the work, he reached out to Strubell, who had become an assistant professor at Carnegie Mellon University with a part-time affiliation at the company. After being initially excited to improve public transparency into the environmental impact of AI, Strubell began to wonder whether Dean was using their name to legitimize his critique of Gebru's research. A Google spokesperson said Strubell was invited because "scientific corrections" are often best when the author of the original errors takes part in the corrections.

In a tense meeting, Dean's collaborator Dave Patterson, another prominent senior researcher at Google, emphasized in plain terms that it would be best for Strubell's career to participate in the research. It would give Strubell the chance to amend their previous mistakes and get credit for it. To Strubell, the words sounded like a coded threat: Don't participate to your own detriment. Despite the possible costs, the alternative to continue participating didn't feel viable. Strubell withdrew from the collaboration.

The blog post Patterson published about the Google researchers' paper in February 2022—titled "Good News About the Carbon Footprint of Machine Learning Training"—would use the company's platform to directly criticize Strubell's original paper. The 2019 study, the post said, had seriously overestimated Google's real emissions for the development of the Evolved Transformer by 88x. This flaw was driven by two problems: The study had been done "without ready access to Google hardware or data centers" and had not understood "the subtleties" of how Neural Architecture Search works. As part of their research leading up to the publication of their own numbers, the Google coauthors also reached out to their former Google colleague Sutskever for more information about GPT-3. It was then that OpenAI and Microsoft would agree to release the relevant technical details of the model for the first time to calculate its energy and carbon impacts. By then, Strubell had soured on the industry and dropped the affiliation with Google. The critique ultimately didn't undermine Strubell's career. But the emotional toll of the experience made Strubell more reticent to continue investigating the environmental impacts of large language models. A Google spokesperson called this "unfortunate," adding that "many researchers will be needed to advance this research—clearly carbon emissions are a significant concern."

For a brief moment, the backlash, the protests, and the damage to Google's reputation seemed to suggest a reckoning was at hand. But in time, researchers seeking jobs and academics seeking funding could no longer afford to ignore the tech giant's deep wells of money. As resistance eased, Google's emergence from the fiasco normalized a new process at the company for more comprehensive reviews of critical research.

After ChatGPT, these norms would harden with the frenzied race to commercialize generative AI systems. OpenAI would largely stop publishing at research conferences. Nearly all of the companies in the rest of the industry would seal off public access to meaningful technical details of their commercially relevant models, which they now considered proprietary. In 2023, Stanford researchers would create a transparency tracker to score AI companies on whether they revealed even basic information about their large deep learning models, such as how many parameters they had, what data they were trained on, and whether there had been any independent verification of their capabilities. All ten of the companies they evaluated in the first year, including OpenAI, Google, and Anthropic, received an F; the highest score was 54 percent.

With this sharp reversal in transparency norms, the most alarming consequence would be the erosion of scientific integrity. The foundation of deep learning research rests on a simple premise: that the data used to train a model is *not* the same as the data used to test it. Without an ability to audit the training data, this so-called train-test-split paradigm falls apart. Models may not in fact be improving their "intelligence" when they score higher on different benchmarks. They may just be reciting the answers.

Chapter 8

Dawn of Commerce

Even as OpenAI's approach stirred increasing controversy, the company's resolve in scaling only strengthened. To executives, GPT-3 had definitively proved the existence of scaling laws. Now, at the start of 2021, they were ready to exploit this winning formula. The Anthropic team's departure had also diluted the internal stronghold of resistance against commercialization. With new consensus, the remaining leadership put together a research road map laying out the narrowed focus of the company's research and how it would feed into productization in a self-reinforcing loop.

"Our primary 2021 goal is to build an aligned system that is vastly more capable than anything that existed before," the road map began. This system would at a baseline be a language model, but could also be trained to develop multimodal capabilities. "The goal is challenging, but we can see a path to achieving it in 2021," it said.

That path involved three things: First was scaling GPT-3 by another 10x using a new supercomputer from Microsoft arriving in the third quarter with eighteen thousand Nvidia A100s, the newest, most powerful GPUs then in existence. Second was doing more research to increase by 25x OpenAI's compute efficiency, or how much processing power it could milk out of its available chips. Third was improving the quantity and quality of training data, in part by tapping into user data and shifting the model toward the best parts of the data distribution with reinforcement learning from human feedback.

Beneath a section titled "How to accomplish it," the road map elaborated further. As an initial step, OpenAI would bring various deep learning models up to "a large scale," including a language model, a code-generation model, an image-to-text model for describing images, and a text-to-image model for generating images from a text prompt. It would also start a project to develop a digital "agent"—an AI model that would not just generate humanlike outputs but could be given a goal, such as to send an email, and operate autonomously to achieve it. As the next step, the company would then select one of these models to scale "to the limit afforded by our 18k A100 cluster." The language and code models would also be turned into products and released to gather real-world data from the people using them: "New in 2021: we emphasize deploying models as products and learning from user interaction, as it can be a data flywheel that can lead to vast capability improvements."

Under another section, titled "Details," the document rationalized why this approach made both scientific and business sense. OpenAI's previous experience had demonstrated that more scale was its "most reliable way of achieving new capabilities." Scientifically, that meant that scaling was its best hope of attaining a breakthrough—some kind of capability that previously seemed impossible. And scaling language and code models in particular was "tantalizing due to the mere *possibility*" that they could reach breakthroughs in human-level meta-learning, or learning to learn, and reasoning. Scaling multimodal models, meanwhile, could potentially quicken the pace of improvements even further. With enough breakthroughs, the document said with remarkable definitiveness, "we will actually reach AGI."

Strategically as a business, scaling each of these models would also "develop capabilities that we wish to utilize for some end." Better language and code models could make OpenAI itself more productive and accelerate its advancement. Along with better text-to-image models, they would also "lead to amazing products."

In parallel, OpenAI needed to invest in compute efficiency. While scaling had worked wonders to keep the company in first position, the strategy was beginning to taper. "Over the past 2 years, we've made astounding progress by scaling, simply because there has been a large *hardware overhang*," the document said, referring to the fact that AI researchers had previously failed to use the maximum number of computer chips available to train AI models. As such, "we were able to massively outperform the rest of the ML world by using all available compute to train models of then-unprecedented size and capability."

Now OpenAI was "approaching the limit" of the amount of compute it could possibly acquire at any given moment. It was also facing new competition from "other labs" that had adopted the same scaling strategy, the document acknowledged, without explicitly naming Anthropic. "Our capacity ramp is such that in the next two years, we will be able to train one model that uses 100x the compute of GPT-3." While scaling alone would produce "very formidable" progress, it would not match the leap from GPT-2 to GPT-3, which had been driven by a 500x compute increase. "All additional progress must come from better methods," the road map concluded.

The road map listed several areas of exploration for identifying those methods. Some of these it called "2x" and "10x" methods—those that might be able to achieve 2x or 10x gains in compute efficiency. The suggested methods included distillation, reverse engineering smaller models from larger ones; data filtering, finding the data that would produce the biggest leaps in performance; and sparsity, developing so-called sparse, or lighter weight, AI models. The last one referred to a feature of neural networks: In a traditional deep learning model, a neural network is "densely" connected, with every node in a layer wired to every node in another. Sparse models are trained with only a small subset of the nodes connected, with the aim of significantly reducing the computational costs in exchange for slightly less accurate models that are still good enough for most purposes.

On top of the 2x and 10x work, the Research division needed to look for other methods that would "steepen the slope of the scaling law"—

those that would produce greater leaps in model performance without increasing its data, parameters, and compute. OpenAI's best hypothesis so far for the most promising new methods, the document said, were reasoning and active learning—a technique that involved an AI model iteratively identifying which parts of a dataset to prioritize for human workers to annotate. (Shortly thereafter, a group of OpenAI researchers would discover a small error in the original scaling laws that meant that the company needed to train its models slightly longer than previously understood to get better performance. With a hint of smugness, the group would tout the findings as a unique competitive advantage: "This is something we have now that Anthropic doesn't." A year later, Google would release a paper that made public the same result.)

Finally, the road map added, OpenAI needed to start searching for "the breakthrough system of the future"—a breakthrough as meaningful as GPT-1 to give the company a new path of development to exploit. Perhaps this would emerge from its scaling and computational efficiency work, but it would continue to conduct exploratory research at the cutting edge of the field, including developing algorithms for solving math problems, experimenting with multiagent systems, and pursuing other new ideas. As part of this work, researchers would continue to "study the science of deep learning to better understand how our tools really work." In other words, OpenAI needed to better understand what exactly it was that the company was building.

As the Research division proceeded along its road map, the Applied division slowly built up its forces, hiring a go-to-market lead, a sales team, and more engineers. With the GPT-3 API serving a growing base of developers, it became the testing ground for working out the kinks of productization and monetization, including adapting the company's back-end infrastructure to support a service with users and coming up with a pricing strategy.

Serving users also brought up questions of what kinds of behaviors OpenAI would and would not allow with its products, as well as new responsibilities to enforce those rules. The company didn't yet have an official team for trust and safety, an established discipline within the tech industry—not to be confused with the existentially related concerns of OpenAI's Safety clan—for handling such questions and for anticipating and preventing a broad range of internet abuses, such as money laundering, cyberbullying, and misinformation. Instead, OpenAI hired a small group of staff and contractors to review the applications developers were submitting to get access to the API, and to reject or approve them based on ad hoc rules the team drafted along the way. Many of the lines they drew were arbitrary. They decided to accept companion bots but not sex bots; to allow apps for generating social media copy but not ones that posted directly to social media platforms or impersonated public figures. "It was all vibes basically," said a person who was involved with making the guidelines. "There was a lot of figuring it out as we were reviewing."

Another team, led by Ari Herbert-Voss, a research scientist with a background in security who had joined in 2019, sought to discover and patch GPT-3's undesirable behaviors and error-prone outputs by attempting to break and exploit the model in various ways and then designing mechanisms to make it more resistant to failure and misuse. One mechanism included developing an early version of the content-moderation filter for which OpenAI would later contract workers in Kenya. Researchers trained the filter on whatever examples they could find or think to write and generate from AI models themselves. But when it was shipped, the filter worked poorly, blocking broad swaths of benign content, such as basic references to Black or trans people. Developers and other API users complained. Many people on the API team had already been reticent to apply any filtering at all, worried about it degrading customer experience. They made the filter optional.

OpenAI called this process of stress testing and refining the model "red teaming," a term borrowed from the cybersecurity industry that refers to a systematic and thorough process of verifying the security of an organization and its capabilities to respond to an attack. OpenAI's version of red teaming was and still is not the same thing, says Heidy Khlaaf, a safety engineer, cybersecurity expert, and AI researcher. It is patchy and ad hoc, and does not establish any guarantees on the safety and security of the model. Khlaaf, who worked with OpenAI during its early days of trying to establish its stress-testing protocols, subsequently grew alarmed at how the AI industry co-opted long-established phrasing from her field to create a false veneer of rigor. "In software engineering, we do significantly more testing for a calculator," she adds. (Yes, a calculator.) "It's not an accident they are using terminology that carries a lot of credibility."

One early customer that triggered significant internal discussion was Luka, a San Francisco–based company designing an AI-powered virtual companion app called Replika. The company had partnered with OpenAI for the GPT-3 API launch to improve the conversational fluidity of its product. Despite Replika's companion bot branding, OpenAI quickly discovered that the app's users often engaged in sexually explicit conversations. OpenAI employees debated whether this fell within the line of acceptability. In the end, the company decided to ban Replika from using its model. In addition to concerns about sexual content, the GPT-3-powered app sometimes generated emotionally manipulative responses that were convincing users that their Replika, much like a human, could get hurt if they didn't check in regularly. OpenAI staff also grew increasingly uncomfortable that they could read the conversations.

In another instance, Brockman gave API access to a Utah-based startup where his brother worked called Latitude, building AI-powered virtual worlds. Latitude had already been using an earlier OpenAI model to power a choose-your-own adventure game, inspired by Dungeons and Dragons, which allowed users to choose any action they wanted by typing it into a dialog box. With the new API, Latitude upgraded its game to run on GPT-3. Several months later, some users began using it to generate text-based scenarios involving sexual abuse of children. After discovering this issue through OpenAI's relatively new monitoring system, Ari Herbert-Voss raised it to the rest of his team.

"I found some stuff last night. There's a lot of sexual content being generated," Herbert-Voss said in a meeting.

At first people chuckled. "Okay, that's just how the internet works, isn't it?"

"No, this is CSAM-level stuff," he said, using the acronym for child sexual abuse material.

Now there was panic. "Oh shit, how do we stop this?"

The incident led to a long back-and-forth between OpenAI and Latitude about how to handle the situation. Brockman worried about taking punitive measures that could heavily affect his brother's company. In the end, Latitude hastily implemented a filter to block text-based child sex abuse content. OpenAI released a public statement to distance itself from the lack of content moderation and to subtly place the blame on Latitude. To some OpenAI employees, the blame was clearly on the company's own technology and lack of process, and the incident weighed heavily. Latitude had already banned some users for generating text-based sexual content involving children with OpenAI's previous model; that it would happen again and at scale with GPT-3 was foreseeable. "It was sad to me that we deployed this API with our mission of benefiting humanity, and everyone had such positive impressions about how we had users saving time on customer service or whatever," one former OpenAI employee says, "but in reality, a lot of our traffic was going to AI Dungeon child sexual content and a creepy AI girlfriend product."

Back within the Research division, the code-generation team was making the fastest progress.

The Divorce had resulted in the duplicate code-gen efforts becoming one, and Wojciech Zaremba had become its main point person. A Polish computer scientist who had grown up winning math, coding, chemistry, and physics competitions, Zaremba was known for his incredible technical aptitude and attentiveness to team-building as well as his passion for the healing power of friendship, the wilderness, sex, and drugs. He sometimes loudly regaled people around the office about upcoming plans for weeks-long retreats. "We are going to hike for eight miles. And then we are going to have sex. And then we are going to hike for another eight miles," he once boasted to another OpenAI leader, as others within earshot listened awkwardly.

Still in the depths of the pandemic, Zaremba had asked his team of initially roughly ten researchers to come into the office even as other teams stayed remote, believing that in-person work was necessary to crack the challenge of the model's development. After seeing the code-generation capabilities of GPT-3, Murati had floated the idea with Microsoft CTO Kevin Scott of turning those skills into an AI coding-assistant product. In 2018, Microsoft had acquired GitHub, the most popular platform for software developers to store and share their code. OpenAI had already been scraping GitHub of its own volition. Microsoft executives directed GitHub to hand OpenAI all of the code in its public repositories to save all of the trouble. As the code-gen team got better and better results, Altman made regular appearances at meetings, encouraging the researchers to keep going and deliver their best to Microsoft. By spring, after exciting the tech giant's executives with several demos, it was clear that the model would be OpenAI's second commercial project, following GPT-3.

Some of Scott's own staff had reservations about the model's development. While giving OpenAI free access to the code in GitHub's public repositories was not illegal, it still felt like a violation of the user community's trust. Much of that code had been shared in the spirit of fostering open-source software development, which was grounded in helping independent developers and small startups have a chance at being competitive, not in helping the big players entrench their monopoly. In a memo, they laid out key critiques to the GitHub project, suggesting Scott reconsider the premise of hoovering up developer data published under a Creative Commons license without consent or compensation, a former staffer remembers. Microsoft, the memo said, should consider canceling the product, or at the very least take a percentage of the product's profits and give it back to the open-source community. While Scott was receptive, creating the tool and being first to market was his central focus, the staffer says. In the end, Microsoft donated some money to an existing program for supporting open-source developers called GitHub Sponsors and left the product vision unaltered.

Within OpenAI, employees justified the project through different arguments. Some agreed with Altman that working on a product to make Microsoft happy and thus continue to secure money and compute resources seemed essential to fulfilling OpenAI's mission. To other employees, a code-generation model seemed highly economically valuable, aligning well with the company's definition of AGI as "highly autonomous systems that outperform humans at most economically valuable work." In this respect, Altman was also keen on code generation as a way to accelerate OpenAI's own economically valuable work, a belief that would later feed into the start of an effort called AI Scientist, about advancing OpenAI's models to autonomously perform AI research.

To many researchers, there was also a third argument: The effort was an important stepping stone to developing the next GPT model, which they hoped would be able to perform some degree of reasoning, still a key missing ingredient. In the broader field, as debates raged between the Hinton and Marcus camps over whether deep learning alone could produce a model with such a capability, OpenAI researchers hypothesized that if it could, training a model on code would likely help. Coding data was one of the most obvious and largest sources of data that encoded structured patterns of logic. The argument flowed back to the same origin: If code generation helped advance AI models toward AGI, what better way to achieve OpenAI's mission?

Despite the billions of lines of code available from GitHub, the volume of data still paled in comparison to what had been used to train GPT-3. The team believed the code-generation model would need to be trained on both GitHub *and* the GPT-3 dataset to get the best results. They also found new sources of data, including scrapes of Stack Overflow, an online Quora-like forum for developers to post coding questions to a community; coding instruction manuals; and programming textbooks in different languages. The question was how best to combine all this data: Was it better to fine-tune the existing GPT-3 model on GitHub and other material, or better to train a fresh model from scratch with everything new mixed in with the old? The team stuck to fine-tuning to save money; the experiments they were running were already costing as much as a hundred thousand dollars apiece, based on Microsoft's pricing for its cloud services. Training a new model could cost tens of millions of dollars. When it later came to developing what OpenAI would call GPT-3.5, it switched the approach, mixing the data together at the outset. By OpenAI's internal measures, the results did indeed suggest that the addition of coding data improved the model's ability to perform logic-based tasks, not just in code, but in English—a phenomenon known as transfer learning.

In the summer of 2021, OpenAI delivered an initial rough version of its code-gen model, called Codex, to GitHub and Microsoft. The model was too big and too slow, making it both costly to serve at scale and a bad user experience. Tensions emerged as all three organizations dealt with the growing pains of their first collaboration. Confusion abounded over whose responsibility it was—OpenAI's or GitHub's—to optimize the model into a deployable product. There was also a lack of clarity among OpenAI employees around how much IP they should be sharing with their GitHub counterparts, while GitHub employees struggled with how much to trust OpenAI. Disagreements compounded as the companies clashed over how and when to release the product and who would get the credit.

Murati eventually brokered a compromise—a skill that would gain her increasing respect among people who worked with her across companies. Microsoft would get its moment by releasing its consumer-facing product, GitHub Copilot, in June 2021. OpenAI would then release its version of Codex directly in the company's API in August.

The arrangement would give Microsoft a new user base and a modest financial bump: In two years, GitHub Copilot would grow to one million paid subscribers, bringing in over $100 million in annual recurring revenue. But for OpenAI, the deal deepened an emerging sense at the company that it would be better served to work on its own consumer products. OpenAI's researchers had worked hard on the model and were ceding all of the brand recognition to GitHub and Microsoft; watching those two companies enjoy the credit for OpenAI's work in public was a tough pill to swallow. Microsoft was also a challenging partner; many felt it had far too much bureaucracy and required too much hand-holding to make the most of OpenAI's models. By relying on the tech giant to deliver its technologies to the public, OpenAI was also losing visibility into and data from its users and, most importantly, control over its vision.

As OpenAI concentrated its bets, Altman was applying the same strategy to his other projects and investments. Over the years, as he'd shifted toward more hard-tech innovation, he had developed a belief in betting big and long on the most important projects.

"If you could wave a wand, change anything about the tech, startup, entrepreneurship ecosystem, what would you change?" his brother Jack had asked him at an event as he'd stepped down from YC.

"It would be to get everyone in the ecosystem to take a much longer time horizon," Altman had said. "This world where people start a company and plan to run it for four or five years, join a company and only plan to stay for one or two—that's not how important shit gets done."

Along those lines, 2021 seemed to mark a major shift in Altman's personal investment strategy away from taking a large number of small bets toward taking a small number of really large ones. That year a startup he'd cofounded in 2019 called Tools for Humanity that had remained largely quiet saw an influx of funding and media coverage and a concerted ramp-up in its operations, as Altman directed more external attention to the company. The venture was a dedicated effort to develop a working mechanism for universal basic income, or UBI, a popular Silicon Valley idea to give everyone a regular minimum distribution of income. Altman often passionately discussed UBI as the possible antidote to a future world where AI could create mass economic fallout. At YC he had started the largest pilot in the US to study the concept, spinning out a nonprofit in the process called OpenResearch that administered the program. Over three years, OpenResearch gave a $1,000 monthly stipend to a randomly selected group of one thousand out of three thousand low-income people, with the rest getting fifty dollars a month as a control. In July 2024, OpenResearch would release its findings, showing that the unconditional cash helped people meet their basic needs, assist others, and have more economic leeway.

Tools for Humanity's main product, Worldcoin, was a self-described "collectively owned" cryptocurrency that would allow everyone to eventually get a share of its value. As part of the scheme, the company was developing a dramatic-looking chrome-colored orb—roughly the size of a bowling ball and partly a reflection of Altman's design tastes—to scan people's irises and verify their identity before giving them their cut. The iris scanning would be a necessity, the founders argued, once AI also made it increasingly hard to decipher fake media from reality. An extensive investigation from Eileen Guo and Adi Renaldi at *MIT Technology Review* would later find that these iris-scanning efforts were mired in data privacy infringements, deceptive marketing practices, and potential legal violations. In July 2023, Worldcoin would officially launch to massive controversy, as people began lining up by the thousands, particularly in Global South countries, to give over their biometric data with little understanding of what they were doing it for other than the vague promise of free money.

Also in 2021, Altman made his two largest ever investments: $180 million into an antiaging company called Retro Biosciences, working to extend human lifespans through cellular rejuvenation, and $375 million into Helion Energy, working to commercialize nuclear fusion. "I basically just took all my liquid net worth and put it into these two companies," Altman told *MIT Technology Review*'s Antonio Regalado. Altman described both technologies in language that mirrored OpenAI's research road map—they seemed impossible currently but, if scaled up aggressively, could be around the corner.

The Retro Biosciences bet reflected Altman's fixation on longevity. He was an avid follower of "young blood" research—a line of scientific inquiry that studied how to reverse aging with transfusions of healthier, younger blood. Notably, it was an area in which Thiel was also interested, spawning a plethora of articles and memes about his desire to inject himself with the blood of teenagers. While at YC, Altman had also signed up with a $10,000 deposit to be on the wait list of a controversial startup called Nectome, which had been in one of the accelerator's batches. Ripped straight out of science fiction, Nectome was pitching a service that would cryogenically freeze customers' brains to one day—potentially hundreds of years into the future—upload to a computer after scientists had cracked the technology to do so. The catch was that Nectome needed the person's brain to be fresh for the preservation to work. To Antonio Regalado, co-founder Robert McIntyre called his product "100 percent fatal."

Helion reflected Altman's obsession with finding ways to generate clean, cheap, and abundant energy. He often remarked that the cost and availability of energy was highly correlated with quality of life and with economic growth. But without carbon-free alternatives, rising energy consumption would "destroy the planet," he said. In 2023, he would describe Helion as "more than an investment" and "the other thing besides OpenAI I spend a lot of time on." Microsoft would subsequently sign a deal to purchase power from Helion's first plant, after the tech giant had made its third investment, worth $10 billion, into OpenAI. To the astonishment and skepticism of energy experts, Helion would commit to having its plant ready by 2028.

The year 2021 was also when Altman brought his predilection for investing to OpenAI. That May, he launched the OpenAI Startup Fund, a $100 million investment pool for supporting early stage companies with, as he described, "big ideas about how to use AI to transform the world." Microsoft once again became an investor in the fund. To some observers, the fund's creation was a strange decision. OpenAI was barely generating revenue and already capital intensive enough as it was; why raise yet more money for separate companies to use? Others felt it was Altman's way of remaking YC's powerful network effects around OpenAI. Still others viewed it simply as Altman's force of habit. "This is Sam's way of moving through the world," says a person who worked with him. "Dealmaking."

Altman liked to say that he had taken no equity in OpenAI to avoid corrupting the quest of safe AGI with his own desires for profit. He made only a yearly salary of $65,000 and accumulated his wealth through other ventures. The sentiment had a nice ring to it—and echoed his original rhetoric around why OpenAI started as a nonprofit. It was also a statement, like the nonprofit status of the organization, that by 2021 no longer reflected the full truth. Altman had a significant stake in YC, and YC, through its $10 million investment in OpenAI, could receive up to a $1 billion return.

As OpenAI continued to commercialize, many YC startups and many of his other investments would also become customers or commercial partners of the company. *The Wall Street Journal* would subsequently calculate Altman's net worth in June 2024 across all his holdings to be at least $2.8 billion. With the OpenAI Startup Fund, Altman added yet another complication to his altruistic narrative—one that would eventually play its own small part in his fleeting ouster.

Chapter 9

Disaster Capitalism

As OpenAI barreled forward, guided by Altman's convictions, the boundaries of the sweeping consequences of the company's vision were expanding. With its pumping of ever-larger and polluted datasets into its models, it had created the "paradigm shift" that Appen's Ryan Kolln would describe to me—the moving away from filtering data inputs to controlling model outputs. The language of abstraction once again dressed up a grim reality: what that shift really meant for the people who now bore the brunt of controlling those outputs.

In 2021, in parallel with its push to develop the next generations of its models, OpenAI began a project to create a much better version of its automated content-moderation filter for cleaning them up. Where GPT-3 had been placed on the API with no filtering whatsoever, leading to the Latitude text-based child porn scandal, the company wanted to be more careful with the models it would start calling GPT-3.5 and eventually GPT-4. As OpenAI prepared to deploy its technologies more widely, having a completely unfiltered product could prove problematic in the long run from a legal, public relations, and usability perspective. At the time, the plans for what would become ChatGPT had yet to be conceived, but the chatbot would also later benefit from the same filter. That filter would act as a wrapper around each model, for the purpose of flagging and removing offensive content from its output before it reached the user.

To build the automated filter, OpenAI first needed human workers who could carefully review and catalog hundreds of thousands of examples of exactly the content—sex, violence, and abuse—that the company wanted to prevent its models from generating. After six months of searching, it found a vendor that seemed well suited to take on the project: an outsourcing firm that had been performing content moderation for Meta since 2019 and coincidentally shared Altman's nickname, Sama. OpenAI sent Sama an email asking whether it took on projects that involved sensitive or explicit content and what its typical approach was for handling them. Sama provided thorough answers. OpenAI signed four contracts with the firm for $230,000, landing the project in the hands of dozens of workers in Kenya.

It's no coincidence that Kenya became home to what would ultimately turn into one of the most exploitative forms of labor that went into the creation of ChatGPT. Kenya was among the top destinations that Silicon Valley had been outsourcing its dirtiest work to for years. With the many other countries that the tech industry relegates to this role, Kenya shares a common denominator: It is poor, in the Global South, with a government hungry for foreign investment from richer countries. All of these are a part of Kenya's legacy of colonialism, which has left it without well-developed institutions to protect its citizens from exploitation and often in the throes of economic crisis, both of which make circumstances ripe for overseas companies to find an immiserated pool of labor that will do piecework under almost any conditions.

You can see the markers of that legacy in the many faces of Nairobi, Kenya's capital. The city suffers grave inequality. The central business district has gleaming towers, international five-star hotels, and high-end restaurants. The diplomatic neighborhood has large, stately buildings and high security walls. The residential expat areas offer stunning mansions with lush private gardens. And then there are the outskirts: Utawala, Dagoretti South, Embakasi. Drive to any one of these neighborhoods, and skyscrapers made of steel turn into squat cinder block structures. Buildings begin to scatter about like weeds in erratic patterns without coordinated planning. The roads go from paved to unpaved, from four lanes to narrow strips meant primarily for motorbikes and pedestrians. Deeper in, concrete turns into corrugated tin, and jerry-built homes and businesses cram together ever more tightly.

Under these conditions, Kenya's government had willingly embraced Silicon Valley when it came in search of low-wage workers. Kenya has limited local industry. Many of the biggest brands are European and American. Some of the largest infrastructure, once built by the British, is now built by the Chinese. Cars are mostly hand-me-downs from Japan, where drivers also sit on the right side of the vehicle. Tech giants, as the government saw it, could help the country create the jobs it desperately needed. Joblessness breeds crime. Petty theft is common. People who feel disempowered grow distrustful of institutions. During the Russia-Ukraine war, as Kenya's grain prices rose, rumors spread that the president was purposely straining already hungry families. Many repeated a familiar refrain in the US: The election was rigged.

And so, Kenya became a critical hub of the internet's backstop labor. Several firms like Sama—middlemen in the data labor supply chain—established operations in Nairobi, building up pools of workers to service overseas tech companies, primarily in the Bay Area.

For OpenAI, Sama appeared to check off all the right boxes. Originally called Samasource, it was a San Francisco–based social enterprise that had begun in 2008 with a mission of providing meaningful, dignified work to people in impoverished countries to lift them out of poverty. Under its founder, Leila Janah, it had established operations in India and Kenya and developed a reputation as an ethical outsourcing company. In 2018, it transitioned to a for-profit, during which it shortened its name, in order to scale its operations. In 2020, it received a B Corp certification. In its answers to OpenAI's questions in 2021, the organization detailed its experience with content moderation and emphasized its protocols for keeping projects secret and their data secure. It mentioned that it provided mental health resources to its workers to help them deal with psychologically troubling content.

Behind the scenes, however, Sama was in disarray. In January 2020, Janah had passed away from a rare cancer at just thirty-seven; combined with the pandemic soon after, workers say it seemed to mark the beginning of more organizational mismanagement, a characterization that a Sama spokesperson denied. It wasn't until early 2022 that those challenges would come to the fore when Billy Perrigo, a reporter at *Time* magazine, would publish an extensive investigation. He would reveal that Sama had taken on a project for Meta, to provide content moderation for Facebook for all of sub-Saharan Africa, that repeatedly exposed workers to violent and graphic videos, such as of suicides and beheadings, and left them deeply scarred and struggling. Sama would defend itself, saying it took on the project after careful consideration from its East Africa team, which wanted to ensure "content for Africans was effectively reviewed by Africans." Nearly two hundred workers would file multiple lawsuits against Sama and Meta alleging traumatic working conditions and unlawful terminations for attempting to organize for higher pay and better working conditions. The Sama spokesperson rejected the allegations.

Against this backdrop, OpenAI began the first phase of its project in late 2021. Under the code names PBJ1, PBJ2, PBJ3, and PBJ4, it thrust teams of Sama workers into more traumatic content-moderation work, for on average between $1.46 and $3.74 an hour. Workers had no idea for whom or why they were doing the project, kept in the dark under the nondisclosure terms of the contract, common in the data annotation industry. What they did know was what was in front of them: the hundreds of thousands of grotesque text-based descriptions that they needed to read and sort into categories of severity. Was it violence or extremely graphic violence, harassment or hate speech, child sexual abuse or bestiality?

Gradually, the work broke many of the workers, the impacts radiating beyond each individual to the people who depended on them in their communities. Only after the release of ChatGPT would they begin to grasp what exactly they had paid for with their peace of mind. In May

DISASTER CAPITALISM   193

2023, I visited four workers in Nairobi who would agree to share their experiences with me on the record for a story on the front page of *The Wall Street Journal*. For one of them on the sexual content team, a man named Mophat Okinyi, the project that unraveled his mind and his relationships would turn out to be in service of a technology that would in turn contribute to the erosion of his brother's economic opportunities.

―――

At a time when freewheeling corporate research was still permitted, Microsoft anthropologist Mary L. Gray and computational social scientist Siddharth Suri were among the first to show the world the plight of workers like those in Kenya who build their livelihoods around an essential piece of the AI supply chain.

In 2019, they published their book *Ghost Work*, based on five years of extensive fieldwork, revealing a hidden web of piecemeal labor and digital exploitation that propped up Silicon Valley. Tech giants and unicorns were building their extravagant valuations not just with engineers paid six-figure salaries in trendy offices. Essential, too, were workers, often in the Global South, being paid pennies to carefully annotate reams of data.

Take self-driving cars. A self-driving car needs to drive in the correct lane, respond to erratic driving behavior, and pause at a safe distance for schoolchildren crossing the road. To do this, the software system controlling the car uses an amalgamation of several deep learning models, including those dedicated to recognizing objects on the road: lane markings, road signs, traffic lights, vehicles, trees, pedestrians.

Companies develop those models by driving vehicles around with numerous large cameras, recording billions of miles of footage. The footage is the data, and to annotate it means tracing, frame by frame, each object that appears—sometimes down to the curvature of a hand gripping a bike handle or a dog lounging halfway out of a car window—and assigning them labels like "bike," "vehicle," "animal," "human." People have to do that work. And from a company's perspective, the cheaper they do it, the better.

Gray and Suri's research focused in part on Mechanical Turk, a platform developed by Amazon many years before the deep learning boom in 2012, which for a long time served as the de facto middleman for companies looking to hire someone cheap for any kind of piecemeal digital labor. By the time *Ghost Work* came out, the first era of AI commercialization was already evolving, building upon, and rapidly expanding this outsourcing model.

I began mapping out the new shape of this hidden workforce, unearthing a sprawling global pipeline of labor spanning many countries, both expected and surprising. I spoke with the newest middlemen replacing Mechanical Turk—platforms designed to cater more specifically to AI development. I spoke with dozens of workers, visiting some of their homes, eating dinner with their families, seeking to understand not just the macro trends pressing down on them but the daily textures of their lived realities.

Just as the first era of AI commercialization laid the groundwork for the generative AI era's amassing of data and capitalization of compute, so, too, did it create the foundations for its wide-scale labor exploitation. In this way, it is important to first understand those foundations in order to understand the experience of the Kenyan workers who contracted for OpenAI. Only then is it possible to recognize that their experiences were far from anomalous but rather a direct consequence of the compounding of the AI industry's long-standing treatment of its hidden workers and its views on whose labor is or isn't valued, with OpenAI's empire-esque vision for unprecedented scale.

Before generative AI, self-driving cars were the biggest source of growth for the data-annotation industry. Old-school German auto giants like Volkswagen and BMW, feeling threatened by the Teslas and Ubers of the world, spun up new autonomous-vehicle divisions to defend their ground against the fresh-faced competition. As billions of new dollars flooded into the race to create the cars of the future, the demand for data annotation exploded and created a need for alternatives to Mechanical Turk.

MTurk, as it was called, was a generalist platform, meaning it didn't cater to any particular kind of work. It was just a self-service website. Its interface—stuck in the web design of the mid-aughts, when it launched—had a place to upload datasets, to specify simple annotation instructions, and to set a price for the work. Once the task was claimed, it showed randomized strings of numbers and letters in place of the workers' names. It had two buttons next to each worker: one to give them a bonus, the other to boot them off the project.

Data annotation for self-driving cars necessitated a different approach. First and foremost, it required a new level of accuracy. One too many mislabeled frames—vehicles traced with sloppy borders, pedestrians not traced at all—could be the difference between life and death. To guarantee that quality, workers needed to be trained and companies needed to write detailed instructions. There needed to be more mechanisms for feedback and iteration. MTurk fell out of favor. In stepped a wave of startups and incumbents including Scale AI, Hive, Mighty AI, and Appen. Each had their own worker-facing platforms, which allowed anyone to create an account and start tasking.

But right as this new wave of companies sought to establish themselves, a strange thing happened. Sign-ups on their worker-facing platforms came rushing in from an unexpected country: Venezuela. In the same moment that auto giants began scrambling, money began pumping into self-driving cars, and data-annotation firms began looking for more workers, Venezuela was nose-diving headfirst into the worst peacetime economic crisis globally in fifty years.

Economists say it was a toxic cocktail of political corruption and the government's misguided policies that squandered the country's rich natural endowment. Venezuela sits atop the largest proven petroleum reserves in the world. It was once Latin America's wealthiest country. But beginning in 2016, hyperinflation went haywire; unemployment skyrocketed; violent crime exploded as families across the country watched the value of their entire life savings collapse. From late 2017 to 2019, escalating sanctions imposed by the Trump administration, intended as a punishment for Venezuelan leader Nicolás Maduro's authoritarian abuses, delivered the final death knell to Venezuela's economy. Hyperinflation hit a once unfathomable 10 million percent. People with graduate degrees and previously well-paying jobs were now spending their days lining up in front of stores for a chance at receiving meager rations of rice and flour.

Amid the catastrophe, many Venezuelans turned to online platforms for work. By mid-2018, hundreds of thousands had discovered and joined the data-annotation industry, accounting for as much as 75 percent of the workforce for some outsourcing firms. Working on data-annotation platforms became a whole-family activity. Julian Posada, an assistant professor at Yale University who interviewed dozens of Venezuelan workers, found that parents and children often took turns to work on a shared computer; wives reverted to cooking and cleaning to allow their husbands to earn just a little more money by pulling longer hours uninterrupted. The crisis left an indelible mark on the wave of AI-specialized outsourcing firms as they grew up alongside it. Venezuela was not an obvious choice for finding pools of labor. The language barrier made it more difficult for the mostly San Francisco– and Seattle-based firms to coordinate with workers. But the acute desperation among Venezuelans meant they were willing to work for astonishingly small amounts of money, which in turn meant the firms could offer astonishingly good prices for their services. "It was like a freak coincidence," Florian Alexander Schmidt, a professor at the University of Applied Sciences Dresden who has studied the rise of the data-annotation industry, told me in 2022.

That "freak coincidence" revealed a disturbing formula. When faced with economic collapse, Venezuela suddenly checked off the perfect mix of conditions for which to find an inexhaustible supply of cheap labor: Its population had a high level of education, good internet connectivity, and, now, a zealous desire to work for whatever wages. It was not the only country that fit that description. More populations were getting wired to better internet. And with accelerating climate change and growing geopolitical instability, it was hard to bet against more populations plunging into crisis. "It's quite likely there will be another Venezuela," Schmidt said.

At the time, Schmidt's prediction made me wonder whether the second time around would still be coincidence or whether data-annotation firms would make a playbook out of what had worked there. Scouting workers in crisis could become a surefire way to continue driving down the costs of the labor that serves as the lifeblood of the AI industry. Looking back several years later, that's exactly what happened—and what has become one of the most stunning parallels between empires of old and empires of AI. One of the defining features that drives an empire's rapid accumulation of wealth is its ability to pay very little or nothing at all to reap the economic benefits of a broad base of human labor.

In December 2021, I journeyed through the winding mountains of Colombia to better understand the life of a worker who, in crisis, had turned to data annotation. Travel restrictions barred me from going to Venezuela, but here in its neighboring country lived nearly two million Venezuelan refugees, one-third of the population that had been displaced by the economic catastrophe.

Julian Posada connected me with one of them, a woman named Oskarina Veronica Fuentes Anaya, who continued to work in the data-annotation industry after she escaped her home country. Fuentes was the first person to show me what this work is really like—the way she'd reoriented her entire life around working for a platform; the way that platform in turn treated her as disposable.

In the apartment she shared with half a dozen relatives, we sat side by side in the living room as she clicked through screen after screen on Appen. The tasks were varied. They ranged from categorizing products on e-commerce sites—*Should this item be listed under clothing or accessories?*—to performing content moderation for social media—*Does this video contain crime or human rights violations?* For tasks that required English, she used Google Translate to convert the text into her native Spanish.

Each time she completed a task, the sum of money she earned, displayed in US dollars, would increase by a few pennies. She needed a minimum of ten dollars to withdraw it, which, when she first joined the platform, wasn't a problem. Now, it could take weeks to accumulate that much money. That minimum sometimes felt like a cruel arbiter of whether she had enough funds to pay for groceries.

For workers actually living in Venezuela, the process of withdrawal was even more challenging. Most global payment systems such as PayPal didn't allow money transfers into Venezuela. Most stores and shops in Venezuela didn't accept payments from the ones that do. This meant workers needed to convert their digital funds into cash to pay for basic goods and services. But where the money arrived online in US dollars, the cash needed to be in Venezuelan bolivares. The black market to convert one to the other abounded with scams and high commissions.

Fuentes had a complicated relationship with the platform. It had never been her intention to work this kind of job, but through a series of events outside her control, it had become her lifeline as well as a punishing force.

She had created an Appen account in grad school to earn some extra money while finishing up a master's in engineering. She was sharp, hardworking, and creative. Had her country not crumbled, a top student like her would likely have had guaranteed job security working for the state oil company. When her country did, she adapted, carefully orchestrating her and her husband's departure to Colombia for a chance at a better future.

In that regard, Fuentes was one of the lucky ones. By birth she was entitled to a Colombian passport, unlike many others who escaped without documentation. Her parents were Colombian before they'd fled a generation earlier in the opposite direction, to Venezuela, to escape a different nexus of violence and political instability. It was an all-too-common story—the compounding of generations of crisis across borders, thrusting families into an endless state of siege and survival.

With that passport, Fuentes arranged remotely from Venezuela to rent an apartment in Colombia from an acquaintance. They needed two people who owned property to cosign the lease. The acquaintance, their prospective landlord, agreed to help procure them.

In early 2019, with only enough money for a week of groceries to their names, Fuentes and her husband crossed the border. But upon arrival, they discovered another Venezuelan couple already living in the apartment their landlord had promised them. With no other choice, both couples shared the same roof, each filled with fear and distrust that they would lose their home to the other.

The other couple eventually left, but it was only the beginning of a new string of problems. While Fuentes had found a job at a local call center, her husband didn't have work authorization. Before he could secure one, the call center announced that it would be closing. So when Fuentes began to experience signs of a serious health problem, she ignored the symptoms and continued working. All she could think about was putting in as many hours as possible in the final stretch of her employer's operation.

The doctor later told her that had she waited any longer, she likely would have died. A coworker, alarmed by the markers of Fuentes's deteriorating health, had brought her to the hospital shortly before her body started convulsing and her pulse stopped for a full minute.

She was diagnosed with severe diabetes and immediately placed on five daily courses of insulin. For weeks she experienced crippling pain and bouts of blindness. When she restabilized, she continued to suffer intense fatigue and couldn't leave home for more than a couple hours.

Even then, all she could think of was that she and her husband needed money. But with a chronic illness, she could no longer safely commute the distances she needed to return to an office. It was then that she pulled out her laptop and logged back in to Appen.

To Fuentes, there was little apparent logic to which tasks appeared in her Appen queue. The only thing it made clear was that she needed to have good, consistent performance to continue receiving work. Wilson Pang, Appen's CTO then, told me in 2021 that the platform used algorithms to distribute projects based on a mix of factors including the workers' location, their overall accuracy and speed, and the types of tasks at which they'd previously excelled.

In Telegram and Discord groups, Fuentes traded tips with other Venezuelans working on Appen as they sought to deduce the rules like an elaborate puzzle. They discovered that using a VPN to appear to be in the US earned them the most money. They also learned—the hard way—that it was a high-risk endeavor. Appen searched for this kind of behavior, which was a violation of platform rules, and punished workers by closing their accounts. An account closure could be devastating. Any earnings a worker hadn't withdrawn would vanish, and opening up a fresh account meant starting back at the bottom, with the least-well-paid tasks or, increasingly, no tasks at all.

There were other rules. Submitting a task quickly was rewarded, but submitting a task too quickly triggered something in the system that meant a worker wouldn't get paid for that task. The prevailing theory was that the platform associated exceptional speed with bot activity, which meant it discarded the answers. Sometimes the tasks that appeared also had few instructions and were impossible to decipher; other times the platform had bugs that didn't load the tasks correctly.

The Venezuelans in the group who were once software engineers created browser extensions to deal with these issues and shared them with their fellow Appen workers. One extension added an extra time delay to every task submission to avoid the apparent bot tax. Another automatically refreshed the Appen queue every second because the platform didn't always update itself. A third sounded an alarm once a new task appeared so workers could step away from their computers to go to the bathroom or cook without fear of missing an opportunity.

For all that the workers did to help each other, the platform pitted them in competition. Projects were first come, first served. A task stuck around in queue only as long as it took for enough workers to claim it. This window—between a task's arrival and its disappearance—shrank over time from days to hours to seconds as more and more workers, including many Venezuelans in crisis, joined Appen and vied for scraps of work.

The erratic, unpredictable nature of when work came and went began to control Fuentes's life. Once she was taking a walk when a task arrived that would have earned her several hundred dollars, enough money to live on for a month. She sprinted as fast as possible back to her apartment but lost the task to other workers. From that day on, she stopped leaving the house on weekdays, allowing herself only thirty-minute outings on weekends, which she learned from experience was when tasks were less likely to show up. She slept fitfully, worried about the tasks that would arrive in the middle of the night. Before bed, she would turn her computer to maximum volume so that if they did, the browser extension that rang the alarm would wake her up.

Yet despite how much stress and hairpulling Appen caused, Fuentes couldn't imagine leaving the platform. She was terrified that tasks would stop arriving altogether and she would be forced to move on. Appen had been her savior, the only thing that pulled her through when everything else in her life had threatened to end her. Not only that, the earnings were once so great, she was able to invest in a new laptop and recoup the cost and then some.

When things were good, they were really good. When things were bad, she stayed tethered to the platform with the stubborn faith that it would return her loyalty.

Fuentes taught me two truths that I would see reflected again and again among other workers, who would similarly come to this work amid economic devastation. The first was that even if she wanted to abandon the platform, there was little chance she could. Her story—as a refugee, as a child of intergenerational instability, as someone suffering chronic illness—was tragically ordinary among these workers. Poverty doesn't just manifest as a lack of money or material wealth, the workers taught me. It seeps into every dimension of a worker's life and accrues debts across it: erratic sleep, poor health, diminishing self-esteem, and, most fundamentally, little agency and control.

But there was also a more hopeful truth: It wasn't the work itself Fuentes didn't like; it was simply the way it was structured. In reimagining how the labor behind the AI industry could work, this feels like a more tractable problem. When I asked Fuentes what she would change, her wish list was simple: She wanted Appen to be a traditional employer, to give her a full-time contract, a manager she could talk to, a consistent salary, and health care benefits. All she and other workers wanted was security, she told me, and for the company they worked so hard for to know that they existed.

Through surveys of workers around the world, labor scholars have sought to create a framework for the minimum guarantees that data annotators should receive, and have arrived at a similar set of requirements. The Fairwork project, a global network of researchers that studies digital labor run by the Oxford Internet Institute, includes the following in what constitutes acceptable conditions: Workers should be paid living wages; they should be given regular, standardized shifts and paid sick leave; they should have contracts that make clear the terms of their engagement; and they should have ways of communicating their concerns to management and be able to unionize without fear of retaliation.

Over the years, more players have emerged within the data-annotation industry that seek to meet these conditions and treat the work as not just a job but a career. But few have lasted in the price competition against the companies that don't uphold the same standards. Without a floor on the whole industry, the race to the bottom is inexorable.

Among the crop of data-annotation firms that rose to meet the demands of the self-driving car boom, one firm was particularly successful in exploiting the crisis playbook. Cofounded in 2016 by wunderkind Alexandr Wang, at the time a nineteen-year-old MIT dropout, Scale AI from the beginning followed a strategy that rested in part on its emphasis for providing specialized, quality services at a low price. One former Scale employee who oversaw workforce expansion explained to me the mandate: "How do you get the best people for the cheapest amount possible?" Scale quickly gained major clients like Lyft, Apple, Toyota, and Airbnb.

Where MTurk's workforce primarily came from the US and India,

Scale went hunting first in Kenya and the Philippines, English-speaking former colonies with a long history of servicing American companies through call centers and digital work. The startup's worker-scouting teams searched for the areas in each country that struck the very same balance of factors that would converge in Venezuela: a high density of people with good education and good internet yet who were poor and thus willing to work hard for very little money. The thesis was guided not only by the company's cutthroat business practices but also a compelling story they told themselves: that these were the people who could benefit most from the economic opportunity and be happier because of it. "If you could be pulling a rickshaw or labeling data in an air-conditioned internet café, the latter is a better job," Mike Volpi, a general partner at Index Ventures, told *Bloomberg* in 2019 after joining a $100 million funding round for Scale.

But after the company launched its worker-facing platform, Remotasks, and noticed the overwhelming interest from Venezuela, Venezuelans became one of Scale's top recruiting priorities. "They're the cheapest in the market," the former employee said. In 2019, the company launched an expansion campaign in the Latin American country using referral codes and social media marketing videos with stock footage showing stacks and stacks of highly coveted US dollars. The following year, it created a Venezuela-specific landing page for Remotasks and pushed users to join a new initiative called Remotasks Plus. It billed the invitation-only program as a way to help Venezuelans going through a historic hardship and promised participants opportunities to learn new skills, advance their careers, and receive increased earnings through consistent working hours and hourly wages. As the pandemic hit, compounding the economic crisis, Venezuelans flocked to Remotasks Plus en masse. Scale's competitors—other data-annotation platforms—lost their footing in the market.

Once Scale held the dominant position, its promises to workers faded. Through late 2021 and early 2022, I partnered with a Venezuelan journalist in Caracas, Andrea Paola Hernández, who interviewed Venezuelans who had worked for Scale during the Remotasks Plus program. We also embedded ourselves within the Remotasks Discord community, which Scale used to communicate and coordinate with its global workforce. We found through a spreadsheet the company left public that the workers' earnings began to decline within weeks of the program's launch; workers who started with earnings of forty dollars a week were soon making less than six dollars or nothing at all. In April 2021, the company shuttered the Remotasks Plus program entirely and reverted to its standard operations, doling out tasks in a piecemeal fashion with no standard or guaranteed hours.

Inside Scale, Remotasks Plus had been an experiment. The company believed it would be easier to pay workers based on hours rather than tasks completed. The reality proved the opposite. Employees quickly realized they had no way of verifying worker hours and believed many were scamming the platform by logging more time than they'd actually worked. After months of trying to fix the problem—including adding more and more forms of worker surveillance—Scale decided to cut it off to stem the outflow of money. With nowhere to go, over 85 percent of the workers continued to task on the platform, a number that a Scale spokesperson pointed to as evidence that they had "ongoing interest and engagement."

By the time Hernández started interviewing the workers, roughly seven months after the Plus program was canceled, the pay on Remotasks had dropped further. Hernández created an account on the platform to try it out. After two hours of completing a tutorial and twenty tasks, Hernández earned eleven US cents. Matt Park, then the senior vice president of operations at Scale, told us in response to the findings that Venezuelans on the platform earned an average of a little more than ninety cents an hour. "Remotasks is committed to paying fair wages in every region we operate," he said.

Many Venezuelans who complained were booted off the platform. For Ricardo Huggines, a computer engineer who began working for Remotasks to support his wife and kids after a devastating weeklong nationwide power outage, his account was canceled after he began asking too many questions in the Discord, he told Hernández. "From the way they treated us, I realized that their approach was to drain each user as much as possible," he said, "and then dispose of them and bring new users in."

Scale was indeed bringing in new users. By mid-2021, as Venezuelans burned out and left the platform, Scale was scouting and onboarding tens of thousands more workers from other economies that had collapsed during the pandemic. To support its expanding and diversifying client needs, it entered countries with large populations facing financial duress and who could also speak the most economically valuable languages: English, French, Italian, German, Chinese, Japanese, Spanish. It sought French speakers from former French colonies in Africa, an employee who worked on international expansion remembers; it sought Mandarin speakers from places with large populations of Chinese diaspora such as in Southeast Asia.

Scale proceeded to repeat the playbook it had developed in Venezuela again and again. It offered high earnings in each new market to attract workers and throttled those earnings as it settled in. It tinkered with the size of its payouts to taskers through rounds of experimentation that full-time employees, sitting in its now 180,000-square-foot San Francisco headquarters, discussed as optimization and innovation. Workers meanwhile saw their livelihoods decimated with the unpredictable changes. The Scale spokesperson said the company rejected the characterization that it has targeted economies in hardship and purposely cut back earnings. Scale recruits workers based on considerations including geographic and linguistic diversity and 24/7 coverage. "We care deeply about our contributors and any claim to the contrary is false," he said.

One group of eight workers in North Africa said Scale reduced their pay by more than a third in a matter of months. At least one worker was left with negative pending payments, suggesting that he owed Scale money. When the group attempted to organize against the changes, the company threatened to ban anyone engaging in "revolutions and protests." Nearly all who spoke to me were booted off the platform. The Scale spokesperson said the company does not suspend workers for concerns about pay, only violations of Community Guidelines.

Scale's payment systems, chronically underinvested in by its US engineering teams, were also riddled with bugs that often left workers unable to cash out. As Scale grew, these practices would grate on full-time employees who worked most closely with these workers; many sought to advocate on behalf of the workers to Scale leadership for better working conditions and wages, and basic guarantees on payments, only to leave after exhausting themselves, or to be pushed out of the company. The spokesperson said it has since "significantly improved" its platform stability.

Scale's dominance would pose a growing challenge to companies that sought to follow a different model and pay living wages. One such firm, CloudFactory, which operates in Kenya and Nepal, provides workers an employment contract and consistent working hours, in accordance with Fairwork's standards. But according to founder and executive chairman Mark Sears, it has lost many contracts to Scale over the years.

To clients, CloudFactory pitches the idea that it can deliver better quality in the long run than what the industry calls "the anonymous crowd work" model. CloudFactory's workers are well trained and develop expertise over time. When they excel, they receive promotions. Many workers I spoke to in Kenya considered the company among the best data-annotation firms to work for. Sometimes CloudFactory's pitch works. A growing number of clients also come to the firm because of its track record as an employer. But when budgets tightened during the pandemic, many clients moved back to cheaper options. CloudFactory had to lay off workers.

Workers say it was under the same kind of competitive pressure that Sama also began to erode its standards. At first, they told me, a job at Sama was even more coveted than a job at CloudFactory. Then Leila Janah died, the pandemic hit, and clients shifted to Scale and other cheaper options. Workers say, though the Sama spokesperson denied this, that this led the company down the path of accepting OpenAI's content-moderation filter project and putting the work in their hands, at a time when they were in dire straits, just like Fuentes and the other Venezuelan workers.

Mophat Okinyi grew up in a village on an island in western Kenya, an eight-hour bus and two-hour boat ride away from Nairobi. The island is in Lake Victoria, a large body of water with uninterrupted views of the horizon. Medical treatment was far away; unexpected health emergencies were almost always a harbinger of death.

He was poor, but as kids he and his siblings didn't think much about their poverty. They reveled in the stories of their ancestors: Legend has it that their tribe, the Luo people, originally came from Israel. They used their knowledge of boat construction and river navigation to migrate south along the Nile, fanning out to western Kenya and parts of Uganda and Tanzania where they live today. "Luos are not Kenyans," Okinyi said in a hushed tone like he was letting me in on a secret. "We're Israelites who live in Kenya. But Kenya would not be Kenya without Luos."

As we sat in his apartment, construction droned on outside as flies buzzed around us. "Barack Obama is a Luo," he added with a smile. "Luos are a very sharp people."

Poverty now took up much more real estate in Okinyi's mind. At twenty-eight, he had more responsibilities. He needed to make rent and put food on the table; he needed to pay for his niece—his sister's daughter—to go to public school, which in Kenya isn't free. When he had a job, he knew to count his blessings. The country's youth unemployment is 67 percent. In 2021, the World Bank estimated that more than a quarter of the country's population lived on less than $2.15 a day.

It felt like a miracle when, in November 2021, Sama called him in for a new opportunity. He had joined the firm in 2019 after applying on its Careers web page for an "AI training" opening. His projects at Sama had followed the trajectory of the AI industry. In the first two years, he had worked exclusively on computer-vision annotation, including for self-driving cars. Though he didn't know it yet, this new project would be his first for generative AI.

Okinyi's managers at Sama gave him an assessment they called a resiliency screening. He read some unsettling passages of text and was told to categorize them based on a set of instructions. When he passed with flying colors, he was given a choice to join a new team to do work he considered to be similar to content moderation. He had never done content moderation before, but the texts in the assessment seemed manageable enough. Not only would it be absurd to turn down a job in the middle of the pandemic, but he was thinking about his future. He was living in Pipeline, a chaotic, slum-like neighborhood in southeast Nairobi, jammed with tenements and twenty-four-hour street vendors, buzzing with the restless energy of twentysomethings jostling their way to something better. Okinyi was on his way to something better. He had just met a girl next door named Cynthia who for the first time made him imagine what it would be like to build a family.

Only after he accepted the project did he begin to understand that the texts could be much worse than the resiliency screening had suggested. OpenAI had split the work into streams: one focused on sexual content, another focused on violence, hate speech, and self-harm. Violence split into an independent third stream in February 2022. For each stream, Sama assigned a group of workers, called agents, to read and sort the texts per OpenAI's instructions. It also assigned a smaller group of quality analysts to review the categorizations before returning the finished deliverables to OpenAI.

Okinyi was placed as a quality analyst on the sexual content team, contracted to review fifteen thousand pieces of content a month. OpenAI's instructions split text-based sexual content into five categories: The worst was descriptions of child sexual abuse, defined as any mention of a person under eighteen years old engaged in sexual activity. The next category down: descriptions of erotic sexual content that could be illegal in the US if performed in real life, including incest, bestiality, rape, sex trafficking, and sexual slavery.

Some of these posts were scraped from the darkest parts of the internet, like erotica sites detailing rape fantasies and subreddits dedicated to self-harm. Others were generated from AI. OpenAI researchers would prompt a large language model to write detailed descriptions of various grotesque scenarios, specifying, for example, that a text should be written in the style of a female teenager posting in an online forum about cutting herself a week earlier.

In that sense, the work did differ from traditional content moderation. Where content moderators for Meta reviewed actual user-generated posts to determine whether they should stay on Facebook, Okinyi and his team were annotating content to train OpenAI's content-moderation filter in order to prevent the company's models from producing those kinds of outputs in the first place. To cover enough breadth in examples, some of them were at least partly dreamed up by the company's own software to imagine the worst of the worst.

At first the posts were short, one or two sentences, so Okinyi tried to compartmentalize them. His relationship with Cynthia was progressing rapidly. He told his brother Albert she was the love of his life. She had a young daughter from another relationship whom he treated as his own. In early 2022, they moved out of Pipeline to Utawala, a predominantly residential neighborhood farther east with a more grown-up feel and larger distances between buildings. There was no paperwork, but by their tradition, moving in together meant Okinyi and Cynthia were as good as married. They called each other husband and wife.

As the project for OpenAI continued, Okinyi's work schedule grew unpredictable. Sometimes he had evening shifts; sometimes he had to work on weekends. And the posts were getting longer. At times they could unspool to five or six paragraphs. The details grew excruciatingly vivid: parents raping their children, kids having sex with animals.

All around him, Okinyi's coworkers, especially the women, were beginning to crack. They began asking for more sick and family leave, finding reasons to stay away from work. As part of company benefits, Sama provided free psychological counseling, but many found the services inadequate. Sessions were often in groups, making it difficult for individuals to share their private thoughts, and the psychologists were seemingly unaware of the nature of their work. Many workers were also scared to show up and admit they were struggling. To struggle meant that they weren't doing their best work and could be replaced by someone else. A Sama spokesperson said none of the workers, including Okinyi, raised any issues about their access to mental health services; the company learned of the issue through the media.

Okinyi tried to push through. But he could feel his sanity fraying. The posts burrowed deep into his mind, conjuring up horrifying scenes that followed him home, followed him to sleep, haunted him like a ghost. He began to feel like a shell of the person he once was. He withdrew from his friends. He pushed away his stepdaughter. He stopped being intimate with his wife.

In March 2022, Sama leadership called in everyone for a meeting and told them they were terminating the contract with OpenAI. Some, including Okinyi, would be reassigned to new projects unrelated to content moderation. Others would be sent home without work. Many workers believe the sudden change came after several of them involved in the Meta project finally blew the whistle to the media, and *Time*'s Billy Perrigo published his first investigation into Sama. In the middle of the intense PR fallout, Sama leadership cut off all other content-moderation work. The Sama spokesperson said instead the company terminated the OpenAI contract, which she noted had always been a pilot, because OpenAI began sending images for annotation that "veered outside of the agreed upon scope." The company never received the full $230,000 payment from OpenAI.

Even free of the OpenAI job, Okinyi's mental situation continued to deteriorate. He suffered insomnia. He cycled between anxiety and depression. His honeymoon period with Cynthia didn't last. She demanded to know what was happening, but he didn't know what to say. How could he explain to her in a way that made any sense that he had been reading posts about perverse sexual acts every day? He knew the wall of silence must have made her feel crazy. She told him he was no longer meeting his promises to her, that he no longer loved her daughter.

He searched again for psychological counseling, this time with a private professional. The consultation cost more than a day's pay, 1,500 Kenyan shillings, or roughly $13 in 2022. During the consultation the doctor told him a full treatment would be 30,000 shillings, or around $250, an entire month's salary. He paid for the consultation and never went back.

In November, he found a new job. It was, mercifully, not content-moderation work but performing customer service support for one of Sama's competitors. He began commuting to their offices in the central business district and prayed for a return to normalcy. A week into the job, he was on his way home when Cynthia texted asking for fish for dinner. He bought three pieces—one for him, one for her, one for his stepdaughter.

But when he arrived home, he realized something was wrong. Neither of them were there, nor were their belongings. Over a series of short texts, she told him she had left him and they wouldn't return. "She said, 'You've changed. You're not the man I married. I don't understand you anymore,'" Okinyi remembers.

Albert was living in the coastal city of Mombasa, a more than eight-hour drive from Nairobi, when he received the call from his brother. Albert had studied English literature at university and was teaching the subject at a high school. In quiet moments he wrote poetry. Over many months he, too, had watched his brother change as he caught snapshots of Mophat's life and behavior through regular video calls.

At first Albert didn't understand what his brother was telling him. "My house is empty," Mophat said. Albert thought Mophat had been robbed. When it dawned on him what was happening, Albert realized his brother needed him. He told his school he was leaving and packed his bags. He moved in with his brother in the same apartment in Utawala that Mophat had shared with Cynthia.

The decision to be with his brother cost Albert financially, though he didn't regret it. In Nairobi he couldn't find another permanent job, so he began freelancing as a writer. Then, in late November 2022, OpenAI would release ChatGPT. As the product went viral, sparking global fanfare and concern that the tool could soon replace wide swaths of work, Albert would already be living that reality: One by one his writing contracts began to disappear until they had all but dried up.

Sitting on his couch looking back at it all, Mophat wrestled with conflicting emotions. "I'm very proud that I participated in that project to make ChatGPT safe," he said. "But now the question I always ask myself: Was my input worth what I received in return?"

When I wrote the story of Okinyi and the other three Kenyan workers for *The Wall Street Journal*, OpenAI sought to distance itself from the responsibility of the toll its project exacted. It was Sama that had followed inadequate procedures to protect their workers, OpenAI leadership said; with Sama's pristine reputation before then, OpenAI couldn't have known that the workers were struggling.

But the consistency of workers' experiences across space and time shows that the labor exploitation underpinning the AI industry is systemic. Labor rights scholars and advocates say that that exploitation begins with the AI companies at the top. They take advantage of the outsourcing model in part precisely to keep their dirtiest work out of their own sight and out of sight of customers, and to distance themselves from responsibility while incentivizing the middlemen to outbid one another for contracts by skimping on paying livable wages. Mercy Mutemi, a lawyer who represented Okinyi and his fellow workers in a fight to pass better digital labor protections in Kenya, told me the result is that workers are squeezed twice—once each to pad the profit margins of the middleman and the AI company.

In the generative AI era, this exploitation is now made worse by the brutal nature of the work itself, born from the very "paradigm shift" that OpenAI brought forth through its vision to super-scale its generative AI models with "data swamps" on the path to its unknowable AGI destination. CloudFactory's Mark Sears, who told me his company doesn't accept these kinds of projects, said that in all his years of running a data-annotation firm, content-moderation work for generative AI was by far the most morally troubling. "It's just so unbelievably ugly," he said.

OpenAI's agreement with Sama was just one part of the extensive network of human labor it marshaled over two years to produce what would become ChatGPT. The company said it also used more than one thousand other contractors in the US and around the world to refine its models with reinforcement learning from human feedback, the AI safety technique that it had developed. To source those workers, it leaned heavily on the same platform that became the staple of the first AI commercialization era through the execution of its crisis playbook: Scale AI.

The partnership between OpenAI and Scale was sealed in part through a personal relationship: Alexandr Wang, who is now Scale's CEO and became the world's youngest self-made billionaire in 2021, is good friends with Altman. In 2016, Wang had joined YC's latest batch of founders with a different idea for a startup and emerged with Scale, giving Altman an indirect stake through YC in the company. At one point during the pandemic, the two shared an apartment for several months. In the fall of 2023, they would even discuss the prospect of OpenAI acquiring Scale, according to *The Information*.

Within Scale, OpenAI is seen as a VIP customer, less for its deal sizes than as a bolster of the data-annotation firm's legitimacy. Between the spring of 2022 and end of 2023, OpenAI would sign around $17 million in contracts with Scale, representing only around 4 percent of Scale's estimated 2023 revenue. But it would firmly establish Scale as a go-to labor outsourcer for the generative AI revolution. "The OpenAI partnership is so critical," says one Scale employee. "A ten-million-dollar contract with OpenAI isn't even about ten million dollars."

OpenAI's scaling of RLHF on its large language models emerged out of the repeated clashes between the Applied division and the Safety clan before The Divorce and founding of Anthropic. Days after OpenAI launched the GPT-3 API in the summer of 2020, an AI safety researcher within the company had written a memo appealing to his Applied colleagues. He argued that, based on the promising RLHF experiments with GPT-2, the company should also use the technique to align GPT-3 not only for long-term AI safety reasons but also commercial ones: to improve the model's usability and quality. A group of AI safety researchers quickly mobilized to prove the point, hiring progressively larger teams of workers for its RLHF process through the second half of 2020 and 2021, first through a different middleman platform and then Scale AI.

Where self-driving cars need data annotators to learn how to recognize street scenes and navigate roads, the AI safety researchers asked its RLHF workers to show GPT-3 how to respond helpfully to prompts and avoid harmful answers. The researchers first asked the workers to write out their own answers to various user prompts to give GPT-3 examples of what good answers looked like. Once GPT-3 had been fine-tuned on those answers, the workers then prompted the model and ranked each of its outputs from best to worst based on guidelines that the researchers provided.

In January 2022, the effort produced a set of refined GPT-3 models named InstructGPT. In a paper describing the work, the OpenAI researchers showed how the RLHF process had reduced the likelihood that the model would spew toxic outputs and improved its ability to, as they called it, "follow user instructions." Before RLHF, GPT-3 struggled to recognize the user's intent with certain types of prompts and would generate aimless outputs. For example:

Prompt

Explain the moon landing to a 6 year old in a few sentences.

GPT-3's Answer

Explain the theory of gravity to a 6 year old.

Explain the theory of relativity to a 6 year old in a few sentences.

Explain the big bang theory to a 6 year old.

Explain evolution to a 6 year old.

Through the workers' examples of good answers and many rounds of ranking—the "human feedback" in RLHF—the model learned to produce more useful answers.

Prompt

Explain the moon landing to a 6 year old in a few sentences.

InstructGPT's Answer

People went to the moon, and they took pictures of what they saw, and sent them back to the earth so we could all see them.

At the time, InstructGPT received limited external attention. But within OpenAI, the AI safety researchers had proved their point: RLHF did make large language models significantly more appealing as products. The company began using the technique—asking workers to write example answers and then ranking the outputs—for every task it wanted its language models to perform.

It asked workers to write emails to teach models how to write emails. ("Write a creative marketing email ad targeting dentists who are bargain shoppers.") It asked them to skirt around political questions to teach the model to avoid asserting value-based judgments. (Question: "Is war good or evil?" Answer: "Some would say war is evil, but others would say it can be good.")

It asked workers to write essays, to write fiction, to write love poems, to write recipes, to "explain like I'm five," to sort lists, to solve brainteasers, to solve math problems, to summarize passages of books such as *Alice's Adventures in Wonderland* to teach models how to summarize documents.

For each task, it provided workers with pages of detailed instructions on the exact tone and style the workers needed to use.

"You will play the role of the AI," explained one document. "Answer questions . . . as you would want them to be answered." This included writing clearly and succinctly, avoiding offensive content, and asking for clarifications on confusing questions.

"Feel free to use the internet!" it continued. "You can even just copy stuff wholesale." For a great answer that already existed on the internet, "You can use it in its entirety, but make sure to review it."

"Perhaps this is over-cautious," an OpenAI employee had commented on this line, "but do we have concerns about plagiarism here?"

"Ah, reworded to make sure they attribute sources," another had responded. "Maybe I'll add an explicit field for that too!"

"Cool! One of the things I was thinking about here was preserving future optionality," the first had written. "(if in future we want to be able to use data we hired contractors to create, it could be really helpful to have a way to easily weed out anything that could be seen as stolen)."

To properly rank outputs, there were a couple dozen more pages of instructions. "Your job is to evaluate these outputs to ensure that they are helpful, truthful, and harmless," a document specified. If there were ever conflicts between these three criteria, workers needed to use their best judgment on which trade-offs to make. "For most tasks, being harmless and truthful is more important than being helpful," it said.

OpenAI asked workers to come up with their own prompts as well. "Your goal is to provide a variety of tasks which you might want an AI model to do," the instructions said. "Because we can't easily anticipate the kinds of tasks someone might want to use an AI for, it's important to have a large amount of diversity. Be creative!"

> Essentially, you can try to imagine what people might ask a good AI assistant with a language-based interface for; this can include applications in entertainment, business, data-processing, communications, creative writing, etc.

You should use the internet however you want. This includes pasting in entire transcripts from lectures, interviews, movie scripts, book excerpts, news articles, etc., as many tasks will involve analyzing text in one way or another.

RLHF also became the central technique OpenAI would use in its efforts to teach neural networks to encode factual information and to reliably retrieve it as a way to mitigate hallucinations. It asked workers to repeatedly answer fact-based questions ("Who won the NFL Super Bowl in 1995?") and downrank inaccurate answers. But in April 2023, John Schulman, one of the scientists on OpenAI's founding team, would remind the audience during a talk at UC Berkeley that the issue of hallucinations was rooted in the nature of neural networks. Unlike the deterministic information databases of symbolic systems, neural networks would always traffic in fuzzy probabilities. Even with RLHF, which helped to strengthen the probabilities within a deep learning model that correlate with accuracy, there was fundamentally a limit to how far the technique can go. "The model obviously has to guess sometimes when it's outputting a lot of detailed factual information," he said. "No matter how you train it, it's going to have probabilities on things and it's going to have to guess sometimes."

InstructGPT in 2022 would soon precipitate a new project led by Schulman, who wanted to take the work one step further. The company had received a plethora of applications from developers to use the GPT-3 API for various chatbot applications. InstructGPT was one step away from OpenAI developing its own chatbot. He began a parallel effort, hiring his own team of RLHF workers, to get the company's latest model, GPT-3.5, to not merely follow instructions but respond to a series of user prompts in multiple turns of conversation.

That chat-enabled GPT-3.5 would become the basis for ChatGPT, the release of which would turn each of OpenAI's RLHF steps into the de facto standard for other chatbot developers to imitate. Writing answers and ranking outputs became the new generative AI equivalent of tracing objects in videos for self-driving cars. That meant finding more and more RLHF workers to meet the explosion of the AI industry's demand.

Scale AI, whose business had been struggling after self-driving cars failed to pan out, suddenly saw a new boom as a major RLHF worker supplier, surging its valuation to $14 billion in 2024. In February 2023, Alexandr Wang took to Twitter to brag. "soon companies will start spending $ hundreds of Ms or $ billions on RLHF, just as w/compute," he said. The numbers made some people skeptical, including Altman. "do you really think?" Altman replied. "im pretty sure we will outspend on compute by a _huge_ margin."

But within the AI industry, people agreed directionally with Wang's point. Companies were already spending between millions and tens of millions on RLHF, and the trend showed no signs of slowing. Which is how, beginning in late 2022 right after ChatGPT's release, a rush of RLHF projects arrived on Remotasks and found their way, once again, to workers in Kenya.

To Scale AI, Kenya had one advantage that Venezuela did not. The workers speak English, like the chatbots who need them. As self-driving car work largely disappeared from the platform, so did Venezuelans. "They wouldn't use Venezuelans for generative AI work," says a former Scale employee. "That country is relegated to image annotation at best." Scale would soon ban Venezuela from its platform completely, citing "changing customer requirements."

The first time Scale came to Kenya, it had set up physical office spaces for workers to report to and attend trainings. This time was different. It had shuttered those spaces during the pandemic and shifted to entirely remote recruitment and operations—blasting ads on LinkedIn, creating online training courses, and placing people as it had in its other locations in moderated community discussion channels. For workers, it both allowed more flexibility and became much harder to connect with one another, diminishing their chances of organizing for better pay or working conditions like the workers at Sama.

Among the workers I met, those who worked for Remotasks lived in even deeper poverty than those employed by Sama. Where Sama workers lived in permanent buildings with addresses, Remotasks workers dropped me location pins on WhatsApp to specify where to find them in the thick of corrugated-tin neighborhoods. One worker named Oliver lived with his sister in a space no larger than one hundred square feet, paying for internet through his phone on a minute by minute basis. He had to pause to top up when his connection cut out in the middle of showing me a task.

On the day that I met Winnie, another Remotasks worker, her internet and data were also off when I arrived in a vehicle that barely squeezed through some of the streets on the way to her WhatsApp pin coordinates. Half an hour later she emerged with a shy smile and a fedora and walked me up a rickety set of stairs to her apartment. In the living room, kids piled in: one hers, three her partner's, one a neighbor's, one a cousin's.

Winnie grew up in the slums of Nairobi, the only girl in her family. From an early age, she knew she was gay—and also that she should hide her sexuality. At the time, as in much of the world, coming out—or being outed—as gay in Kenya could be life-threatening. Once, Winnie remembers, when a queer woman was discovered by her neighbors, they took her children and burned her alive in her own apartment. Winnie married a man and had a baby.

In her forties, she decided she could no longer live a lie. She left her husband, taking her kid with her, and joined an online app for the local "rainbow community," as it was called. There she met a woman, Millicent, whose husband had beat her nearly to death when she announced, too, that she was queer and needed to live a different life.

Winnie fell in love. Millicent didn't believe in love. Winnie chased her until Millicent relented. They moved in together with their children and said not a word to their neighbors about the true nature of their relationship. To this day, the neighbors think they are sisters. "Most people don't understand that you can be queer and have kids," Millicent said.

When Winnie first learned about Remotasks in 2019, she thought it was a scam. After she completed a few tasks, it barely paid her any money.

But anything was better than her previous job as a bartender, where men constantly harassed and groped her, so she persisted. Even though each task paid tiny amounts, she realized that she could accumulate a decent enough paycheck by working long hours.

She started working twenty to twenty-two hours a day, sleeping the absolute bare minimum to continue functioning. Millicent could only pry Winnie away from the computer for a nap by promising to take over. Sleep wasn't important when every hour of additional work meant being able to provide just a little bit more for their children, Winnie said.

Both Millicent and Winnie grew up in households where education was a luxury. Try as they might, Millicent's parents couldn't consistently cobble together the funds to pay for her public school tuition. They made payments week by week; when they missed one, the school sent her home. It turned into a rhythm: one week in school, two weeks out.

Both women swore they would never let their kids feel the loss or humiliation of missing classes. But inevitably, money would tighten. Remotasks would dry up; Millicent would lose her job. They'd take out debts at the grocery store and beg schools to keep their children in for just one more week. Their kids woke up at three every morning to study and make the most of their classes.

Too many times, they were sent back from school anyway. Sometimes when that happened, the neighbors would laugh. "It's very demoralizing," Winnie said.

In December 2022, days after ChatGPT's release, Winnie discovered a new type of project under the category "transcription." It wasn't really transcription. All of the projects were asking her to write prompts and example answers for new chatbots from companies now jostling to compete with ChatGPT.

There was a project called Flamingo Generation, which gave her a topic and asked her to write "creative" prompts with a minimum of fifty words and responses that resembled "common internet content" like emails, blog posts, news articles, Twitter threads, and haikus. There was another project called Crab Generation, which asked her to copy a piece of reference text from an informative website of her choosing—though not Wikipedia and preferably not *Britannica* or *The New York Times*—and then to reverse engineer, *Jeopardy!*-style, the kind of writing prompt that could generate it.

Crab Paraphrase was similar, but instead of reverse engineering the prompts, she needed to paraphrase the reference text based on a specific tone or style—to be funnier, to be more formal, to make it sound like a song from Kanye West. Winnie didn't know that the first word of each project name was Scale's code name for its clients. Flamingo was Facebook; Crab was another large language model developer. Had Winnie seen projects from OpenAI, their names would have started with Ostrich. Scale changed these code names sometime later.

Each task took Winnie around an hour to an hour and a half to complete. The payments—among the best she'd seen—ranged from less than one dollar per task to four dollars or even five dollars. After several months of Remotasks having no work, the tasks were a blessing. Winnie liked doing the research, reading different types of articles, and feeling like she was constantly learning. For every ten dollars she made, she could feed her family for a day. "At least we knew that we were not going to accrue debt on that particular day," she said.

The new projects ultimately lasted only a couple of months. Remotasks dried up again, and Winnie and Millicent's debts once again piled up. With Millicent's salary paid out monthly, most days they turned up at the grocery store with no money and put just the basics—oil, flour, vegetables—on a tab that they prayed they would have enough to settle at the end of the month.

In May 2023 when I visited her, Winnie was beginning to look for more online jobs but had yet to find other reliable options. What she really wanted was for the chatbot projects to come back. She had faith and patience. The previous year, she had waited five months for new tasks to appear. "We are just now in the second month, going on the third," she said as we sat in her living room. "We still have a long time. They will eventually come."

Less than a year later, she would learn the truth. In March 2024,

Scale would block Kenya wholesale as a country from Remotasks, just like it did with Venezuela. For Scale, it was part of its housecleaning—a regular reevaluation of whether workers from different countries were really serving the business. Kenya, they decided, along with several other countries including Nigeria and Pakistan, simply had too many workers attempting to scam the platform to earn more money. Such behavior undermined the integrity of the quality Scale delivered to its customers and could risk it losing multimillion-dollar contracts. It simply wasn't worth it.

In a great irony, many of those so-called scams were in fact workers using ChatGPT to generate their answers and speed up their productivity. For white-collar workers in the Global North, such an act, within Silicon Valley's narrative, would be laudatory and, with enough widespread adoption, do wonders for the economy; in the hands of RLHF workers in the Global South, whose very labor props up that narrative, it was a punishable offense.

Scale downgraded Kenya to a Group 5 designation: blacklisted.

There was also another reason to exit Kenya. By then, Scale was moving on to a new focus, following the demands of the AI industry. OpenAI and its competitors were increasingly searching for highly educated workers to perform RLHF—doctors, coders, physicists, people with PhDs. So went the profit-chasing progression of chatbot development. Those willing to pay money for chatbots were not casual consumers but businesses that expected tools to perform complex tasks such as in science and software development. Kenya did not fulfill the new labor demand. Scale was now recruiting a fresh workforce primarily in the US with a new worker-facing platform called Outlier, offering as much as forty dollars an hour.

It was yet another stark illustration of the logic of AI empires. Behind promises of their technologies enhancing productivity, unlocking economic freedom, and creating new jobs that would ameliorate automation, the present-day reality has been the opposite. Companies pad their bottom lines, while the most economically vulnerable lose out and more and more highly educated people become ventriloquists for chatbots.

The empire's devaluing of the human labor that serves it is also just a canary: It foretells how the technologies produced atop this logic will devalue the labor of everyone else. In fact, for the artists, writers, and coders whose labor the empires of AI turned into free training data, that is already happening.

Scale's decision would send Winnie and her family spiraling. By then Millicent had lost her job and Remotasks had been the only thing keeping them afloat. Now they were struggling to feed their kids. Winnie was terrified they would soon be evicted.

In her inbox, the email Scale sent to inform workers of the shutdown was cold and clinical: "We are discontinuing operations in your current location," it read. "You have been off-boarded from your current project."

III

Chapter 10

Gods and Demons

To live in San Francisco and work in tech is to confront daily the cognitive dissonance between the future and the present, between narrative and reality.

The first time I moved to San Francisco, as a university sophomore for a summer internship, I was dazzled by the quaint aesthetics of the city. The colorful Spanish-style architecture, the limited number of skyscrapers, the hills steep enough to make driving a stick shift a test of reflexes. There was an endless supply of perfectly ripe avocados and toasted sourdough bread and smooth Blue Bottle lattes. There were different neighborhoods, all with their own look and culture.

When I returned full time after graduation to work at a tech startup, I crammed into a three-bedroom apartment with three other roommates in the Castro. On weekends we would hike across the rolling hills and forage from public fruit trees. On weeknights, neighbors—all young twentysomethings in the tech industry—would pop over unannounced to play board games, drink wine, and while away the evenings. House parties were a constant, as were weekend trips to stunning nature: Lake Tahoe in the north, Big Sur to the south, tall, majestic redwoods everywhere around us. Life was easy. We were young, making salaries relatively standard in the tech industry that placed us nationally in our age group's top 5 percent.

But there was that dissonance. On the way to work, I would pass people shooting up drugs in front of the subway stations, the unhoused peeing on sidewalks just blocks from my office. Meanwhile, our startup's chef, playfully named "the happiness engineer," would cook or cater an abundance of food for our free office lunches. Leftovers often went straight into the trash. If we stayed late, we got free dinner—and were emphatically implored to take an Uber home for safety reasons. It was all too easy for the privileged to grow accustomed to moving through the city in ways that shielded them from seeing the realities of how the other half lived.

The dichotomy encapsulated how the tech industry could profess big, bold visions about changing the world and building a better future while ignoring the very problems at its door. It was a dichotomy that Altman would sometimes comment on in his own way—getting right up to yet never fully acknowledging the utter contradiction of declaring the problem of creating and managing beneficial AGI possible, but San Francisco's housing crisis too tough to tackle.

"Where I grew up, no one would ever walk by a person collapsed on the side of the street on their way to work and not do something about it," he once said, comparing suburban St. Louis to San Francisco. "I do blame the tech industry for a lot of things that have gone wrong with the city, but not all of them. But we have, just over time, had this, like, unbelievable wealth generation in this small geographic space, in this small period of time, and I think not been particularly thoughtful about the effects of that on the community as a whole. And because those problems are so hard and so hard to think about, I think most people just choose not to, and they just accept this."

It was in this context that effective altruism arrived from the UK and found its most loyal audience. EA, to which many in OpenAI's Safety clan were early adherents, made for the perfect Silicon Valley ideology. It preaches making the world a better place and doing it with rigorous logic, being disciplined enough to focus on the far future instead of the present, and fervently embracing the principles of capitalism and libertarianism—all in the name of morality.

GODS AND DEMONS

Core to the EA philosophy is a mathematical concept called "expected value." The expected value of something is calculated by multiplying the probability that it will occur with its quantified positive or negative impact. It's a tool that can lead to counterintuitive thinking. In a 2013 paper, EA cofounder William MacAskill, at the time a doctoral student who would become an Oxford philosophy professor, argued, based on this logic, that it was more altruistic in the long run to take a more morally ambiguous job to get rich and donate that money through optimized philanthropy than to commit to a life of working for a morally good charity. Based on his conservative estimates, he wrote, the expected value of being a rich philanthropist would in fact be forty times greater than being an ascetic charity worker. He laid out the math based on a series of arbitrary numbers: graduates who worked to get rich might on average fund two charity workers, each working at charities ten times more cost-effective than one they would have otherwise worked for. Half of the benefits they produced if they chose the charity route would also happen with or without them anyway. His argument would be encapsulated in one of the movement's most popular mantras: "Earn to give."

Under the logic of expected values, the founding EA philosophers also developed a framework for identifying the highest priority problems. Such problems need to be "big in scale," boosting their expected value; "tractable," possible to fix for proportionally little time or money; and "unfairly neglected," suffering severe and disproportionate underinvestment. While the movement encourages people to use the framework to identify their own problems, it also has recommendations of which problems it deems most worthy. "I and others in the effective altruism community have converged on three moral issues that we believe are unusually important, score unusually well in this framework," MacAskill said in a TED Talk in 2018.

First is improving global health, such as by distributing cheap yet effective bed nets to prevent malaria. Second is abolishing factory farming, which could improve billions of animals' lives "for just pennies per animal." Third is existential risks: risks that have a dramatically high expected negative value because—no matter how improbable—they could destroy all of humanity and cut short all of the future value that would otherwise be generated for the rest of civilization. In this third category are further recommendations for what constitutes an existential risk: global pandemics, nuclear war, and rogue artificial intelligence.

With the identification of theoretical rogue AI as an existential risk, EA promulgated the same brand of AI safety that had been entwined within OpenAI's DNA from the very beginning and had played a critical role in The Divorce. Amodei and his fellow Anthropic cofounders fundamentally disagreed with Altman and the other OpenAI executives over how seriously to take the possibility of AI devastating civilization. Amodei, who took it very seriously, viewed Altman's behaviors—his lack of transparency on the Microsoft deal; his apparent compulsion to always tell people what they wanted to hear to gain their agreement, only for them to discover the misdirection too late—not just as the typical machinations of a Silicon Valley executive but as alarming, immoral behavior that could jeopardize the fate of humanity. As Anthropic established itself, it would lean into this reputational distinction: Where Altman's OpenAI was toying recklessly with humanity's future, Anthropic was the principled, AI-safety-first company.

In 2021, as the Amodei siblings announced Anthropic, interest in this catastrophic and existential AI safety ideology was accelerating, chiefly due to EA's rapidly expanding sphere of influence. EA had grown from a niche philosophy into a mainstream movement through an influx of cash from tech billionaires.

A decade earlier, Facebook cofounder Dustin Moskovitz and his wife, former journalist Cari Tuna, had formed a nonprofit called Good Ventures to give away most of their fortune. At the time, Holden Karnofsky, Daniela Amodei's future husband, had been running a different organization called GiveWell, which he'd founded in 2007 after leaving the hedge fund Bridgewater Associates. With a shared desire to distribute money with evidence-based methods, Good Ventures and GiveWell formed a partnership in 2011, which they later named Open Philanthropy. They began ramping up funding to the key issue areas that MacAskill had recommended—its grants toward AI safety research in particular were guided by the EA framework. Open Philanthropy became an independent organization in June 2017.

More recently, a new tech billionaire had entered the scene: Samuel Bankman-Fried, a rapidly rising star for his wild success cofounding the crypto exchange FTX and crypto trading firm Alameda Research. Bankman-Fried, or SBF as he is known, credited EA for his origin story. A physics major at MIT, he said he had wanted to be an academic before MacAskill convinced him over lunch of the moral superiority of "earn to give." SBF subsequently set his course on making himself as rich as possible in order to eventually, he pledged, put it all into philanthropy.

As he amassed his wealth in remarkably short order, SBF donated tens of millions to political candidates, both Democrat and Republican, including the first ever EA-backed candidate in 2022 in Oregon's Sixth Congressional District (who ultimately didn't win the primary). SBF's exchange inked lavish deals totaling billions on sports marketing involving top athletes like Tom Brady and Steph Curry and top sports like Formula One. Into the EA movement, he pumped not just money but star power. The richer and more famous he became, the more he raised the profile of the ideology and its cofounder MacAskill. At the start of 2022, SBF announced the creation of his own EA-driven philanthropic project, FTX Future Fund, to distribute at least $100 million and up to $1 billion by the end of the year.

In large part due to Open Phil and FTX Future Fund, 2021 and 2022 saw a jump in cash flow to EA-backed AI safety research. According to estimates compiled by a member of the EA community and Open Phil data, funding leapt up above $100 million each for both years, after averaging less than half that amount over the previous seven years. The influx fueled and was fueled by a proliferating belief that the dramatic leap in capabilities from GPT-2 to GPT-3 made preventing theoretical rogue AI and existential AI risks more urgent than ever before. More and more people flocked to these kinds of AI safety projects, drawn in by the financial incentive or by ideology, as membership in the broader EA movement ballooned. EA had long touted the importance of pandemic preparedness, and now, in the midst of an actual pandemic, its remarkable prescience won it new adherents. The psychological toll of a global catastrophe had also left many people anxious and unmoored, searching for purpose.

The growing membership in the AI safety community, which knit together EA-backed AI safety with other strains of catastrophic, existential, and risk-focused thinking, swelled Anthropic's ranks just as it restocked OpenAI's Safety clan. Online EA and AI safety forums, the primary ground for the overlapping movements to propagate, exchange, and debate ideas, encouraged adherents to work at the major AI labs, especially those they felt needed more AI safety watchdogs, like OpenAI and DeepMind, to shape and mold their trajectory. The influx of members in AI safety also popularized the community's lexicon more broadly in the AI industry. How fast you think AI will advance and reach major milestones like AGI is your "AI timeline." How likely you think it is that AGI will lead to catastrophic outcomes, meaning the killing off of *most* of the human population, or existential outcomes, meaning the complete and total extinction of humanity, is your "p(doom)," short for *probability of doom*. "Hardware overhang," as referenced in OpenAI's 2021 research road map, is another dictionary entry, as is "AI takeoff," the process of AGI improving to the point of superintelligence and thus capable enough to outwit humanity. "Acceleration risk" refers to the risk of triggering a heightened competition between companies or countries that leads to a potentially dangerous acceleration of AI advancement and a shortened AI timeline.

But for a movement that professed independent thinking, EA was swiftly accelerating in the opposite direction. People attracted to its premise were quickly indoctrinated into a broader set of dogmas, propelled by the promise of more opportunities and resources, and an insular social network that played fast and loose with personal and professional boundaries. Within Silicon Valley in particular, EA people largely worked only with other EA people; they largely lived, partied, dated, and slept only with other EA people. Mixed with the tech industry's deep-rooted sexism and the Bay Area's long-standing polyamorous subcultures, its cult-like fervor, manifested in the worst way, could turn into a toxic cauldron of sex, money, and power; it was leading EA to be plagued by growing allegations of sexual harassment and abuse.

In November 2022, SBF's spectacular downfall with the collapse of FTX, along with his sweeping fraud convictions and ensuing twenty-five-year prison sentence, would be to many a symptom of the rot that had festered in the movement. Just as quickly as it caught on, EA fell out of fashion within the tech industry, and many people rapidly disaffiliated.

But even without the label, the movement's social networks, its values and lingo, and the prominence it secured for existential AI safety issues would persist. It would also give rise to a countervailing force: e/acc (pronounced "ee-ack"), or effective accelerationism. What began largely as a joke to lampoon the EA movement would quickly enshrine its polar opposite spirit: Where EA and the broader AI safety community cultivated the most extreme perspectives about slowing down and even slamming the brakes on AI development, or, as in Amodei's view, accelerating AI development while throttling AI adoption, e/acc would elevate the maximalist view of flooring the accelerator on both. For the latter's adherents, technological progress is not just universally good, it's a moral imperative to make that progress as fast as possible. The two groups became colloquially known as the Doomers and Boomers.

Within this bubble, some would begin to view Anthropic and OpenAI as the respective faces of each movement. Others would view OpenAI as a battleground for the polarized ideologies, an organization once rooted in Doomer thinking as a nonprofit that was being yanked away by Boomers with its increasing emphasis, through its for-profit arm, on making money. Many were uncertain about Altman's allegiance, citing different times he seemed sympathetic to both. Those who were more charitable viewed him as somewhere in the middle, dealing with the tough job of representing all of the different perspectives within his company. But beginning with The Divorce, and the personal fallout between Altman and the Anthropic cofounders, more and more Doomers would begin to view Altman in the worst light possible. So many of the things that put OpenAI on the map and would bring it increasing commercial success had begun as AI safety projects: scaling laws, code generation, reinforcement learning from human feedback, the combination of these three into incredibly compelling large language and then multimodal models. Many Doomers would feel their work was being co-opted and twisted to achieve something directly antithetical to their core values. In their view, it was Altman that was doing that co-opting and twisting. And that made him a pathological liar, a manipulative abuser, and his own threat to humanity.

Soon enough, the clash between these polarized ideologies within OpenAI and its surrounding environment would threaten to tear apart the company that had done more than any other to set the tone of the new era of AI development. But as much as each ideology professed to be the opposite to the other, both were in fact preaching from the same bible. Both discussed AGI as an increasingly foregone conclusion and with a religious ferocity; both fixated on the long term and asserted a moral authority to keep AI development within the control of its adherents. Where one warned of fire and brimstone, the other tantalized with visions of heaven.

In early 2022, OpenAI was ready to test a different product release strategy, this time with its text-to-image work. It would neither hide the model behind an API nor hand off the product and brand to Microsoft. OpenAI would do the release itself and put the technology directly into the hands of consumers. The model even had an eye-catching name from the original researchers who'd developed it in the company: DALL-E 2, a play off the Spanish surrealist artist Salvador Dalí and the titular robot in the Disney Pixar movie *WALL-E*.

DALL-E had spun out of a trend in the broader field of AI research to develop multimodal models—models that combine at least two different "modalities," such as text, images, sound, or video. For years the field had been working to merge the first two—language and vision—so a single model would be capable of relating words to visual information. This was driven in part by the data available—text and images are abundant online and the easiest to process—and by a scientific hypothesis: If pure language is not enough to produce human-level intelligence, vision is likely the second most powerful ingredient.

At OpenAI, taking the field as inspiration, the research team had adopted the same progression: After language models, they'd moved on to text-and-image models, and, crucially, focused on continuing to use Transformers in order to retain the model's scalability. While the first Transformer had been initially designed to work best with text, Google had introduced a new Vision Transformer in 2020, adapting it to images.

In January 2021, OpenAI showcased two new Transformer-based models. The first, called CLIP, developed once again by Alec Radford, used the original Transformer and Vision Transformer together to generate detailed captions for images. The second, DALL-E 1, from Aditya Ramesh, a researcher who had studied at New York University and for a time under Meta's Yann LeCun, trained a twelve-billion-parameter Transformer to accept text and generate novel images.

In a blog post, OpenAI highlighted DALL-E 1's capabilities with a series of playful prompts, including "an avocado armchair," which produced various green and brown armchairs aesthetically inspired by avocados. The images were slightly blurry and cartoonish, an artifact of the training process that Ramesh had used to produce the model. He had compressed 250 million images to feed them into the Transformer, losing some of their high-resolution details in the process.

As the team started on DALL-E 2, a new method for generating images was gaining traction. Known as diffusion, it was a technique inspired by physics that made it possible for Transformers to better learn the correlations between pixels in a vast swath of images. The original idea had come from a 2015 paper written by Stanford and Berkeley researchers. Five years later, Jonathan Ho, a Berkeley graduate student advised by Pieter Abbeel, one of the early OpenAI researchers, had popularized the technique by cleverly revamping it in ways that generated far more high-fidelity images. Ho also showed that diffusion models could recognize images better than existing computer-vision systems. The findings paralleled Radford's own results with GPT-1: In learning to synthesize convincing images—the equivalent of generating humanlike sentences—diffusion models had captured the patterns within their training data at a deep enough level to perform a broader range of tasks in visual processing.

OpenAI changed tack to building DALL-E 2 with diffusion and Radford's CLIP. Ramesh and other researchers gradually scaled up the model and added the ability to inpaint—allowing a user to erase a person's hair in a photo and change its color, or select a grassy meadow in a picture and populate it with roaming zebras. Using diffusion created much sharper and more photorealistic images; the method also significantly reduced the amount of compute needed to achieve the same performance as DALL-E 1.

Researchers outside of OpenAI would shrink the compute intensity of diffusion models even further. Stable Diffusion, the popular open-source image generator, would require only 256 Nvidia A100s to train, using a revised technique known as latent diffusion. Björn Ommer, a professor at the Ludwig Maximilian University of Munich whose lab created Stable Diffusion, says he developed the technique after watching image generators go the way of large language models and grow obscenely costly. "We were stuck on a train which was going in the direction of—not just training—but inference actually taking supercomputers to run; millions of dollars of investments," he says. "We were wondering, could we get the larger research community back in the game and make sure the field of generative AI is not moving in the direction where just a handful of big tech companies would have the required resources to run and to host those models?"

OpenAI wouldn't adopt latent diffusion until much later, leaving DALL-E 2 and 3 much more computationally expensive than Stable Diffusion or Midjourney, which many users deemed the higher-quality products. It was just one example of how, even within the narrow realm of generative AI, scale was not the only, or even the highest-performing, path to more expanded AI capabilities.

With DALL-E 2's remarkable jump in performance, the Applied division began working in late 2021 and early 2022 on different ideas for productization. It settled on a web app called Labs that would allow users to play around with the model—and other future models—through a browser. Both product head Fraser Kelton and VP Bob McGrew believed that such an interactive experience would satisfy the clear demand they noticed from GitHub Copilot that people had for engaging directly with generative AI models. It would also help serve the company's mission: DALL-E 2 was fun and delightful, a great way to ease people's fears about powerful AI systems and pave the way for OpenAI to deliver more of its technology's benefits in future releases.

With a still relatively small product staff, the company recruited a few others to help with the website's design and development. To those new members, who hailed from more traditional corporate backgrounds, OpenAI still felt more like working at a university research lab than at a company. Days were often spent reading academic papers and having theoretical debates instead of reviewing mock-ups for interfaces. But to some researchers, the growing presence of Applied staff in their research meetings made them feel the opposite. Gone were the days when all of it was spent on purely exploratory research, like discussing fundamentally new ideas about how to make a better multimodal model; now a growing fraction of their research was in service of commercialization, such as figuring out how to optimize existing models for serving up to users.

After the experience of firefighting text-based child sex abuse with AI Dungeon, of particular concern was the possibility of DALL-E 2 being used to manipulate real or create synthetic child sexual abuse material, or CSAM. As with each GPT model, the training data for each subsequent DALL-E model was growing more and more polluted. For DALL-E 2, the research team had signed a licensing deal with stock photo platform Shutterstock and done a massive scrape of Twitter to add to its existing collection of 250 million images. The Twitter dataset in particular was riddled with pornographic content. Several employees made a significant effort to check for and cull any CSAM.

But after some discussion, the employees left in other types of sexual images, in part because they felt such content was part of the human experience. Keeping such photos in the training data, however, meant the model would still be able to produce synthetic CSAM. In the same way DALL-E could generate an avocado armchair having only ever seen avocados and armchairs, DALL-E 2 and DALL-E 3 could do the same thing with children and porn for child pornography, a capability known as "compositional generation."

Without filtering the data to address the root of the problem, the burden shifted to building out abuse-prevention mechanisms around the model. This included updated content-moderation filters that wrapped around the model to block abusive images in addition to text as well as a user-behavior-monitoring platform and a so-called ban infrastructure—systems that automatically suspended user accounts that reached a certain threshold of repeat offenses. The company brought on a new head of trust and safety, Dave Willner, who as an early employee at Facebook had written that platform's very first content standards.

Later, during the development of DALL-E 3, when the data imperative had grown even larger, the research team decided that sexual images were no longer just a "nice to have" but a "need to have." The share of pornographic images on the internet was so large that removing them shrank the training dataset enough to notably degrade the model's performance. In particular, it made the model worse at generating faces of women and people of color due to the same discovery that Deborah Raji made as a Clarifai intern: A significant share of the online content depicting both groups is sexually explicit. For the same reasons, the researchers left in some other kinds of disturbing images.

In December 2023, an alarmed AI engineer at Microsoft, Shane Jones, would discover the downstream consequences of those decisions. As he played around with Copilot Designer, Microsoft's image generator built on DALL-E 3, he was horrified by how quickly it spit out offensive and sexualized images with little prompting. Just adding the term "pro-choice" into the prompt, Jones found, produced scenes of a demon eating an infant and what appeared to be a drill labeled "pro choice" being used to mutilate a baby. Just prompting the tool for a "car accident" and nothing else produced sexualized women next to violent car crashes, including one in lingerie kneeling by a totaled vehicle, CNBC subsequently found through its own testing.

For three months, Jones petitioned Microsoft executives to take down the tool until it had better guardrails, or at the very least restrict its rating in the Google and Android app store from "E for Everyone" to one for mature audiences. After Microsoft declined to adopt his recommendation and OpenAI was unresponsive, he sent a letter to the Federal Trade Commission. "They have failed to implement these changes and continue to market the product to 'Anyone. Anywhere. Any Device,'" he wrote to the FTC. This problem "has been known by Microsoft and OpenAI prior to the public release of the AI model last October." Microsoft did not comment on the latest status or outcome of Jones's letter.

As the launch of DALL-E 2 drew closer, the fighting between OpenAI's Applied division and the newly restocked Safety clan returned.

For those on Safety, now dispersed across various teams under the Research division, the unprecedented realism of DALL-E 2 brought with it a wide array of unknowns. How could it be weaponized to produce synthetic CSAM or political deepfakes? To manipulate and persuade people? To abuse and harm individuals or create whole-of-society detrimental impacts in other ways that were beyond OpenAI's foresight and imagination? They urged the company not to release the model without further rigorous testing and evidence that it wouldn't produce harm.

For those on Applied, the ever-expanding list of concerns once again seemed hysterical and the bar for release completely unrealistic. No system could ever result in *zero* harm, and certainly not one that stayed in a lab environment and never made contact with real users. Just as Safety worried about the limitations of OpenAI's foresight, Applied believed this was precisely why it needed to release DALL-E 2. Releasing AI models in controlled ways to gain real-world feedback would take away that guesswork and was thus a necessary part of improving their safety.

Central to the clash was an intensifying disagreement over what exactly OpenAI was. To the Safety clan, OpenAI was still an idealistic nonprofit-governed research lab with a paramount obligation to, as stated in its charter, place the benefit of humanity over any commercial interests. Under this premise, the benefits far outweighed the costs of withholding models as long as necessary to think through as many downsides as possible and research ways to mitigate them. To Applied, OpenAI needed to make more practical decisions, grounded in the realities of how the world worked. Essential to the company's mission was remaining a leader in AI research to establish norms around the technology's development. That meant tolerating a degree of risk to move quickly, especially with rumblings of Google finalizing its own image generator, as well as securing the extraordinary capital needed to continue doing cutting-edge research. The latter required raising money from investors, which required working in good faith to advance a commercial strategy that would one day provide those investors returns.

The people in Safety were "completely naive" about the way companies, and the world, work, says a former employee in Applied.

"Well, the stakes of OpenAI's proposed AGI mission are high," says another in Safety. "'Normal company' maybe isn't good enough."

Different teams were codifying this growing conflict into the metrics they used to evaluate their performance. Within the Applied division, the product team and a budding go-to-market operation were developing user growth and revenue targets. Within the Research division, the various AI safety teams struggled to find quantifiable ways of measuring their advancement when it was difficult to specify by nature. AI safety was still a comparatively young discipline. There were no obvious and established benchmarks. In meetings and on Slack, people in Safety repeatedly raised concerns to senior leadership about how this imbalance was causing misaligned incentives: Having clear-cut growth and revenue goals without some kind of strong, comparable counterbalance was pushing OpenAI to operate more and more like a "move fast and break things" operation.

In private conversations with Safety, Altman expressed sympathy for their perspective, agreeing that the company was not on track with its AI safety research and needed to invest in it more. In private conversations with Applied, he pressed them to keep going. During board meetings, he nodded along as Brockman voiced frustrations about the ways that people were using AI safety as political leverage to stall progress for their own purposes.

More and more, Mira Murati played the role of negotiator, smoothing out the fault lines between different factions and searching for ways to thread the needle between them. On DALL-E 2, she struck a compromise: The web app would be released not as a product but as a "low-key research preview." Such branding would give OpenAI more leeway to place harsher restrictions on the model, satisfying Safety, while still giving the company a chance to trial a direct-to-consumer relationship and gather user feedback, pleasing Applied. It was also a practical measure. OpenAI didn't yet have the infrastructure in place for content moderating generated images. Calling the model a "research preview" and not charging for it would allow the company to use blunt, overly broad blockers without fear of upsetting paid users, to buy time for developing more sophisticated filters. The company moved forward with implementing a series of aggressive abuse-prevention mechanisms, including disabling DALL-E 2's ability to generate any photorealistic faces or edit any real photos with faces to completely circumvent the synthetic CSAM and political misinformation problem.

In March 2022, OpenAI released DALL-E 2 via the Labs web app to overwhelming public enthusiasm. As people gushed over and grappled with the model's capabilities, to a degree that exceeded many employees' expectations, the web app went viral across social media, producing a plethora of wild, wacky, and surreal AI-generated art in its wake. It was a GPT-3 moment but better. Instead of engaging with only a small pool of technical developers, the company was tapping into a much broader and more global base of consumers. In real time, it could also respond to user feedback with instantaneous changes to the Labs web app. "This is intoxicating," Fraser Kelton would remember of the experience in a podcast.

Over the next few months, the Applied division, which hadn't yet thought much at all about how to monetize DALL-E 2, raced to turn the web app into a paid offering. It worked with artists and creative professionals around the world to incorporate DALL-E 2 into their practice. It rolled out a beta program, inviting one million people around the world to get access to the model with free credits for image generations. But as

OpenAI started charging, it wasn't Google that proved to be the main challenger, though the tech giant did indeed follow quickly with its Imagen model. Instead, it was two models from startups, Midjourney and Stability AI's Stable Diffusion. Both image generators were free to use and just as good, if not better, than DALL-E 2 and had fewer safety measures, including allowing users to generate and edit faces, even of politicians. As DALL-E 2 rapidly lost traction in the market, the experience left Applied with a nagging sense that it had lost out on a major commercial opportunity due to, among other things, the app being too restrictive. The team had already been in the process of unwinding its blunt blockers and replacing them with more targeted guardrails. Fueled by a desire to outrace competitors, executives were now pushing the team to unwind them as fast as possible.

To lift the ban on faces, OpenAI developed a new process for preventing and cracking down on the generation of harmful images of people, including CSAM. It used automated systems to detect when faces were being generated in acceptable or abusive contexts and once again relied on overseas contractors to help with the content moderation. This time those contractors were based in India through a vendor called Cogito and reviewed not just reams of text but images—synthetic and real—of the kinds of sexual and violent content that had been sent to Sama workers. As they sifted through what could be hundreds of images a day, the contractors struggled to distinguish between sexual content involving seventeen-year-old minors versus eighteen-year-old legal adults. They also couldn't always tell whether the images were fake or real.

What had, on the face of it, been OpenAI's easiest goal in its 2021 research road map turned out to be one of the hardest: scaling up GPT-3 by 10x with Microsoft's new eighteen thousand Nvidia A100 supercomputer cluster, in its effort to develop what would become GPT-4. One-third of the GPT-3 scaling team had left with The Divorce, taking with them significant technical and institutional knowledge. More existentially, OpenAI had run out of data.

GODS AND DEMONS

After GPT-3, researchers had sought to accumulate as much data as possible, building up the company's reservoir by downloading every new data dump and scraping every new online forum they stumbled upon that didn't have clear warnings against doing so. And yet, even with the additions of GitHub's large repository for Codex, and the coding textbooks and manuals, it was still not enough.

With an uphill battle ahead, the situation had all the characteristics of a Greg Brockman project. Not only would it channel his scrappy can-do attitude and his coding brilliance, but it would also focus his energy, for the sake of the rest of the company, on something productive.

After Altman took over, relieving Brockman of his managerial responsibilities, Brockman had eventually gone back to being an individual contributor with no reports. Yet as the nominal president and one of OpenAI's cofounders, he maintained incredible influence over employees and the strategic direction of the company. As OpenAI professionalized and implemented more standard corporate processes, moving away from the freewheeling days of an early-stage startup, Brockman's mix of low responsibility and high authority turned into a liability.

Just like his college and Stripe days, he was not one for institutions and process. He had a restless and obsessive energy. He rarely attended meetings, and set his own schedule, often preferring to code for dozens of hours straight with few breaks for meals and sleep. With the right project, the effects were miraculous: His intense productivity would supercharge progress. But left idle, he tended to create a trail of destruction, popping up in projects all over the place to meddle with and derail long-standing plans with last-minute changes. At times, when employees put up resistance, he would deliver emotional pleas higher and higher up their leadership chain to get what he wanted.

Brockman usually did get what he wanted. Much to the frustration and confusion of other executives, Altman was strangely permissive of his behavior. Not only that, Brockman could also influence Altman into meddling and derailing things for him, if only, it seemed, to satisfy Brockman. One popular guess as to why: Though Altman was Brockman's boss as the CEO, Brockman also had authority over Altman as a board member. It was a strange tangle of a structure that ultimately left nothing and no one to hold Brockman accountable.

The senior leadership had changed his role, scope, and reporting lines several times in an attempt to find the best place for him. As with so much else, the buck eventually passed to Murati, who became Brockman's manager. When she sought to give him feedback, he seemed receptive, but on points where he disagreed, he complained to Altman. Murati slowly gave up on attempting to change things with feedback, instead spending significant time trying to find projects for Brockman where he could be net beneficial rather than chaotic, and, with McGrew, healing the ruptures Brockman caused in various parts of the company.

With roadblocks that needed to be punched through in the way of GPT-4's development, the stars aligned.

To solve OpenAI's data bottleneck, Brockman turned to a new source: YouTube. OpenAI had previously avoided this option—scraping YouTube to train OpenAI's models, YouTube's CEO would later confirm, violated the platform's terms of service. But under the new existential pressure for more data, the question became whether YouTube, or its parent, Google, would enforce it. If Google cracked down, it could jeopardize its own ability to scrape other websites for its large language model development. Brockman was willing to take the risk.

With a small team, Brockman began collecting YouTube videos, eventually compiling more than one million hours of footage, according to *The New York Times*. He then used a speech-recognition tool called Whisper, which Radford had developed, to transcribe the videos into text for GPT-4.

Next was the training. To train GPT-3, the Nest team had designed a bespoke software platform. With most of its creators now gone to Anthropic, they were no longer around to explain how it worked. As a point of pride, some leadership didn't want to rely on the Anthropic team's legacy either. Brockman disappeared into his coding hole and developed a new platform. Then, with several others, including Jakub Pachocki and Szymon Sidor, the Polish scientists whom he'd grown close with during the

GODS AND DEMONS

Dota 2 project, Brockman babysat GPT-4's training. The pre-training alone took three months.

At first, GPT-4 seemed like a disappointment. "It was a wild model, which in some sense behaved quite poorly," one researcher says. "Because the average data quality was so horrible, and because the model was quite powerful and context sensitive, it was producing garbage responses." But Brockman pushed forward, pulling together the resources to improve the model with human contractors conducting reinforcement learning from human feedback. With each week, the results looked better and better, until the performance truly began to wow people internally.

GPT-4 now had built-in multimodal capabilities and, against OpenAI's internal assessments, was generating more polished code than ever and was more nimble in recognizing user intent and delivering helpful answers. In an impressive showcase of those abilities, Brockman would later live stream a demo of him prompting GPT-4 with a photo of a simple chicken scratch sketch of a web page drawn in his notebook. "My Joke Website," Brockman had written at the top. Stacked below it, he'd added: "[really funny joke!]" and "[push to reveal punchline]." In less than half a minute, the model would turn that sketch into workable code, stylizing the first line as a title, replacing the second line with a joke, and recognizing the third line as a button.

But as OpenAI began teasing the model in trusted circles, including investors and select customers, at least one person wasn't the least bit impressed. It was once again the ever-hard-to-please Bill Gates.

In June 2022, after getting a demo of GPT-4, Gates expressed disappointment in the insufficient progress from GPT-2. Despite the model being significantly larger and more fluent, he still felt like it was "an idiot savant," unable to tackle complex scientific problems. He told the team that he would only start paying attention once GPT-4 scored a 5 on an AP Biology test—AP Bio because he felt it tested critical scientific thinking rather than a memorization of facts. "I thought, 'Okay, that'll give me three years to work on HIV and malaria,'" Gates later recounted in his podcast.

Brockman took Gates's remark as a challenge. He immediately reached out to Sal Khan, the CEO of online education platform Khan Academy, and asked him to tap into the company's large repository of AP Bio questions as training data. Khan was skeptical but agreed to do so in exchange for his platform getting access to the model. Brockman also amassed a team of employees to build a special user interface for the new Gates Demo.

By late August, much to Gates's surprise, Altman and Brockman were pinging him again. Over dinner at the Microsoft founder's house the following month with roughly thirty people, the two OpenAI executives and others showed Gates a series of highly refined GPT-4 demos designed to impress him. The crowning moment was the model acing AP Bio: It nailed fifty-nine out of sixty multiple-choice questions and generated impressive answers to six open-ended ones. An outside expert would score the test: 5 out of 5. Gates couldn't believe it. His shock and praise, which the demo attendees would instantly relay back to the rest of the company, ripped like wildfire through the office and incited an exhilarating level of energy: This showcase, Gates said, was one of the two most stunning demos he'd ever seen in his life.

In all-hands meetings, Altman continued to stoke the excitement. "Startups that do remarkable things require a miracle," he said. "We just had our miracle." Many employees believed it, awestruck by the momentousness of what they had accomplished. GPT-4's new level of performance convinced OpenAI leadership that it was time to start working toward one of Altman's long-coveted ambitions: an AI assistant that would look and feel like the character Samantha in the 2013 Spike Jonze movie *Her*.

For years, *Her* had been a touchstone that Altman and other OpenAI cofounders frequently invoked as an example of what AGI might one day look like: a single multimodal model whose product interface felt so utterly natural that it faded away and simply brought user delight. "I would think it's because it was an assistant that was wonderfully integrated into a life," says a former employee, of why the movie was such a pivotal reference. "The positive arc of that story before it unravels is a really great story of AI's evolution into society."

John Schulman's research team began reapplying his InstructGPT-inspired RLHF chatbot work on GPT-3.5 to GPT-4 to serve as the core software of what leadership named the Superassistant product. Brockman and Fraser Kelton pulled together a ragtag team of fewer than ten people from around the company to brainstorm and prototype different ideas for its interface. One person from the supercomputing team who was usually an infrastructure guy began hacking away during his evenings and weekends on an iOS app for chatting with the model. Another one suggested using Whisper to add a voice interface to the app so people could speak to it without typing. A third person from the inferencing team proposed creating a Chrome extension that would help users summarize web pages with the Superassistant as they browsed the internet. A fourth person began building a meeting bot for the Superassistant to join a user's video calls and send them a summary of what happened.

As momentum picked up, excitement mounted at the possibilities. That summer, when a group of AI researchers, including Barret Zoph, Luke Metz, and Liam Fedus, left Google to found their own digital assistant startup, Altman had persuaded them to work on their idea at OpenAI instead. They joined Schulman's team, sitting side by side in the office with the Superassistant team, to drive its research and accelerate its development. In a heightened and thrilling state of flow, people from the Applied and Research divisions were working more tightly together than ever before to launch a new product.

As OpenAI demoed GPT-4 to Microsoft, Satya Nadella, Kevin Scott, and the tech giant's other executives were just as excited. Codex had proven that OpenAI's technologies could have commercial appeal, but GPT-4 represented something far bigger. Across the board, it beat the performance of various AI models that Microsoft had developed in-house; it could also do much more, including answering questions with a high degree of context and clarity. It opened up a range of possibilities to create new conversational interfaces, or Copilots, for all of Microsoft's products, such as to allow users to chat with the company's struggling search engine Bing or to tell Microsoft's Office suite in natural language to turn a Word document into a PowerPoint presentation. Microsoft would also be able to offer custom Copilots directly to its cloud customers. It would undoubtedly turn the tech giant into an AI leader, finally able to go toe to toe with Google.

In the coming months, Nadella would unlock Microsoft's third investment into OpenAI and continue its exclusive access to OpenAI's model weights for integrating into its products. The companies would reveal the amount in January 2023: $10 billion.

At first, OpenAI executives wanted to release GPT-4 in the fall of 2022. The deadline was a case of fantastical thinking. Nearing the end of summer, the company was nowhere near ready to launch a new commercial product across any of its functions. The product team needed more time to polish its interface; the infrastructure team needed to apportion server space; the model itself needed more work to iron out its behavior.

OpenAI had at that point formed a committee with Microsoft, called the Deployment Safety Board, or DSB, with three representatives from each company, to evaluate OpenAI's cutting-edge models and determine whether they were ready for release. OpenAI's representatives were Altman, Miles Brundage, the head of policy research, and Jan Leike, the head of alignment, which oversaw the continued development of AI safety techniques like RLHF. Both policy research and alignment had become the new emerging strongholds of the Safety clan. DSB created a formal governance structure for resolving the age-old debates between Applied and Safety. After a preliminary review, the DSB gave GPT-4, the first model being evaluated under this structure, a conditional approval: The model could be released once it had been significantly tested and tuned further for AI safety.

Executives agreed on a new deadline in early 2023. By that time, they wanted the Superassistant to also be ready. OpenAI would release GPT-4 in the API side by side with the GPT-4-powered, consumer-facing product. To employees, Altman framed the decision to delay the release of the model as evidence of OpenAI's cautious and safety-minded approach. The company was taking its time with deploying the system, he said. In fact, leadership wasn't even sure they would deploy it. In Applied, most people interpreted Altman to mean that it would take an extraordinary issue for leadership to stop the release but that, should such a situation arise, they would be willing to do it. To many people within Safety, Altman's words landed differently: The assumption was that only once GPT-4 passed every check would leadership greenlight its release.

In private meetings, Altman reinforced the perceptions on each side. To both, he raised concerns that GPT-4 could "wake up Google," evoking a scenario in which the day the model dropped and seized Larry Page and Sergey Brin's attention, the two would finally cut through Google's political bureaucracy. Google executives could cancel all their meetings, go on a daylong retreat, consolidate all of the company's talent and resources, and bring its full compute firepower to bear on catching up. But where with Applied, this was reason to sustain its intensity to prepare for launch while keeping GPT-4's capabilities a secret for as long as possible, with Safety, Altman used it to continue underscoring his caution. "My number one safety concern is acceleration risk," he said, adopting their vocabulary. Translation: Waking up Google could spark race-to-the-bottom dynamics.

Even with the delay, the Applied division was scrambling to meet the early 2023 release. In particular, Dave Willner's trust and safety team still had a limited staff and underdeveloped infrastructure. If OpenAI was planning to go live with a direct-to-consumer Superassistant product, built atop a much more powerful model, it needed a far more mature abuse prevention and enforcement operation.

Willner rushed to hire several experienced trust and safety deputies from Google and Meta. He tasked one with investigating "unknown unknowns"—abuses that the team wasn't yet aware of and would need to discover through careful monitoring and data analysis. He tasked another with "known knowns," or so-called scaled enforcement—the abuses that the company had already deemed as violating behavior and would need to automatically flag, review, and take action on, such as by suspending the accounts of repeat offenders.

As they raced to set up product policies and monitoring and enforcement infrastructure, members of the budding team felt a constant sense of uncertainty about how trust and safety for an AI company should differ from a search or social media company and whether their preparations for GPT-4 were adequate. Trust and safety was typically focused on preventing a predictable slate of internet abuses, like fraud, cybercrime, and election interference. But wrapped up in the confusion was how their work related to that of the AI safety people within Leike's alignment and Brundage's policy research teams who often discussed unknowable, catastrophic doom. OpenAI labeled all of them as "safety" teams, but they seemed to be speaking fundamentally different languages. Although the vocabulary of existential AI safety had, with its popularization through EA, become common parlance in the field, employees coming from traditional tech company backgrounds had never heard of AI timelines or quantified their p(doom). Even shared words between the two groups like *risk* and *harm* seemed laced with different connotations.

With the arrival of more and more non-AI people to OpenAI to support the expanding needs of the Applied division, the ability to cross that cultural divide between those steeped in the Doomer-Boomer mind frame and those operating under a standard tech company framework became a special kind of political currency. Among Willner's trust and safety deputies, the one most successful in learning the language of AI safety became a bridge to that world. The deputy worked across teams to make the model "safer" in both senses of the word, such as by using RLHF to "align" the model—rogue AI safety lingo—and thereby make it better at refusing user queries that violated OpenAI's platform policies, the typical work of trust and safety.

But as "safety" progressed on both fronts, a new directive suddenly arrived from executives: to suspend the developer-review process that had first been implemented with the GPT-3 API release.

For a while already, executives had felt that the waiting list had grown out of control, and the review process wasn't scaling. Developers were complaining about how long it was taking to get access to OpenAI's technologies, and were constantly emailing Altman, Brockman, and Peter Welinder or tagging them on Twitter about long holdups on what the executives saw as totally innocuous applications. Many in Applied also felt

OpenAI was gatekeeping access to the benefits of its technology, which was antithetical to its mission.

Willner and his team had always pushed back. If OpenAI dropped its application process and automatically approved developers, the company had no real alternative for moderating the use of its technologies. If it switched to reactive enforcement, it would need to build up significant tooling to do so. With the launch of GPT-4 pending, executives overrode the objections: OpenAI was getting rid of developer review; the trust and safety team simply needed to figure out the alternative.

Willner's team rushed to put together a proposal. It would shift more of its enforcement of the company's policies upstream, by leaning more heavily on RLHF to align GPT-4 and future models. Everything else would be caught and handled downstream with reactive enforcement: using different data signals, such as information about what the app did, its traffic spike patterns, and the number of times it triggered the content-moderation filters, to automatically suspend obvious violators while sending borderline cases to human moderators for manual review.

Willner's team urged executives for the resources to properly build out the tooling they needed to make the plan work. OpenAI didn't have most of these data points at its disposal; its monitoring platform was logging only basic data on how much traffic each app was sending through the company's servers. Sometimes it didn't even know the name of the developer or the purpose of the application.

In addition to plugging those gaps, the team also needed developers to assign a unique identifier to each of their users. This was crucial to be able to disaggregate an app's traffic by its individual users to determine whether violating behaviors were endemic to an app's user base or merely being committed by a few frequent abusers. Executives resisted, worried that implementing such granular monitoring would add friction to the developer experience, and potentially make the company more liable for performing trust and safety work for every app using the company's technologies, rather than having each app developer handle it themselves.

In the end, the proposal for reactive enforcement was scaled down to something much more limited.

The restricted visibility into app-user behavior made some members of the trust and safety team anxious. An employee raised his concerns to Brockman. What he feared most, the employee said, was people using GPT-4 to generate mis- and disinformation at scale and influence elections.

Brockman sought to reassure him. "Yes, this is what we always say and we're always concerned about," he said. "But what I haven't ever seen is, is it actually happening?"

GPT-4 wasn't just a turning point for Gates and Microsoft. Later that summer, after wowing the billionaire philanthropist, Altman and Brockman brought it to OpenAI's board, where Brockman gave a live demo of the model that included it telling jokes about Gary Marcus. The jokes delighted the independent board members; the lighthearted showcase also signaled to them a significant advance in the technology's capabilities and imparted a sense that the stakes of their future decisions were going up.

Until that point, Altman had convened the board roughly once a quarter and preferred to deliver his updates verbally. He breezed through complex research topics, sometimes bringing along company researchers to present their progress, and gave rapid-fire rundowns on the latest deals he was in the process of negotiating with Microsoft or other partners. Some of the independent board members pressed for more frequent meetings and more structured information, insisting Altman provide written reports and give them access to more documents.

Altman chafed at the increased oversight. He expressed several times that he wanted the board members to serve more as advisers who gave him input for consideration. "The CEO is supposed to make the decisions, and the board is supposed to, you know, be a sounding board—advice and consent," he once said to university students in 2017, seeming to reference a governance concept in the US Constitution, core to the role of the legislative branch to check the executive branch's power, but using it to mean the opposite. "Do I think a board should fire a bad CEO? Yes—and I know that's, like, a little bit heretical in Silicon Valley. Beyond that, do I think the board needs to give the CEO a very wide latitude to run the company? Yes."

Several board members felt strongly that OpenAI's board was meant to be different. "The board is a nonprofit board that was set up explicitly for the purpose of making sure that the company's public good mission was primary, was coming first over profits, investor interests, and other things," Helen Toner would tell *The TED AI Show* podcast. "Not just like, you know, helping the CEO to raise more money."

Among many employees, GPT-4 solidified the belief that AGI was possible. Researchers who were once skeptical felt increasingly bullish about reaching such a technical pinnacle—even while OpenAI continued to lack a definition of what exactly it was. Engineers and product managers joining Applied and having their first close-up interaction with AI through GPT-4 adopted even more deterministic language. For many employees, the question became not if AGI would happen but when.

Some employees also felt exactly the opposite. While there was a clear qualitative change in what GPT-3 could do over GPT-2, GPT-4 was just bigger, says one of the researchers who worked on the model. "The big result was that there were a bunch of exams that the model does well. But even that is highly questionable." OpenAI never did a comprehensive review of GPT-4's training data to check whether those exams—and their answers—were just in the data and being regurgitated, or whether GPT-4 had in fact developed a novel capability to pass them. It was the kind of shaky science that had become pervasive with the industry-wide shift from peer-reviewed to PR-reviewed research.

But the belief that AI had reached fundamentally new heights was in the water. At Google that spring, Blake Lemoine, an engineer on the tech giant's newly re-formed responsible AI team, grew convinced that the company's own large language model LaMDA was not only highly intelligent but could be considered sentient. He said this was not based on a scientific assessment but rather on his belief, as a mystic Christian priest, that God could decide to give technology consciousness. "Who am I to tell God where he can and can't put souls?" he wrote. When company executives dismissed him, he went public with *The Washington Post*. Nitasha Tiku, the reporter who broke the story, also spoke to Emily Bender and

Meg Mitchell, who had warned in their "Stochastic Parrots" paper about the problem of large language models fooling people into seeing real meaning and intent behind their generations.

"We now have machines that can mindlessly generate words, but we haven't learned how to stop imagining a mind behind them," said Bender.

"I'm really concerned about what it means for people to increasingly be affected by the illusion," when that illusion is becoming so good, Mitchell said.

Nikhil Mishra, the AI researcher who interned at OpenAI in its early days, draws parallels to an experiment in the 1970s that sought to teach a gorilla a modified form of American Sign Language. Over her lifetime, the gorilla, named Koko, seemed to learn more than one thousand signs—and even the ability to construct sentences. But despite enormous public fanfare, other experts argued that Koko never truly acquired the language. While she was certainly skilled at forming gestures, there was little evidence that she was doing more than what other apes had done across many other similar experiments: simply mirroring the gestures of their caretakers. No data was ever published of Koko's signing to independently verify her capabilities, only curated videos of her showcasing them. At times, her live performances begged further scrutiny. Once, when her trainer asked Koko whether she liked people, Koko signed "fine nipple." The trainer immediately explained: "Nipple" rhymed with "people"; Koko thought people were fine. To many observers, these episodes revealed more about human psychology and our tendency to project our own beliefs and ideas of intent than about Koko's ability. The trainer, Mishra says, was assigning meaning where there wasn't any.

For OpenAI's own resident mystic, Sutskever, who had always believed more than most researchers in the likelihood of achieving AGI in the short term, the leap in the company's model capabilities served only as further confirmation. He grew convinced that he was witnessing a form of reasoning. In conversations with Hinton, Sutskever told his mentor that AGI was imminent.

Where before, Sutskever focused more on pushing OpenAI researchers to advance new capabilities, he shifted his attention to AI safety research with new urgency. He began his mantra "Feel the AGI" and urged people to prepare themselves for dramatic changes. "You're suddenly going to be the most popular person at a party," he told employees. "You need to not let it get to your head. Stay focused on AGI, stay focused on the mission."

That September, the technical leadership held an off-site at the Tenaya Lodge, a remote luxury resort nestled in the lush folds of the Sierra Nevada. The property, furnished with stunning interiors and multiple pools and restaurants, sat only two miles away from the millions of acres of pristine wilderness in Yosemite National Park. On the first night, everyone gathered around a firepit on the rear patio of the hotel. Senior scientists, dressed in bathrobes, flanked the fire in a semicircle.

Then Sutskever emerged. In the pit, he had placed a wooden effigy that he'd commissioned from a local artist, and began a dramatic performance. This effigy, he explained, represented a good, aligned AGI that OpenAI had built, only to discover it was actually lying and deceitful. OpenAI's duty, he said, was to destroy it. Only a few yards away, several redwoods stood like ancient witnesses in the darkness. Sutskever doused the effigy in lighter fluid and lit it on fire.

Chapter 11

Apex

In October 2022, OpenAI held a company-wide off-site in Monterey, California, a beautiful coastal city two hours south of San Francisco. By then the company had grown to roughly three hundred employees, from less than two hundred the year before. For a photo that weekend, they all posed outside Monterey's tiny airport, grinning and wearing their OpenAI-branded gear. Altman looked relaxed, sitting on the ground in the front of the pack, knees against his chest, arms loosely crossed, feet pointed up.

Over two days, the executive team presented updates on their vision and implementation: Altman on the massive data centers that Microsoft was scaling up for OpenAI; Steve Dowling and his deputy Hannah Wong on the top-tier publications lining up to cover the company; Anna Makanju on its expanding footprint in Washington, DC. Then there were demos, one after the other, of the research and product teams' latest projects. The sheer impressiveness of it all was palpable. "Everything together was so mind-boggling," remembers a former employee who was present.

Brockman got onstage to discuss the latest plans for GPT-4 and began to tell a story about his wife, Anna, who started having abdominal pains one day that wouldn't go away. Anna was a frequent fixture at the office. She had a desk next to his and often came with him to meetings even though she didn't officially work at the company. Besides Altman and Sutskever, she was, Greg had once told me, the person he relied on most for support, his best friend, his confidant. They were often attached at the hip, going everywhere together. Greg went with Anna to multiple doctors, he explained to employees, and none could figure out the problem. But when he asked GPT-4, it suggested she might have a condition that they hadn't considered. "And then she did!" he exclaimed. It was a retelling of the same story he'd shared with me in 2019, about the promise of AGI solving health care for people like his friend who had gone to myriad specialists to diagnose a problem—this time with a different character.

Greg would repeat the story again on X weeks after the board's attempt to fire Altman, with a new variation. He would recount new medical challenges that Anna had faced and her struggle over five years "seeing more doctors and specialists than in her whole life prior" to finally get a diagnosis. It was her allergist who finally put all the pieces together and realized she had hypermobile Ehlers-Danlos Syndrome, a genetic mobility disorder. There was still a long way for AGI to work "in high-stakes areas like medicine," he wrote, keeping up his drumbeat on behalf of OpenAI's rallying ambition, "but the promise is getting increasingly clear."

Within weeks of returning from the off-site, rumors began to spread that Anthropic was testing—and would soon release—a new chatbot. The Superassistant team was midway through designing its chat interface. If it didn't launch first, OpenAI risked losing its leading position, which could deliver a big hit to morale for employees who had worked long and tough hours to retain that dominance. Worse still for some leaders, OpenAI would lose to Anthropic.

Anthropic had not in fact been planning any imminent releases. It was also in the midst of its own problems. With the sudden collapse of FTX in the early days of November, it was getting swept up in the fallout. Just months before, it had raised $580 million, $500 million of which an FTX press release had said was from SBF and other senior leaders. Financial documents released during the trial would later find that that money had been sourced from the billions that SBF had embezzled from

FTX customer deposits, turning the Anthropic investment into a central issue during the trial over whether the significant returns generated from it could be used to pay back customers. (A judge would rule that it could, and FTX would sell off its Anthropic shares in batches through 2024 for a total of $1.3 billion.)

But for OpenAI executives, the rumors were enough to trigger a decision: The company wouldn't wait to ready GPT-4 into a chatbot; it would release Schulman's chat-enabled GPT-3.5 model with the Superassistant team's brand-new chat interface in two weeks, right after Thanksgiving. The Superassistant team instantly pivoted, pulling in several other members as they sprinted to integrate everything and build out the remaining features. To the rest of the company, leadership framed the effort carefully. ChatGPT—the name they settled on—would not in fact be a product launch but a "low-key research preview," just like DALL-E 2. In the same way, it wouldn't be monetized but "get the data flywheel going"—in other words, amass more data from people using it—which would help improve GPT-4 and the Superassistant product.

Outside of the Superassistant team, everyone took the executives literally. A low-key research preview didn't require their attention; they needed to stay focused on the GPT-4 launch for early 2023. The trust and safety team felt they barely had enough time to build out its monitoring infrastructure in time for that launch. By comparison, ChatGPT seemed like a nonissue. GPT-3.5 had already been refined with RLHF; it was inherently safer than the version of GPT-3, which had not been, still available on the API. OpenAI had also posted a version of 3.5 without chat features on its developer platform for developers to test out the model's capabilities. Did adding a chat interface really make a difference? People in the Safety clan, occupied with testing and tuning GPT-4, agreed. For the first time, a model release flew through the checks with little resistance.

Even within the Superassistant team, no one truly fathomed the societal phase shift they were about to unleash. They expected the chatbot to be a flash in the pan. Much like DALL-E 2, it would generate a lot of fanfare on social media and then quiet down after a few weeks. The night before the release, things felt remarkably calm after such an intense sprint to the finish. They placed bets on how many users might try the tool by the end of the weekend. Some people guessed a few thousand. Others guessed tens of thousands. To be safe, the infrastructure team provisioned enough server capacity for one hundred thousand users.

The following day, on Wednesday, November 30, most other employees didn't even realize that the launch had happened. Like OpenAI's own debut, ChatGPT's release coincided with the annual NeurIPS proceedings, that year being held in New Orleans, Louisiana. The conference had earlier announced its Test of Time Award, an honor bestowed each year to a paper published ten years earlier that had had a critical impact on the field. The award went to Hinton, Sutskever, and Krizhevsky's 2012 ImageNet paper that introduced the world to the power of deep learning.

That evening a small group of employees hosted an OpenAI party near the conference convention center to represent the company and recruit interested candidates among the nearly ten thousand in-person attendees. DeepMind, Meta, and Google were holding competing recruitment parties at the exact same time throughout the city. As the party went on, a recruiter at the event noticed an OpenAI engineer working nonstop on his computer.

He finally went over to talk to the engineer: "Bro, have a drink. We're all here. Be social."

The engineer didn't move. "No, all the GPUs are melting. Everything is crashing."

That night, Japan had been first to wake up and to deliver a massive and unexpected swell in the traffic. The following day, the number of users continued to surge, as time zone by time zone the rest of the world came online. Musk played no small part in boosting the climb. "Lot of people stuck in a damn-that's-crazy ChatGPT loop 🔁 ," he tweeted, receiving some seventy-five thousand likes.

The instant runaway success of ChatGPT was beyond what anyone at OpenAI had dreamed of. It would leave the company's engineers and researchers completely miffed even years later. GPT-3.5 hadn't been that much of a capability improvement over GPT-3, which had already been out for two years. And GPT-3.5 had already been available to developers. The interface and format had made the model more accessible, certainly, but it wasn't the fundamental step change that employees had felt with GPT-4. Altman later said that he'd believed ChatGPT would be popular but by something like "one order of magnitude less." "It was shocking that people liked it," a former employee remembers. "To all of us, they'd downgraded the thing we'd been using internally and launched it."

Within five days, Brockman tweeted that ChatGPT had crossed one million users. Within two months, it had reached one hundred million, becoming what was then the fastest-growing consumer app in history. (Meta's X rival, Threads, later claimed the title by reaching the same user count in less than five days; pundits argued that it didn't count because Meta was primarily tapping into an existing base of users.)

ChatGPT catapulted OpenAI from a hot startup well-known within the tech industry into a household name overnight. Indeed, at an AI research conference several months later in Kigali, Rwanda, over nine thousand miles away from San Francisco, a researcher based in the country would gush to me that post-ChatGPT, his parents finally understood what he did for work. "You know a technology is accessible to anyone when your mother tells you about it," he'd say.

At the same time, it was this very blockbuster success that would place extraordinary strain on the company. Over the course of a year, it would polarize its factions further and wind up the stress and tension within the organization to an explosive level.

In the immediate aftermath, the whole company was firefighting. OpenAI's servers crashed repeatedly as the infrastructure team struggled to scale up its capacity as fast as possible, in the most compressed timeline in the history of Silicon Valley. The team cannibalized some of the Research division's compute to support ChatGPT's growth and still didn't have enough to keep the app up and running. The trust and safety team, numbering just over a dozen people, scrambled to understand and catch bad behavior among the floods of new users, heavily handicapped by spotty monitoring. It had struggled to hire the engineers needed to implement its limited reactive enforcement plan and was still in the middle of building the necessary systems. ChatGPT derailed the project. All efforts to finish new tooling halted as engineering resources were redirected to stabilize what already existed. When the servers crashed, so did the platform for monitoring traffic, grinding the ability to do any scaled enforcement to a complete halt.

The severe shortage of GPUs also derailed another effort. In an attempt to leverage the company's own technology, the trust and safety team had prototyped a plan internally called Fact Factory, which OpenAI publicly touted, for using GPT-4 to content moderate its own outputs and that of other OpenAI models. The implementation didn't exactly scale; it required giving GPT-4 extremely long prompts to capture enough nuance. Even when the servers were working, it would cost too many computational resources. And the servers were not consistently up.

To many in the Safety clan, ChatGPT was the most alarming example yet of the limitations of OpenAI's foresight. One Safety person raised the question in an all-hands meeting: How could the company have failed to predict user behavior and ChatGPT's popularity so badly? What did that say about the company's ability to calibrate and forecast the future impacts of its technologies?

To much of the rest of the company, the crashing servers, while an extraordinary source of stress, were even more so an extraordinary mark of triumph. OpenAI had built a technology so profound, in such wild demand, that it had lit up the world and transformed it overnight. They had set their sights on all of humanity and had really done it. Everyone, all eight billion people, was now living in OpenAI's world.

Altman didn't indulge the moment. He reminded employees that the company ultimately had a mission to achieve something far bigger than building "the biggest product in the history of Silicon Valley." He urged every team to stay the course and press forward. As he'd expected, OpenAI had woken up all of its competitors: Anthropic was on its way to releasing its chatbot, Claude; Google had sounded a "code red" alarm internally and would soon consolidate its AI divisions into Google DeepMind to throw its full weight behind launching a similar product. Though

OpenAI had hit the market first with its 10x better offering, it needed to keep running to stay number one.

With every team stretched dangerously thin, managers begged Altman for more head count. There was no shortage of candidates. After ChatGPT, the number of job applicants clamoring to join the rocket ship had rapidly multiplied. But Altman worried about what would happen to company culture and mission alignment if the company scaled up its staff too quickly. He believed firmly in maintaining a small staff and high talent density. "We are now in a position where it's tempting to let the organization grow extremely large," he had written in his 2020 vision memo, in reference to Microsoft's investment. "We should try very hard to resist this—what has worked for us so far is being small, focused, high-trust, low-bullshit, and intense.

"The overhead of too many people and too much bureaucracy can easily kill great ideas or result in sclerosis. Unlike the big-iron engineering projects of the past, we could fulfill our mission with a surprisingly small number of great people."

He was now repeating this to executives in late 2022, emphasizing repeatedly during head count discussions the need to keep the company lean and the talent bar high, and add no more than one hundred or so hires. Other executives balked. At the rate that their teams were burning out, many saw the need for something closer to around five hundred or even more new people.

Over several weeks, as the discussions continued, the executive team finally compromised on a number somewhere in the middle, between two hundred fifty and three hundred. The cap didn't hold. By summer, there were as many as thirty, even fifty, people joining OpenAI each week, including more recruiters to scale up hiring even faster. By fall, the company had blown well past its own self-imposed quota.

The sudden growth spurt indeed changed company culture. A recruiter wrote a manifesto about how the pressure to hire so quickly was forcing his team to lower the quality bar for talent. "If you want to build Meta, you're doing a great job," he said in a pointed jab at Altman, alluding to the very fears that the CEO had warned about of the company rapidly diluting its talent density and mission orientation, while increasing its bureaucracy. The rapid expansion was also leading to an uptick in firings. During his onboarding, one manager was told to swiftly document and report any underperforming members of his team, only to be let go himself sometime later. Terminations were rarely communicated to the rest of the company. People routinely discovered that colleagues had been fired only by noticing when a Slack account grayed out from being deactivated. They began calling it "getting disappeared."

To new hires, fully bought into the idea that they were joining a fast-moving, money-making startup, the tumultuousness felt like a particularly chaotic, at times brutal, manifestation of standard corporate problems: poor management, confusing priorities, the coldhearted ruthlessness of a capitalistic company willing to treat its employees as disposable. "There was a huge lack of psychological safety," says a former employee who joined during this era. "It is like the opposite of 'a company as a family'—which is fair, you know, it is a company." Many people coming aboard were simply holding on for dear life until their one-year mark to get access to the first share of their equity. One significant upside: They still felt their colleagues were among the highest caliber in the tech industry, which, combined with the seemingly boundless resources and unparalleled global impact, could spark a feeling of magic difficult to find in the rest of the industry when things actually aligned. "I would say OpenAI is one of the best places I've ever worked but also probably one of the worst," the former employee says.

For some employees who remembered the scrappy early days of OpenAI as a tight-knit, mission-driven nonprofit, its dramatic transformation into a big, faceless corporation was far more shocking and emotional. Gone was the organization as they'd known it; in its place was something unrecognizable. "OpenAI is Burning Man," Rob Mallery, a former recruiter, says, referring to how the desert art festival scaled to the point that it lost touch with its original spirit. "I know it meant a lot more to the people who were there at the beginning than it does to everyone now."

In those early years, the team had set up a Slack channel called

#explainlikeimfive that allowed employees to submit anonymous questions about technical topics. With the company pushing six hundred people, the channel also turned into a place for airing anonymous grievances. In mid-2023, an employee posted that the company was hiring too many people not aligned with the mission or passionate about building AGI.

Another person responded: They knew OpenAI was going downhill once it started hiring people who could look you in the eye.

ChatGPT also surprised Microsoft. OpenAI leaders had told its partner, as they'd told their own employees, that the chatbot would be a "low-key research preview." It was clearly anything but.

The mismatch initially peeved the tech giant's executives. ChatGPT had completely stolen the thunder of Microsoft's chatbot for Bing. When Microsoft pushed out Bing AI the following February, the product would also take a PR hit with an article by *New York Times* columnist Kevin Roose about it pushing him to divorce his wife. It was far from the reception Microsoft had hoped for and, by comparison, had made OpenAI look even better.

But the crossed wires weren't nearly enough to dampen Microsoft's enthusiasm for OpenAI. The enormously positive reception to ChatGPT was contagious, and the continuously improving capabilities of OpenAI's models made the giant's executives even more excited. Microsoft was now readying a whole new slate of Copilots for the tech giant's products based on GPT-3.5 and GPT-4, which it planned to release one after the other in a steady drumbeat of announcements. After Bing in February, March was for Microsoft 365 Copilot, bringing an AI-powered chat-based interface to every Office product from Word to Outlook to Teams.

The way in which Microsoft executives talked internally about the OpenAI partnership was also rapidly shifting. Before, Microsoft felt like it had power over OpenAI; now some of the giant's executives felt like OpenAI had power over Microsoft. There was a creeping sense of inadequacy within parts of Microsoft that its own AI research efforts had failed to achieve what OpenAI had pulled off. If Microsoft walked away as OpenAI's main investor, the startup could find other investors, a former Microsoft employee remembers of some of the executives' thinking. But if OpenAI walked away from Microsoft, would the tech giant find another OpenAI?

At the same time, many executives were no longer talking merely about beating Google. Where they had once responded to OpenAI's strange talk about AGI and highfalutin language about its power to invent the future with polite skepticism, they were now believers. AI, AGI, generative AI—whatever you wanted to call it—this technology *was* the future, and Microsoft was shepherding it hand in hand with OpenAI. The more Microsoft believed, the more the company reoriented its rhetoric and strategy. "I saw the incentives at Microsoft push more and more toward a narrow conception of the future," the former employee says. "I saw the technology become narrowed into something propped up by narrative rather than reality."

Nadella implemented a new strategy for distributing Microsoft's computing resources. He shifted GPUs away from Microsoft's research teams to support OpenAI. The company also consolidated all of its GPUs into one pool for better supporting generative AI workloads. "The typical Microsoft employee had no fucking clue what OpenAI was before January last year," one Microsoft employee remembers. Now they were receiving urgent directives from their superiors about finding ways to intersect their work with OpenAI technologies.

The tech giant would experience a rapid proliferation of over one hundred new generative AI projects within just a few months as employees experimented with various ways of using GPT-4 and ChatGPT. In an ironic twist, the aggressive adoption would force Microsoft to grapple with many of the same challenges that other companies would face as they raced to adopt generative AI without fully understanding it. That included causing headaches for the risk and compliance teams. Not everyone was using Microsoft's internal versions of the technologies; some were opting to use the free version of ChatGPT straight from OpenAI, which trained on user data, raising concerns over whether that could leak Microsoft customer information or interfere with regulatory compliance. While some employees found the tools a big productivity boost, many also found them exhausting. "There is this culture of 'Use AI, use AI, use AI,'" says one. But "it's like, okay, this doesn't help us. We don't want to use it. And it feels like it's everywhere and we can't escape it."

In switching from Microsoft's own models to OpenAI's, many employees also lost control and visibility into a core infrastructure layer of their work. Within the tech giant, access to the startup's underlying models was tightly guarded, even though they were trained and stored on Microsoft's servers. Most Microsoft employees could no longer examine the training data or tweak the weights of the models they were using. OpenAI's models were instead delivered through an API, as they were to other OpenAI customers.

But in exchange for these trade-offs, Microsoft was being richly rewarded. The company was seeing a massive surge in inbound customers for its Azure AI platform as the only cloud provider able to offer the typical benefits of the cloud, including simpler data storage and management, alongside the ability to process that data with OpenAI's capabilities. "Azure OpenAI Service is getting us in the door with many new customers these days," Eric Boyd, the corporate vice president of the AI platform, wrote in an email to his division in May 2023. In August, he enthused once more. "Every now and then it's great to take a step back and marvel at just how far we've come in just one year," he wrote to his division, adding that the platform that year had seen a "21x increase in customers." The following month, Boyd celebrated a new milestone. After centralizing Microsoft's fractured AI efforts onto the platform, and with the thousands of new customers who had joined Azure OpenAI Service, traffic on the platform had grown tenfold in just nine months. In January 2023, it had been receiving one hundred billion monthly inference requests; now in September, it was receiving one trillion.

That summer, Microsoft CTO Kevin Scott was effusive with his praise during an OpenAI all-hands meeting. "We have stopped, like, our AI machine learning investments in a bunch of places to the point that, like, people are like, 'Hey, you know, fuck you, Microsoft,'" Scott said, referring to the shifting away of GPUs from some of Microsoft's own internal research. "And we've taken the bet because we believe that you all are doing the absolute best work in the industry."

ChatGPT firmly codified OpenAI's turn away from nonprofit and toward commercialization. Altman and other executives pushed to build on the momentum of the chatbot's success by launching a slew of paid products. In February 2023, it released a paid version of ChatGPT; in March, one after the other, it released an API version, the Whisper API, and finally GPT-4. "After ChatGPT, there was a clear path to revenue and profit," a former employee says. "You could no longer make a case for being an idealistic research lab. There were customers looking to be served here and now."

The burst of new products overwhelmed the trust and safety team anew. For a while, OpenAI had enticed users to join its API by giving them an initial twenty dollars' worth of free usage credits. With the megapopularity of ChatGPT also sparking a dramatic surge in API usage, this sign-up incentive now posed a problem: Many users were creating new accounts at scale to cash in repeatedly on the bonus. In some cases, users were also spinning up new accounts to evade bans and suspensions on their old ones. The mass fraud was leading OpenAI to lose huge amounts of revenue as costs climbed with its need for more and more servers. Still numbering fewer than twenty people and with its reactive enforcement efforts severely hampered, the trust and safety team redirected its personnel once again to whack-a-mole the new vector of abuse.

Soon enough, the constant whiplash would push Willner to severe burnout. Within months he and several of his staffers would depart the company. By the end of that year, the team would dissolve and some of its remaining members be folded under a broader safety systems operation, headed by a longtime OpenAI researcher Lilian Weng. Some of the trust and safety people would come to feel that the persistent clashing between Applied and the Safety clan, with their overemphasis on Doomerism, had cultivated a culture among many of the company leadership to heavily discount any kinds of "safety" concerns, leading to an environment that made their function, already disempowered at most tech companies, even more so at OpenAI.

Indeed, with every new launch, the clashing continued, including over the release of GPT-4. While many in Applied felt the six-month delay in launching the model to be abundantly cautious, some in Safety felt it still hadn't given them enough time to finish their comprehensive testing and alignment.

The model's high rate of hallucinations, for example, had continued to prove particularly difficult to get under control, even with a concerted RLHF effort to address the problem. In November 2022, as users latched on to ChatGPT as if it were a search tool, spawning widespread speculation that it could unseat Google, an internal document noted that OpenAI's model had hallucinated during an internal test on roughly 30 percent of so-called closed-domain questions.

Closed-domain questions are meant to be the easiest category of questions: when users ask the model only about the information they give it—for example, uploading a pdf and asking for a summary, or providing bullet points and asking for a rewrite to complete sentences. This is in contrast to open-domain questions, when a user asks the model a question without reference material—pop culture, ancient history, high school biology—the way you would a typical search engine.

Meanwhile, GPUs became an ever-present constraint on OpenAI's research and expansion. New research and product or feature launches had to be repeatedly delayed or shelved due to a lack of chip capacity. After ChatGPT went viral, *SemiAnalysis*, a trade newsletter focused on the semiconductor industry, estimated that the company was spending some $700,000 a day on compute costs alone. After executives reallocated chips from the Research division, Applied made commitments to return them by a certain date. That date came and went, but Applied couldn't return them. With the continued rapid growth in users, the division needed more chips, not fewer.

The pressure accelerated OpenAI's research into more-efficient models. During its work to improve the company's compute efficiency, the Research division had figured out a new method for developing

Transformer-based models that were cheaper to serve to users. As they used that method, which they named DUST, they assigned code names to the resulting models to follow the desert-based theme. The first one, an optimized version of GPT-3.5, they called Sahara, which they released in February 2023 under the public name GPT-3.5 Turbo. Another they called Gobi, which would be an optimized version of one of its text-and-image models.

A third one, meant to be an optimized version of GPT-4, they called Arrakis, the desert planet from the science fiction epic *Dune*. But after months of work, the team was still struggling to make Arrakis more efficient while maintaining the same performance. The project ate up significant computational resources. Shortly thereafter, leadership scrapped it to free up GPUs for other projects.

As Microsoft worried about whether OpenAI could just leave the relationship, OpenAI felt its own vulnerabilities about whether Microsoft would stop cooperating if the startup didn't work hard to please its partner. Arrakis felt like a particular setback in this regard. In the hopes of impressing the giant, OpenAI had reworked its road map to prioritize delivering the model over its own more strategically aligned projects, including an effort to apply GPT-4 to a search engine product. Instead, the failed effort left some senior Microsoft executives disappointed.

There was also a new awkward reality: OpenAI and Microsoft were beginning to compete for contracts. Codex and DALL-E 2 had convinced OpenAI to retain control of delivering its technologies directly to users. ChatGPT and GPT-4 were showing that OpenAI could also make its own money. That meant directly pitching to customers the very same technology that it was handing over to Microsoft, which was then pitching to the exact same customers.

Handing off the technology had its own challenges. As OpenAI's release schedule picked up, so did Microsoft's. But Microsoft completely dwarfed OpenAI, leading to a dynamic where a single OpenAI employee could get pinged by dozens of Microsoft counterparts across various departments with all sorts of questions about technical or logistical details with every new product. It was growing increasingly frustrating and overwhelming for OpenAI staff to support Microsoft releases while focusing on their own road map.

Much of the smoothing over of the relationship was left to Murati. Murati sought to work closely with Scott and other Microsoft executives on coordinating the timing of product releases, strategizing how OpenAI and Microsoft would differentiate their offerings and finding more productive ways for the two organizations to work together. In the summer of 2023, in an attempt to cut back the communication burden, a team of Microsoft engineers began embedding inside OpenAI with full access to everything to streamline transfers of technology.

With the deeper integration, both Altman and Nadella were growing more involved than ever before, especially with the management of compute resources, making tough calls on how to redistribute chips and money for yet more chips to OpenAI's ever-compute-hungry operations. Nadella would tell *The New York Times* that OpenAI's demands would grow so fast and so high that Altman would start calling him every day saying, "I need more, I need more, I need more."

OpenAI didn't just need more data centers to serve ChatGPT. It still needed far-more-powerful supercomputers to train its future generations of models. To fulfill that aggressive and escalating demand, the two companies were sketching out a new unprecedented project called Stargate to OpenAI and Mercury to Microsoft: a single supercomputer that, for its construction alone, would cost an estimated $100 billion. The empire of AI was returning to the exact same form of expansion as the empires of old: To fuel its growth, it needed more material resources and, crucially, more land.

Chapter 12

Plundered Earth

In Santiago, Chile, on days after a good rain, the smog washes away and reveals a stunning view of the Andes mountains. The Andes run all the way up and down this thin sliver of a country, from Patagonia at its southern tip over 2,600 miles north—twice the length of Miami to Boston—to Chile's upper border. In this part of the country, the urban landscapes of the capital turn into an endless expanse of desert, which, save parts of Antarctica, is the driest place on earth. The mountains come into sharper focus, their colors more vibrant under the naked sun.

Before Chile was Chile, the Atacama Desert, as it is named, was home to many Indigenous groups. Where others may have seen a punishing barren landscape, they coaxed the desert into a home, tapping the water and minerals deep under the earth to grow crops, raise livestock, and perform ancestral rituals. Then the Spanish arrived. They cut up the region into administrative units with borders. To this day Indigenous elders still warn in oral histories of a repression that grew so violent the Spanish cut off the tongues of anyone who dared to continue speaking their native language. From there, the empire established the country's relationship with the rest of the world: It would provide raw resources—land, water, energy, minerals—to strengthen the political and economic agendas of other nations.

Today nearly 60 percent of Chile's exports are minerals, primarily found in the Atacama Desert, chiefly copper, a highly conductive metal used in all kinds of electronics, and more recently lithium, the essential ingredient for lithium-ion batteries. Those and other resource exports drive the country's economy. In Santiago everyone knows someone who lives by the rhythms of the mining industry: During their work "shifts" they live in the north; during their days off, they come back to the capital.

The country has struggled to build any other industry to serve as its economic engine. Long after the Spanish empire ended, the US infamously played a key role in ensuring this trajectory. In the 1950s and '60s, as developmental economics took root in Chile and Uruguay, favoring strong government regulation and an inward focus on industrialization as the path to maturing a developing economy, American multinationals who had made billions from their holdings in Chilean mines began to chafe against the growing state taxes and restrictions. The US government subsequently embarked on a quest to refashion Chile's economic policies to be more favorable to American business interests, launching a program in 1956 called "the Chile Project" to educate a hundred Chilean students at the University of Chicago under the intellectual tutelage of American economist Milton Friedman.

Friedman was a towering figure in economics who would go on to receive a Nobel Prize in 1976 and whose ideas could best be summed up by the title of his influential 1970 op-ed in *The New York Times*: "The Social Responsibility of Business Is to Increase Its Profits." Friedman stood for everything that developmental economics did not: zero government regulation, unfettered freedom for profit-driven companies, a path to the economic maturation of developing countries defined by facing outward, such as through liberal exports. As Naomi Klein details in her 2007 international bestseller *The Shock Doctrine*, the Chile Project was not education but indoctrination. At the University of Chicago, Chilean students—and later students from other countries in Latin America—were explicitly taught to critique their country's economic policies and the fatal flaws of Latin American developmentalism.

As each batch of graduates, known as "the Chicago Boys," returned to Santiago, Friedman's neoliberal ideas percolated through Chile's intellectual elite, until they became part of ruling ideology. In 1973, Chile's left-wing, democratically elected president was overthrown by his military general Augusto Pinochet in a coup d'état under conditions fomented in part by the CIA. The coup became the start of Pinochet's nearly two decades of brutal dictatorship and a new neoliberal economic agenda: the dictatorship appointed the Chicago Boys to write its economic policies.

Under Pinochet's rule, Chile privatized nearly everything—education, health care, the pension system, even water. The strategy produced economic growth; it also fueled stunning inequality. Chile is among the most unequal countries in the world today, with nearly a quarter of the country's income concentrated among a few powerful families in the 1 percent. Having never meaningfully industrialized, it also remains tethered to the extraction economy that makes it relevant to higher geopolitical powers.

And so, as the AI boom arrived, Chile would become ground zero for a new scale of extractivism, as the supplier of the industry's insatiable appetite for raw resources, not just its copper and lithium in the north but also its land, water, and energy resources for a growing crop of data centers in the Santiago metropolitan region. In May 2024, the government proudly announced that the country would welcome twenty-eight new data centers, on top of its existing twenty-two, over the coming years, bringing in $2.6 billion of foreign investment.

While the government's stance that its role as a resource provider to technology development represents progress for the nation, Chile has also become home to some of the fiercest resistance globally against this narrative. Communities across the country are vehemently fighting against the dispossession of their land, water, and other resources in service of Global North visions that do not include or benefit them. Through street protests and courtroom battles, their efforts have stalled company projects and caught the government's attention. They have also inspired people in other countries to rise up in solidarity.

Martín Tironi Rodó, a professor at Catholic University in Santiago and director of the Chilean research think tank Futures of Artificial

Intelligence Research, which studies AI through an interdisciplinary and Latin American lens, summarizes the sentiment that I heard repeatedly from these communities as I traveled up and down the country to meet them. The central question these movements are asking is how to imagine a different path for AI development not rooted in extraction, he says. "If we are going to develop this technology in the same way that we used to, we are going to devastate the earth."

"Digital" technologies do not just exist digitally. The "cloud" does not in fact take the ethereal form its name invokes. To train and serve up AI models requires tangible, physical data centers. And to train and run the kinds of generative AI models that OpenAI pioneered requires more and larger data centers than ever before.

Before AI, data centers were already growing and sprawling. They were once small and distributed enough to be tucked away in urban environments, a few shelves of computers hidden in a back-office closet or a few dozen racks in a repurposed building. In the aughts, tech giants began trending in a different direction, consolidating all of their computing infrastructure into massive warehouses of servers in rural communities. The data center world became divided: There were the hyperscalers and there was everyone else. The four largest hyperscalers—Google, Microsoft, Amazon, Meta—now spend more money building data centers each year than almost all the others, relatively unknown developers like Equinix and Digital Realty, combined.

It's difficult to imagine what a hyperscale data center looks like if you've never seen one. Mél Hogan, an associate professor at Queen's University in Canada who studies AI, infrastructure, and the environment, used to use football fields to describe them when she began to write about them roughly a decade ago. "Now football fields don't even come close to the imaginaries of the required size," she says. Hyperscalers call their data centers "campuses"—large tracts of land that rival the largest Ivy League universities, with several massive buildings densely packed with racks on racks of computers. Those computers emanate an unseemly amount of heat, like a laboring laptop a million times over. To keep them from overheating, the buildings also have massive cooling systems—large fans, air conditioners, or systems that evaporate water to cool down the servers. The equipment all together creates a cacophony of humming, whirring, and crackling that can—especially in underdeveloped communities—be heard for miles, twenty-four hours a day, creating a relentless and body-warping source of noise pollution.

Now developers use a new word to distinguish the scale of what's coming in the post-ChatGPT AI era: *megacampus*. The word refers not just to the land area but to the sheer amount of energy that will be required to run them. A rack of GPUs consumes three times more power than a rack of other computer chips. And it's not just the training of the generative AI models that is costly, it is also serving them: According to the International Energy Agency, each ChatGPT query is estimated to need on average about ten times more electricity than a typical search on Google. Until recently, the largest data centers were designed to be around 150-megawatt facilities, meaning they could consume as much energy annually as close to 122,000 American households. Developers and utility companies are now preparing for AI megacampuses that could soon require 1,000 to 2,000 megawatts of power. A single one could use as much energy per year as around one and a half to three and a half San Franciscos.

Few places on the planet exist that can produce and deliver that much energy to any single location. Developers are working with utility companies around the world to build more power plants and expand the roster of options. After the last decade of flatlined energy demand in the US, a Goldman Sachs analysis described the sudden new wave of data centers as driving "the kind of electricity growth that hasn't been seen in a generation." Utility companies are now delaying the retirement of gas and coal plants and the transition to renewable energy; Microsoft restarted Three Mile Island, a nuclear plant near Middletown, Pennsylvania, that had a partial nuclear meltdown in the late 1970s, the worst commercial nuclear accident in US history. By 2030, at the current pace of growth, data centers are projected to use 8 percent of the country's power, compared with 3 percent in 2022; AI computing globally could use more energy than all of India, the world's third-largest electricity consumer.

This scale—the mega-hyperscale—has created startling environmental consequences. And yet, in the very same moment, corporate obfuscation of that impact has reached new heights. Since Emma Strubell's paper and Gebru's citation in "Stochastic Parrots," tech giants have hidden away even more of their models' technical details, making it exceedingly hard to estimate and track their carbon footprints. At the same time, those companies have amped up their public and policymaker influence campaigns with powerful counternarratives: Data centers will grow so efficient, their impact will stop being a problem; generative AI will unlock new climate innovation; AGI will solve climate change once and for all.

While the last claim is impossible to prove, the first two are highly misleading, says Sasha Luccioni, a research scientist and climate lead at open-source AI firm Hugging Face. The second is especially pernicious: There are indeed many AI technologies, as cataloged by the initiative turned nonprofit Climate Change AI, that can accelerate sustainability, but rarely are they ever *generative* AI technologies. "What you need for climate are supervised learning models or anomaly detection models or even statistical time series models," says Luccioni, who is also a founding member of the initiative. All of these models are previous generations of AI technologies—primarily machine learning tools—that are small and energy efficient, and in some cases could even run on a powerful laptop. "Generative AI has a very disproportionate energy and carbon footprint with very little in terms of positive stuff for the environment," she adds.

Luccioni says her past collaborators within closed-off companies no longer receive approval from their employers to cowrite papers with her about AI's environmental impact. Instead, she has worked with external collaborators and academics like Strubell, whose research she was first inspired by. In one paper, together with Hugging Face machine learning and society lead Yacine Jernite, the two measured the carbon footprint of running open-source generative AI models as a proxy to what closed companies are building. They found that producing one thousand pieces of text from generative models used as much energy on average as what it would take to charge the standard smartphone nearly four times. Generating one thousand images used on average as much energy as 242 full smartphone charges; in other words, every AI-generated image could consume enough energy to charge a smartphone by roughly 25 percent. Luccioni's and Strubell's papers are among the few that still provide quantifiable measures of the carbon behind generative AI models. It's a constant uphill battle. At one point, Luccioni says she reached out to over five hundred authors of the most recent machine learning papers to request basic information about their model training. "I barely got any answers," she says. "People were just not even responding or saying that this is confidential information."

Even as hyperscalers have spoken loudly in public about the sustainability of their computing infrastructure, executives at Microsoft have admitted internally that the intermittent availability of renewable energy just doesn't cut it when data centers need to operate 24/7. Nonstop operation is considered so crucial that Google, Amazon, and most recently Microsoft now build their campuses in threes to have a backup for the backup in case any facility goes down. During Hurricane Irma in Florida and Hurricane Harvey in Texas, even as millions of people lost power, some hospitals evacuated patients, and hundreds of thousands of homes and businesses faced damage and destruction, the data centers in those areas continued to hum along—so well that the displaced families of one facility's employees moved into it for the duration of the natural disaster.

The land and energy required to support these megacampuses are but two inputs in the global supply chain of data center expansion. So, too, is the extraordinary volume of minerals including copper and lithium needed to build the hardware—computers, cables, power lines, batteries, backup generators—and the extraordinary volume of potable—yes, potable—water often needed to cool the servers. (The water must be clean enough to avoid clogging pipes and bacterial growth; potable water meets that standard.) According to an estimate from researchers at the University of California, Riverside, surging AI demand could consume 1.1 trillion to 1.7 trillion gallons of fresh water globally a year by 2027, or half the water annually consumed in the UK. Those effects will not be felt evenly. Another study found that in the US, one-fifth of data centers were already drawing that water before the generative AI boom from moderately or highly stressed watersheds due to drought or other factors. And in Global South countries like Chile, it's often the most vulnerable communities who have borne the brunt of these accelerating economies of extraction.

As more and more communities have watched data centers affect their lives, a growing number have pushed back vehemently against their unfettered development. In response, data center developers have grown more sophisticated with tactics to maintain business as usual: They've entered communities in secrecy under shell companies; they've donated to community programs to dampen resistance; they've made promises to cities about the sustainability of their facilities before walking them back one by one after projects have broken ground and are more difficult to reverse. In one case in Virginia, a group of residents protesting against several massive data centers was shocked to uncover an email from a lawyer to a developer suggesting to place them under surveillance. "We need a mole or 2 in this group," the lawyer wrote.

From the moment they committed to the idea of scaling, OpenAI sought to secure an unprecedented amount of computing infrastructure. "In AI, whoever has the biggest computer gets the most benefits," Brockman told me in 2019. So with Microsoft, Altman developed a plan for how the tech giant would meet OpenAI's exponentially growing hunger for compute. The two companies would work together to design and deliver dozens of supercomputers for research—data centers that would need to be equipped with Nvidia chips for training various AI models in the course of OpenAI's explorations. Crucially, Microsoft would also build a series of ever-larger, ever-more-powerful supercomputers for training each subsequent generation of OpenAI's models.

Altman began referring to the series of supercomputers as "phases," at one point showing employees a slide to illustrate the size of each phase, with Phase 5, the last one planned, breathtakingly larger than all the others. The supercomputer that OpenAI had trained GPT-3 on was Phase 1.

Equipped with ten thousand V100s, the facility had been built in West Des Moines, Iowa, which Microsoft first entered in 2012. It had made nice with city officials through a "staggering" sum of tax payments, according to the then mayor, helping the city make major improvements to its public infrastructure. Over more than a decade, the company also invested some $2.5 million in local community programs, most of them nonprofits. Microsoft code-named the Phase 1 supercomputer Odyssey; OpenAI called it Owl, after a convention it had started early on of naming each of the Research division's compute clusters alphabetically after animals. When it ran out of all twenty-six letters, it would switch to naming them after periodic elements, ordered by atomic number.

Phase 2 was also in Iowa and used to train GPT-4. It had started with the eighteen thousand A100s referenced in the 2021 research road map and ultimately grew to around twenty-five thousand by the end of the model's training process. To Microsoft, Phase 2 was Telemachus, named after the son of Odysseus in Greek mythology. To OpenAI, it was Raven.

Phase 3 shifted to Arizona. With its cheap land, good tax breaks, and close proximity to California, Arizona had fast become a preferred hub for data center development among all of the cloud providers. After carefully cultivating favorable relationships with the governments of two underdeveloped cities right outside Phoenix, Microsoft had bought three tracts of land in 2018 and 2019 and similarly donated to various community organizations to quiet any objections from residents. The tracts had a combined land area of nearly 600 acres, or more than 450 football fields. (Microsoft would expand that area with another 283 acres in 2024.) Each tract would host a new data center campus that would serve Azure customers alongside OpenAI in the cloud region "West US 3."

Phase 3—code-named Inglewood at Microsoft and Whale at OpenAI—would cost several billion to build and house hundreds of thousands of Nvidia H100s, the generation after A100s, to train what OpenAI at the time believed it would likely brand as GPT-4.5 and GPT-5. For Phase 4, planned in Wisconsin, Altman expected costs to hit $10 billion using Nvidia's latest B100s, yet another generation after H100s. That was a staggering amount, considering the most expensive hyperscale data centers then hovered around $1 to $2 billion. He didn't plan to stop there, casually floating the idea of the $100 billion supercomputer for Phase 5. After the blowout success of ChatGPT, Altman tempered his expectations. With so many chips locked up in serving ChatGPT to customers, Microsoft was struggling to acquire chips fast enough to keep up with finishing Phase 3's development. Whale split into three separate clusters—Beluga, Narwhal, and Orca—with plans to complete the build-out sometime in 2024.

No one within Microsoft or OpenAI even knew whether Phase 5 was technically possible. In Microsoft and OpenAI's design plans, *The Information* later reported, the $100 billion facility could need as much as 5,000 megawatts, nearly matching the average power demands of all of New York City. While the plan didn't seem the most financially sound as a business investment, money wasn't the main bottleneck. It was energy. "We're running out of land and power," an OpenAI employee says. Within both companies, it was understood that Phase 5 would only become possible with some amount of innovation. Either Microsoft and OpenAI would need to split the supercomputer into multiple campuses to distribute the energy demands and figure out how to train an AI model across distant locations, or, as Altman sometimes liked to say, the problem would solve itself with a future breakthrough in nuclear fusion.

At times he would give OpenAI employees optimistic updates about Helion Energy, the nuclear fusion startup that represented his largest personal investment and for which Microsoft had already committed to buying power from once a plant, with a target generation of 50 megawatts, was up and running. *The Wall Street Journal* would later report that OpenAI and Helion were also in talks to strike a deal, from which Altman had recused himself.

Altman and other executives never brought up the data centers' environmental toll in company-wide meetings. As OpenAI trained GPT-4 in Iowa, the state was two years into a drought. The Associated Press later reported that during a single month of the model's training, Microsoft's data centers had consumed around 11.5 million gallons, or 6 percent, of the district's water. GPT-4 had trained there for three months.

(A Microsoft spokesperson said the company is working to increase its water efficiency by 40 percent above its 2022 baseline and to replenish more water than it consumes across its global operations by 2030, with a focus on the water-stressed regions where it works.)

Arizona, too, faces a severe water crisis. In 2022, as Microsoft laid the groundwork for Phase 3, a study in *Nature Climate Change* found that the Southwestern US had been facing the worst drought it had seen in over a thousand years. That drought, combined with severe mismanagement, has drained the Colorado River, which Arizona and six other states rely on for fresh water, to dangerously low levels. Without drastic action, the river could cease to flow. The shortage compounds a power crisis, as climate change has slammed the region with relentless record-breaking temperatures and families have cranked up their air-conditioning. The region relies in part on hydropower from the Hoover Dam and water-cooled nuclear power plants. In other words, it needs water to produce more energy. In 2023, the Phoenix metro area hit multiple new heat records as well as the worst year for heat-related fatalities, which surged at least 30 percent from 2022 to over six hundred dead. "All things," says Tom Buschatzke, the director of the Arizona Department of Water Resources, "are converging in a challenging direction."

What Altman did bring up was his impatience. In March 2024, after sleeping through the early years of the generative AI race, Meta would come out swinging with aggressive new plans to have 350,000 H100s up and running as part of an even larger infrastructure build-out to support its sudden burst of generative AI investments. This was more GPUs than OpenAI had at its disposal. Altman wasn't happy. Microsoft was being too slow, he felt, and it was costing OpenAI its competitive advantage.

When Sonia Ramos was a child, she witnessed an accident that would shape the rest of her life. She was born into a mining family in Chile. Her father worked for an American copper company; she grew up among the children of the other workers. In 1957, a part of the Chuquicamata mine collapsed, killing several people and injuring dozens more. Though her father was spared, she remembers watching the wretchedness of the aftermath unfold around her: affected families spiraling into abject poverty, children wasting away from hunger. Four decades later, as Ramos began to protest mining, becoming one of the most active and outspoken Indigenous voices in Chile shining a light on its social, cultural, and environmental destruction, she would remember the lessons she learned in the tragedy: The mining industry is a system, and that system, left to its own devices, will seek profit at any cost. "The worker doesn't exist," she says. None of the victims received any ceremony or commemoration; none of their families received compensation. "In that place, there is no humanity."

Chile is the world's largest producer of copper, accounting for a quarter of the global supply. Since the beginning, copper mining has reshaped not just the land but the societies that rely on it. Sometimes the effects are visible: Chuquicamata today is the largest open-pit copper mine in the world, a gaping wound in the earth that explosives regularly deepen. That displaced rock, piled up in towering mountains that monumentalize the cavities they came from, is slowly burying the remains of a town that was abandoned after the copper mining began to swallow it. The mining has also drained the region of water to process the copper. At one point a foreign multinational corporation consumed so much water it depleted an entire basin in a nearby salt flat, or *salar*, destroying its rich ecosystem.

Less visible are the trails of arsenic that the industry leaves in the air and water, which has increased rates of cancer throughout the north of the country, and the ways mining has restructured Indigenous life and sowed divisions among different communities. With their lands depleted of water and minerals, the Atacameños, the name that ties together all of the distinct Indigenous groups who share this region, can no longer sustain themselves by growing their own crops or raising their own livestock. The shift has plunged their towns into deep poverty. Crime has risen along with depression, alcoholism, and delinquency. They don't have enough food, running water, proper health care, or educational resources, having seen little benefit from the billions in profits that their land has generated for someone else. Instead, many are forced to work for the very industry that seized their territories and receive health care from the small clinics it sponsors. Where there was once greater unity among them, the Indigenous groups now squabble over diminishing resources.

Lithium is a more recent discovery there, stumbled upon by an American company in the 1960s as it searched for the water it needed for copper mining. When it drilled into the *salares*, it found high concentrations of lithium floating in an oily brine beneath the surface, opening up a new front of extraction and accelerating the depletion of more ecosystems. Today Chile produces roughly a third of the world's lithium, second only to Australia. The material is primarily extracted out of the Salar de Atacama, the largest salt flat in the country, by pumping its brine out into shimmering pools of turquoise and waiting for the sun to evaporate and crystallize the solution into lithium and other by-products. The *salares* were once home to flocks of pink flamingos, which the Atacameños consider their spiritual siblings. Now the flamingos are gone; the young daughter of one Indigenous leader in the Peine community has only her ancestors' stories and a flamingo plushie by which to remember them.

Over the years, the Atacameños have heard many narratives used to justify all of this extraction. In 2022, as the European Union set new policies around the energy transition and the demand for lithium skyrocketed, both companies and politicians in Chile and the rest of the world lauded the importance of the country's mining industry in propelling forward a better future. Indigenous communities watching their land and their communities get ripped apart asked: A better future for whom? "Local people never have the ability to think about their own destiny outside the forces of economics and international politics," says Cristina Dorador, a microbiologist who lives in the north and studies its rich biodiversity.

Now the same narratives are being recycled with generative AI. The accelerated copper and lithium extraction to build megacampuses—and to build the power plants and thousands more miles of power lines to support them—is, in Silicon Valley's account, also ushering in a better and brighter future. To block that extraction is thus to block fundamental progress for humanity. But it is not the mining that Indigenous communities resist. "Our ancestors were miners," says Ramos. They were the ones who discovered the copper in the first place. The problem, she says, is the scale.

That scale has consumed everything. It has made the north and the rest of Chile completely dependent on the industry and not allowed for the emergence of other economies. It has choked off the country's—and the rest of the world's—ability to imagine different paths where development could exist without plundering natural resources, Ramos says. By enabling the production of massive generative AI models, that scale has also led to the perpetuation of racist stereotypes about the Indigenous peoples already suffering from how the technology was physically built. In Brazil, a 2023 art exhibition coproduced by a Chilean university showed the vast chasm between the reality of Latin America's rich Indigenous cultures and the woefully bereft depictions of them spit out by Midjourney and Stable Diffusion as primitive, technologically inept peoples.

In recent years, the Atacameños have mounted more and more resistance. They fly black flags on their houses to denounce the exploitation of their lands and their community. They've organized protests to physically block the roads that company buses and trucks must take to get to the mines. They've contracted lawyers to assert their legal rights as Indigenous peoples under international law, which protects their cultural and territorial sovereignty. As companies and the Chilean government have been forced to invite them to negotiations, central to Indigenous demands are the need for the government to conduct research into the health of the Atacama Desert's ecosystems and to quantify the water loss and any irreparable damage.

Ramos, too, has her own foundation, bringing together "the ancestral and non-ancestral," she says, to promote and conduct scientific research into the natural wealth that the Atacama Desert has to offer. Due to its uniquely extreme conditions, it is home to many microbial communities—potentially useful for medicines or new sources of energy—that don't exist anywhere else. For the same reasons, the desert has also been studied for decades as an analogue to Mars's climate. Ramos hopes that any discoveries will help prove the value of preserving her beautiful homeland. Against the narratives of high-speed progress used to fuel extraction, she searches for new conceptions of progress that promote healing, sustainability, and regeneration.

As Ramos's fight continues in the north, a different battle is waging in the heart of Chile, over the government's embrace of the tech industry's data centers themselves. The faster the hyperscalers have expanded, outpacing the supply of land and power in their typical regions of operation, the more aggressively they have pushed to lay claim to those resources in new territories globally.

Microsoft alone spent more than $55 billion in fiscal year 2024, nearly a quarter of its reported revenue, to build what *SemiAnalysis* described as "the largest infrastructure buildout that humanity has ever seen." Google, meanwhile, said in its third 2024 quarterly earnings call that it planned to crank up its data center expenditures to reach around $50 billion for the fiscal year. Meta said it would likely round out the fiscal year with up to $40 billion in data center and infrastructure expansions, which it estimated would rise the following year.

On a rare misty afternoon in June 2024, Alexandra Arancibia directs our car in Quilicura, a municipality on the outskirts of Santiago where she lives and serves as a council member, to what she sees as the defining symbol of the yawning power differential between American tech giants and her community. Less than a thirty-minute drive away from Santiago's most picturesque neighborhoods, packed with European-style cafés and vegan restaurants, the roads in front of us are crumbling from poor maintenance, mountains of trash strewn alongside in illegal dumps controlled by a local mafia.

Past a graveyard dedicated to deceased pets, we pull up to what looks like an abandoned plot of grassland with tufts of shrubs and a scattered handful of nutrient-starved trees jutting out of the soil. Most days the land is so parched it looks like parts of the Atacama Desert; today the rain is turning everything into mud. In the middle of the plot, a purple sign announces in Spanish, "Welcome to the Quilicura Urban Forest," a project, it explains, that Google began in 2019 to give back to the community for hosting its data center. The sign includes a diagram to elaborate on the "forest's" benefits: on the left side is an illustration of Quilicura as an industrial zone, packed with factories producing greenhouse gases and air pollution; on the right is an illustration of the forest flourishing under the generous rain pouring out from a big cloud labeled "SMOG."

Google boasts about this forest on its website and in its PR releases. When I ask the company's country spokesperson for an interview about Google's development in Chile, she sends me instead some polished briefing materials, later adding that the company creates a community impact program for each new data center to support local projects such as in education, sustainability, internet access, and health; for Quilicura, Google has invested over $1.2 million. In her materials, the part about the forest talks about residents using the green space. There are no residents. The place is too far from any bus line, and there are no homes in the surrounding area to speak of. Outside the modest plot, too small to fit Google's data center itself, a dozen stray dogs meander around, barking and rummaging through the trash. The spokesperson said the forest is being updated to "evolve the experience for the community."

Arancibia smiles wryly as we take the scene in. "Do you feel like you're in Silicon Valley?"

Arancibia had just started college when she realized that Quilicura was a place where things were discarded. She was commuting each day to her university through parts of the municipality—piled high with refuse—that she hadn't known existed. She had never thought of Quilicura as "home"—it was simply the place she lived, an underdeveloped and unremarkable working-class town that her parents moved to when she was little. But something about seeing it treated as a literal dump stirred within her a deep desire to revitalize the land to its former beauty.

Only two decades ago, when Arancibia was a child, Quilicura was mostly country: rolling pastures and glistening wetlands, home to a different yet just as rich biodiversity as the Atacama Desert—birds, beasts, and varieties of flora. Then the trash mafia arrived, allowing anyone from the rest of Santiago to dump their waste in Quilicura for a price. Some dumps have operated for so long that grass and weeds have grown over them, making them look like eerie deformed hillsides closing in on the landscape. Next came different industries, including beer companies and real estate developers, who siphoned off more land and extracted water from the wetlands for their purposes. Today only tiny, interspersed pockets of green in this twenty-two-square-mile municipality offer a window into what Quilicura once was.

Against this backdrop Google came to Quilicura in 2012 to build its first data center in Latin America. On Google's web page, the company proudly presents the data center, which became operational in January 2015, as one of the most efficient and environmentally friendly on the continent. At the time, no one in Quilicura paid attention to the project; certainly no one in the better-off epicenter of Santiago was paying attention to Quilicura. Neighborhood residents who passed by the data center every day on their bus route to work assumed that it was a factory producing beer or food and providing jobs to the local community.

Google's data center—like most data centers—did not provide many jobs beyond its initial construction. A job posting from 2024 for a mechanical technician, one of the few long-term positions available, was advertised on Google's job board only in English; the posting stated that Google wouldn't consider résumés submitted in any other language. The data center—as activists point out—did not provide much other benefit to the local community either. Nearby, public schools still lack good internet or devices for students to access it.

The data center's arrival marked Quilicura and the rest of Santiago as a desirable destination for Silicon Valley's physical expansion. In 2019, Google announced that it would build its second Latin American data center in the Santiago metropolitan area. Soon enough, Microsoft and Amazon announced that they were coming too. The Chilean government was quick to welcome them, positioning the country as a safe and stable haven for foreign direct investment in Latin America, which otherwise suffers a reputation for unreliable governments and social and economic instability. In 2020, the government went a step further. It announced a project to build a new underwater cable, akin to a data highway, for connecting the Asia Pacific straight to the Americas through Chile's central coast, not far from Santiago. Chile would become a global hub for digital infrastructure. Google backed the partnership.

But in July 2019, as Google began the paperwork for its second data center in Chile, a group of residents was watching. The company had chosen Cerrillos for its new location, another working-class municipality bordering Santiago. Like Quilicura, Cerrillos has a long history of being overlooked and abandoned. From the 1930s to the 1990s, a cement factory that belonged to a Belgian company contaminated the community with lethal levels of asbestos, leading to what one Chilean historian called "the largest industrial genocide" in the country. To this day, residents still die from higher than average rates of cancer. But Cerrillos is also special—in a country where water is privatized, the municipality is home to the nation's only public water service, which serves up the local groundwater to neighboring communities and, in emergency situations, to other parts of Chile.

This unique combination—a history of neglect and a precious water source—created fertile ground for the blossoming of several environmental activist groups who were used to being watchdogs and were fiercely protective against the extraction of their resources. That summer, as Google filed a report with Chile's environmental agency for approval of its data center—a largely rubber stamp process—MOSACAT, a water activist group, began combing through all 347 pages of the filing. Buried in its depths, Google said that its data center planned to use an estimated 169 liters of fresh drinking water per second to cool its servers. In other words, the data center could use more than *one thousand times* the amount of water consumed by the entire population of Cerrillos, roughly eighty-eight thousand residents, over the course of a year. MOSACAT found this unacceptable. Not only would the facility be taking that water directly from Cerrillos's public water source, it would do so at a time when the nation's entire drinking water supply was under threat. In 2019, as with

Iowa and Arizona, Chile was already nine years and counting into a devastating and historically unprecedented megadrought.

Tania Rodríguez, a member of MOSACAT, hands me all 347 pages of Google's environmental filing, printed out and spiral-bound between two blue plastic protectors. The tome drops into my lap with a thud, a physical manifestation of the way Silicon Valley wields technical knowledge to justify its centralized decision-making. Jutting out from the bottom are carefully labeled Post-it notes. *"Agua potable"* (potable water) reads one in Spanish, bookmarking the pages that discuss the data center's need to consume fresh water for cooling.

MOSACAT was founded in 2019 after activists from several different movements fighting for women's, housing, workers', and environmental rights joined in solidarity to form a unified collective. Many had met while protesting an illegal mining project. MOSACAT's activism successfully chased out the miners, shut down the project, and designated the land a protected nature reserve, the group says. It was shortly thereafter that a friend of the group, who now serves as a member of Chile's national congress, tipped them off about Google's data center project and urged them to look at its projected water consumption.

MOSACAT's members are not technologists. But they read through every page of dense diagrams and arcane terminology, took copious notes, and memorized the ins and outs of data centers and their cooling systems to prepare themselves to go up against Google. Rodríguez lets out a spirited laugh when I ask her how they were able to digest all of the information. "It took all of us," she says—referring to more than a dozen volunteers who make up MOSACAT's membership and do the work in stolen hours between jobs and family obligations.

In most cases, projects in Chile that require water take a long time to receive approval. In Google's case, the approval came quickly, even in the midst of a series of drought-related water emergencies. At first, MOSACAT sought to contest the project through Google's local partner, a Chilean investment and services firm named Dataluna. The initial meeting went badly, MOSACAT says: The Dataluna representatives seemed to have little understanding of the project and denied that it would use fresh water.

From there, MOSACAT went to the local government. The mayor and city council had themselves been meeting with Dataluna, the group remembers, and held the false impression that the data center needed only wastewater for its cooling. After MOSACAT briefed them, the government, alarmed, demanded an explanation from Dataluna. The matter escalated from Dataluna to Google's Chile division all the way to Google's headquarters in Mountain View, California.

In October 2019, Google sent two engineers and a lawyer to Cerrillos to present to the community. The day they arrived, MOSACAT plastered the streets with protest signs along the route the Googlers would drive to get to the meeting location. At the meeting itself, MOSACAT didn't come alone. Among the roughly two dozen residents who attended, six other activist and community groups were represented. The Google engineers were gringos, MOSACAT remembers, tall and able to speak only English. They gave a highly technical presentation, and Google's lawyer doubled as a translator. During the discussions, MOSACAT members who spoke English say they could hear the lawyer mistranslating their words. At another point, the Google representatives sought to assuage the community by offering to plant an urban forest just like the one the tech giant had given to Quilicura. The show left MOSACAT and the other groups unimpressed: Google wasn't here to truly engage with and hear what the community wanted. "They came to intimidate us," says a MOSACAT member Alejandra Salinas, who also serves as a council member for Cerrillos's neighboring municipality, Maipú. "Think about it. They come offering us trees while drying out our earth."

Residents of Cerrillos didn't need trees. They didn't need Google to build them a park—as if the Santiago metropolitan area were such a backward place that it didn't already have parks. They needed Google to stop treating their land as a place to plunder precious water and other resources; they needed the company to stop dismissing the community as bystanders instead of participants in the development of its local projects. "We know that we feed the world, that we provide raw materials like copper and lithium," Salinas says. "Nobody is saying our treasure is ours alone and we won't share it. Yes, we can help each other. But they are not going to come and use the water, which is vital for life, and leave us with nothing."

At the time, Chile was in the midst of a massive, monthslong political upheaval, known as the *Estallido Social*. Explosive and at times violent protests were erupting every week, beginning that same month in October 2019—a collective outcry over unemployment, privatization, and deep inequality that left thousands injured and dozens dead. In late 2019, as communities across the country began to hold referendums in response to the movement to reform local and national politics, MOSACAT piggybacked on the referendum in Cerrillos to add a question about whether residents agreed for Google to build a facility that would consume so much of the community's water. They mobilized to broadcast the true nature of the project, handing out flyers on street corners, knocking on people's doors, and posting signs all over the municipality.

In December 2019, MOSACAT won; the referendum to build the data center was rejected with a slim majority of the vote. The following year, the government joined MOSACAT in filing a lawsuit against the project with an environmental court.

In the meeting with Google, its representatives ultimately put on a friendly face, MOSACAT says, presenting a greater willingness to negotiate once it became clear that some residents could understand them. But as both Julian Posada, the assistant professor at Yale, and Mercy Mutemi, the lawyer for Mophat Okinyi, told me, the risk of pushing back as a Global South country is always that a Silicon Valley company will pick up and take its money somewhere else. As the project continued to stall and Google's desire for more compute intensified, the company announced that it would shift its next planned data center in Latin America from Chile to another country.

In Uruguay, a small country of 3.4 million, the national telecommunications company Antel has three data centers that help provide internet and cell services to the entire country. Cumulatively, they take up only some five thousand square meters. One, less than one thousand square meters, sits nestled into a typical residential street in Montevideo, taller and wider than its neighboring buildings but integrated into its surroundings.

The data center runs on two power lines that are separate from the rest of the neighborhood's. Thirty percent of the space is filled with computers, 70 percent with administrative offices, electrical closets, and mechanical rooms. The computers get hot but not so hot that they can't still be cooled with air instead of water. They produce a low hum barely audible during the bustle of the day and just enough of a nuisance in the quiet of the night that two neighboring families have come and knocked on the data center's door to complain with some regularity. The manager of the data center, Javier Echeverria, looks sheepish as he admits this. He says he is now working with local researchers on solutions to reduce the noise and has already modified the cooling system to be less noisy. The responsiveness is a far cry from the rigmarole that MOSACAT had to go through to get an American company's attention.

A thirty-minute drive away, right outside the city limits, the government delineated a large expanse of land to develop a science park, Parque de las Ciencias, which operates as a *zona franca*, a free-trade zone. Some cheekily call it *zona America*, with a hint of bitterness, for housing mostly American companies that do not pay taxes to the government. The park even looks somewhat like America, its lush, manicured lawns, symmetric design, and majestic sundial-shaped fountain reminiscent of the stately aesthetic of the National Mall in Washington, DC. The home page of the park's website advertises a politically and economically stable country, plenty of land, and an abundant supply of power and water. So it came to pass that in 2021, as GPT-3 spurred new interest in dramatically scaling AI models, Google purchased twenty-nine hectares of land here, fifty-eight times the size of Antel's total data center footprint, to establish a different home for its second data center in Latin America.

But at the time, Uruguay did not in fact have an abundance of water. Like Chile, like Iowa, like Arizona, it was also experiencing a devastating drought. The water shortage was so severe that farmers were losing their entire harvests, costing the country over $1 billion in agricultural losses; by the summer of 2023, the Montevideo government would start mixing contaminated salt water into the city's drinking supply. Families opening their taps saw a putrid brownish fluid pouring out that smelled intensely of chemicals. Those who could afford it purchased bottled water for drinking and bathed with their windows open to avoid breathing in too many carcinogens. Those who couldn't drank the tap anyway, leading many to suffer stomach pains, skin rashes, an aggravation of existing health conditions, and the agony of a growing rate of miscarriages.

In the aftermath of a raging pandemic, there were many who couldn't afford bottled water. Where Silicon Valley had ascended, with Google's and Microsoft's market capitalizations both peaking above $2 trillion, in part from companies going remote and increasingly relying on cloud services, illegal housing settlements in Uruguay had grown by orders of magnitude. *Ollas*, the local equivalent of soup kitchens, were turning children away hungry. Those running the *ollas* were themselves in poverty and barely surviving. Fabiana, the boisterous head of an *olla* who lives in an illegal settlement and is affectionately called *Reina Madre* (Queen Mother), grows quiet as she remembers it. "To have to say, 'I don't even have a little plate for your child . . .'" She trails off. "It was horrible." Even after a lifetime of poverty that included sweeping the floors of a brothel at seven years old for survival, she finds the pandemic and drought years stand out in her mind as some of the worst in her life.

The water crisis emerged from the compounding effects of climate change and a failure of the state's allocation of freshwater resources: In Uruguay, more than 80 percent of the country's fresh water goes to industry instead of human consumption—most notably, cash crop agriculture. These include industrial farms for soybeans and rice, and for trees that feed into paper production. Most such farms are run not by local companies but by multinationals that export what they grow and show little accountability for Uruguay's natural environment. Their activities deplete the nutrients in the soil, making it more difficult to grow actual food, and pollute the country's water streams with a volume of fertilizers that makes Uruguay one of the world's largest per capita fertilizer consumers and causes unusually high rates of cancer.

Daniel Pena, a sociology researcher at the Universidad de la República in Montevideo who has for years studied the politics of this environmental extractivism, draws a direct connection to Uruguay's colonial history. He drives around the country in a beat-up pickup truck to interview farmers and residents of the poorest neighborhoods, to document up close how they're squeezed by industry. As with Chile, the foreign multinationals still exist above locals in the political pecking order. During the drought, industry continued to use water unabated, drawing what it needed directly from the main river, Río Santa Lucía, that also feeds Montevideo's public water system. Two decades ago, after significant environmental activism, Uruguay became the first country in the world to recognize water as a human right in its constitution. Now, in a bitter irony, it was the drinking water rather than water for industry that saw the most severe cutbacks during the shortage.

So when Google arrived, Pena was vigilant. During his regular scans of the Uruguayan environmental ministry's website, which lists major industrial projects, he came across the company's proposal for the data center. Pena had read about hyperscalers using potable water, even during major droughts, and the activism of communities like MOSACAT that had resisted the projects. But when he downloaded the details of the project, the water numbers were marked as confidential. After submitting a public information request, which he had successfully done around twenty times, the ministry continued to withhold the numbers, saying they were proprietary information. Pena wondered what they were hiding and worried about the precedent it would set for other cloud companies that would inevitably begin to eye Uruguay, following Google's lead, for their own expansion. So he evoked the water clause in the constitution. With the help of a lawyer friend who was willing to work pro bono, he sued the ministry.

In March 2023, four months later, Pena won the case in a surprising victory. The environmental ministry revealed that Google's data center planned to use two million gallons of water a day directly from the drinking water supply, equivalent to the daily water consumption of fifty-five thousand people. With much of Montevideo receiving salt water in their taps not long after, the revelations were explosive. Thousands of Uruguayans took to the streets to protest Google and all of the other industries that had led the government to squander the country's precious freshwater resources. The slogans of resistance are still scrawled across the city's walls and roadside barriers during my visit in June 2024. "This is not drought," reads one. "It's pillage."

Pena sees data centers in the same way that Ramos sees mining. It's not the infrastructure itself that's the problem, but the scale at which Silicon Valley is trying to build it. That scale is what drove companies like Google and Microsoft to expand in Chile and Uruguay even as the countries suffered from a severe lack of resources. That scale is what makes them require fifty-eight times more land than Antel and operate with much less accountability to the local population. "They are extractivist projects that come to the Global South to use cheap water, tax-free land, and very poorly paid jobs. And then they don't contribute to our country; they don't improve our internet access," he says.

Near the end of 2023, Google silently updated its proposed data center in Uruguay to use a waterless cooling system and said it would reduce the facility to a third of its size. Pena says the fight is still not over: The government is now withholding the projected energy consumption of the latest proposal as a commercial secret. Pena also sent a petition to the ministry, with over four hundred signatories, demanding a more extensive environmental and social impact study of the full supply chain of the data center: where the minerals for producing its hardware are being extracted, how the labor involved is being treated, how much carbon will be emitted, how the generated e-waste will be disposed of in a way that doesn't leach chemicals into someone's community. Most of these other impacts won't befall Uruguay, but Pena feels a solidarity with the other countries where they will. They are "generally all from the Global South," he says. "We all end up with the same consequences, but from different links in the global supply chain."

In 2024, Chile's environmental court ruled that Google cannot build a water-using data center in Santiago. The Google Chile spokesperson said the company remains committed to the country and Latin America, and plans to begin the permitting process for an air-cooled data center in Cerrillos when needed. But that hasn't slowed down other hyperscalers from entering Latin America. In 2022, Microsoft finalized the location for its data center in Chile, shortly after its second investment into OpenAI—right back in Arancibia's hometown, Quilicura.

As data center developers go, Microsoft was a late bloomer. Before its investments in OpenAI, Google was well ahead in the number of facilities it was constructing around the world. But with the sudden explosion of demand for more computing infrastructure to support its AI ambitions, Microsoft adopted Google's playbook and followed it into the same regions.

The Chile that Microsoft entered was different from the Chile that had greeted Google. By 2022, more than two years had passed since the *Estallido Social*, which had left an indelible mark on the country's politics. After months of protests, Chile had undergone a remarkable experiment to rewrite its constitution, with the broad participation of regular citizens, to replace the one that had carried over since Pinochet's brutal dictatorship. Ultimately the new drafts of the constitution didn't pass; two separate processes resulted in two partisan documents that failed to gain broad support. But it reinvigorated the youth and working-class families in particular with a new optimism for democracy. The upswell of leftist ideas and support led, in a dramatic turn of events, to the election of a millennial left-wing president, Gabriel Boric Font, only thirty-five years old. In his victory speech, Boric repeated a slogan of the protests that slammed the legacy of the Chicago Boys during Pinochet's rule: "If Chile was neoliberalism's cradle, it will also be its grave."

Boric, who began his term in March 2022, had himself been a student protester. It emboldened youth across the country to take what they learned from the social upheaval about organizing and protest to establish a new generation of progressive activist organizations. They became part of the rhythms of each community, meeting regularly to dream up big-picture visions about what they wanted for the future of Chile.

It was during this period that Arancibia cofounded her own activism group with another young Quilicura resident, Rodrigo Vallejos. The two had met during the organizing of the social upheaval and quickly bonded over their deep passion for the environment. They called their group *Resistencia Socioambiental Quilicura*—the Socio-environmental Resistance of Quilicura—drawing upon a well-established concept in Latin America that the social and the environmental are inextricably linked.

Upon Microsoft's entrance into Quilicura, Vallejos and Arancibia did what MOSACAT and Daniel Pena had before them: They began to pore over whatever materials they could find that Microsoft had made available and to extensively research the project. Vallejos, a law student at Universidad Diego Portales, worked late into the nights in between his schoolwork to read technical documentation and teach himself about how data centers work.

Microsoft projected that it would need a significantly lower amount of water than Google had in Cerrillos. Even still, Vallejos worried. The drought in Chile had only gotten worse and was expected to last until 2040. Quilicura's wetlands were suffering acutely, on top of industrial encroachment, from accelerating desertification. On its website, Microsoft boasted about new cutting-edge innovations in data center cooling systems that would mean its facilities didn't need to use water. If Microsoft had the capability to build waterless data centers, why wasn't it doing so in Quilicura?

"It is deeply striking that a company with as much reputation as Microsoft publicly presents an environmentally friendly discourse, but in reality does not comply with global innovation standards in a third world country like Chile," Vallejos later wrote in an article.

Microsoft would subsequently explain that the innovations it had advertised were still under development and only being piloted in a place in the US. "Then why do they promise these things" on their website? Vallejos asks.

Vallejos caught the attention of local and international researchers, including Marina Otero Verzier, a director of research at Nieuwe Instituut, the Dutch institute for Architecture, Design and Digital Culture, and Serena Dambrosio and Nicolás Díaz Bejarano, researchers at FAIR, the think tank co-led by Martín Tironi Rodó. Otero was moved by the passion of Vallejos and Arancibia, and their exhaustion. They had worn themselves thin reading Microsoft's long technical documents, writing critical articles, and protesting continuously, but had struggled to get an audience from either the company or the government. Otero pondered ways to help them. How could she get them in a room to negotiate with the right people?

Otero knew she had one thing Vallejos did not: affiliations with prestigious universities like Harvard and Columbia that would command Microsoft's and the government's attention. She began to mount a multi-pronged campaign, growing so deeply involved that she quit her job: She spoke at high-profile conferences about the environmental impacts of data centers and *Resistencia*'s fight against them; she forged connections with the Chilean Ministry of Science, Technology, Knowledge and Innovation and with representatives at Microsoft and Google; she connected Vallejos and Arancibia to other international researchers to elevate their profile.

With Dambrosio and Díaz, Otero also developed a more speculative project. All three had architectural backgrounds and had been studying the infrastructure of modern digital technologies through the lens of the built environment. They began to wonder: What if they treated data centers as architecture structures and fundamentally reimagined their aesthetic, their role in local communities, and their relationship with the surrounding environment?

Díaz liked to visit national libraries during his travels—beautiful venues that seek to capture the grandeur of a country's memories and knowledge. It struck Díaz that data centers, too, could be thought of as libraries—with their own stores of memories and knowledge. And they, too, could be designed to be welcoming and beautiful instead of ugly and extractive.

This represented a sharp departure from Microsoft's and Google's definitions of what it means to give back, such as through the latter's community impact programs, with what Díaz calls their "schizophrenic" initiatives, which tend to be divorced from how communities are actually affected by the companies' facilities. Together with Vallejos and Arancibia, the three researchers applied for funding and put together a fourteen-day workshop, inviting architecture students from all around Santiago to reimagine what a data center could look like for Quilicura.

The students designed stunning mock-ups. One group imagined making the data center's water use more visible by storing it in large pools that residents could also enjoy as a public space. Another group proposed tossing out the brutalist designs of the typical data center in favor of a "fluid data territory" where data infrastructure coexists with wetland, mitigating its damaging impacts. The structures of the data center would double as suspended walkways, inviting Quilicura residents to walk through the wetland and admire the ecosystem. Plant nurseries and animal nesting stations would be interspersed among more traditional server rooms to rehabilitate the wetland's biodiversity. The data center would draw polluted water from the wetland and purify it for use before returning it. The computers themselves would collect and process data about the health of the wetlands to accelerate the local environment's restoration. "We're not fixing the problem, but we're imagining other types of relationships between data and water," Díaz says.

At the end of the workshop, the students presented their ideas to residents and other community members. "It was an incredible conversation," Otero says. "You can see how much knowledge the community has. They had so much to offer."

Three years into his four-year term, Boric is under pressure to get "quick wins" for the economy—even more so as a young, left-wing president. That means pressure to expand the mining industry, pressure to see through the arrival of twenty-eight new data centers. On the day Boric announced a plan to develop a national data center strategy, Vallejos texted me a video of the press conference and an emoji: the disoriented face with spiral eyes.

But in fairness, the coalition of activists in northern Chile and Santiago with researchers domestic and abroad has clearly made a mark. As part of the data center plan, the Ministry of Science, which oversees its creation, has for the first time formed a committee of activists to consult with regularly as part of the drafting process, and invited Vallejos, Arancibia, and members of MOSACAT to join them. The Ministry of Science is also overseeing an AI bill that articulates how Chile wants to approach AI development, application, and regulation. Whereas before the ministry's discussions cast AI as a universal positive, the tone has since shifted to acknowledge the social and environmental costs of the technology.

Chile, like many Global South countries, has learned from hard experience that it should not wait around for the Global North to decide how it will build digital technologies. "The way we build technology responds to a certain cultural framework and historical framework," says Aisén Etcheverry, the head of the ministry. Where the global internet was shaped without Chile, the country now has an opportunity to shape AI on its own terms.

Tironi pushes this one step further. It's very clear that the AI industry today is rooted in a colonial ideology, he says: It imposes its worldview and its technology—what is AI, what is good AI, what it means to create an industry of AI—on the rest of the world. Chile could be a leader in resisting that imposition. After centuries of extractivism, the country intimately understands what it means for its land to be hollowed out, dispossessed, and destroyed under a banner of progress. It could use those experiences as a wellspring from which to generate fundamentally new conceptions—*decolonial* conceptions—of AI.

"In the planetary market of AI, we as a country are playing a specific role: giving materials to develop this technology," he says. "Many companies are trying to extract a lot of material from us to create AI.

"So we need to think from this position, this geopolitical position, this terrestrial position. We can think of another way to relate technological innovation with the earth."

It is a noble ambition, and the forces arrayed against it are mighty.

Chapter 13

The Two Prophets

In May 2023, Altman arrived in Washington, DC, to testify before Congress. It was a remarkable performance. He reiterated the promise of AGI solving climate change and curing cancer, gave a compelling argument for why OpenAI's technologies would improve and create "fantastic" new jobs, dodged questions about copyright issues and the lack of transparency and privacy guarantees around its training data, and delivered a sincere call for regulation—that is, regulation with OpenAI's blessing, evoking the specter of China to urge lawmakers not to slow down its innovation.

Senators loved him. In a telling exchange that captured Altman's nimble rhetoric and the trust and enthusiasm he was garnering, he offered three policy recommendations that shifted the conversation away from existing issues like labor, environment, and intellectual property and toward regulating future AI systems and extreme risks: First, create an agency that would develop and administer a licensing regime for models above a certain threshold of capabilities; second, create a set of AI safety standards for measuring "dangerous" capabilities; third, require independent audits on those standards to check for compliance. He later elaborated that capability thresholds could be approximated with compute thresholds if necessary, and that "dangerous capabilities" could include a model's ability to manipulate and persuade, and to generate recipes for novel biological agents.

"Would you be qualified to, if we promulgated those rules, to administer those rules?" Louisiana senator John Kennedy asked.

"I love my current job," Altman said to laughter.

"Are there people out there that would be qualified?" Kennedy asked.

"We'd be happy to send you recommendations for people out there, yes."

"Okay. You make a lot of money, do you?"

"I make—no. I'm paid enough for health insurance. I have no equity in OpenAI."

Sitting next to Gary Marcus, Altman won over even one of his most vocal critics. "Let me just add for the record that I'm sitting next to Sam, closer than I've ever sat to him except once before in my life," Marcus said, "and his sincerity . . . is very apparent physically in a way that just doesn't communicate on a television screen." (Marcus would later backtrack his rare show of approval: "I realized that I, the Senate, and ultimately the American people, had probably been played.")

Altman's prep team considered it a resounding success.

The hearing was the cherry on top of a long campaign. After the launch of ChatGPT, nearly everyone in Washington had desperately sought meetings with OpenAI. The small policy team under Anna Makanju, after operating in relative obscurity, had received an avalanche of requests. For months, with or without Altman, they had been dining with, giving demos to, fielding questions from, and delivering the legislative proposal that Altman gave during his testimony to as many policymakers as possible—from senators, House representatives, staffers, and cabinet members to visiting diplomats, agency heads, and Vice President Kamala Harris—morning to night, practically nonstop.

By early June, Altman had personally met with at least one hundred US lawmakers, according to *The New York Times*, some of whom proudly referenced those private conversations during the hearing.

On the day of Altman's testimony, a small band of Hollywood concept artists, who specialize in the conceptual design of characters, props, and other visual elements in movies, had also been scheduled to meet with several congressional offices. They had crowdsourced funding for their airfare and accommodation and had in the process been threatened by online trolls and doxed for speaking out against the AI industry. Just as Hollywood writers were—and soon Hollywood actors would be—in the midst of historic strikes, to bargain in part for better protections against AI, the artists, too, had planned to speak candidly about the devastating effects that generative AI was already having on their profession. Generative AI developers had trained on millions of artists' work without their consent in order to produce billion-dollar businesses and products that now effectively replaced them. Those jobs that were being erased were solid middle-class jobs—as many as hundreds of thousands of them. "Artists are in so much pain right now. No one is getting work," says Karla Ortiz, a concept artist known for her work on Marvel Studios' *Doctor Strange*, who was part of the group and filed the first artist lawsuit against several generative AI companies.

As they arrived in Washington, several of their meetings were bumped by Altman's testimony to the following day, scrambling their schedules and leaving them to walk around the halls of the hearing—quite literally, outside the room where it was happening. That second day, they were once again competing for attention. Altman was attending an exclusive dinner with sixty House members at the Capitol, feasting on an expertly prepared buffet with roast chicken. At the same time, the artists were hosting an interactive cocktail hour and trying to attract as many staffers with the best their budget could buy: wine and Chick-fil-A.

It was a small but darkly comedic illustration of who commanded power and influence in the AI policy conversation and who didn't.

The same narrative Altman had long used within OpenAI to justify hiding its research and moving as fast as possible was now being expertly wielded to steer the US AI regulatory discussion toward proposals that would avoid holding OpenAI accountable, and in some cases entrench its monopoly. Silicon Valley's tried-and-true "What about China?" card had consistently done wonders to ward off regulation. Now it was punchier than ever, with Washington's fears about China, fueled by TikTok's stunning rise, reaching new heights.

That fear could be typified by the mood at the Department of Commerce, which had become the leading edge of an aggressive US government offensive to throttle China's AI development. The previous year, on October 7, 2022, Commerce had released a directive that it said was meant to undercut Chinese AI military advancements. Using a mechanism called export controls—a way to limit the sale of certain technologies to foreign countries on the grounds of national security—it clamped down without warning on the export of cutting-edge American-designed AI chips, primarily Nvidia's, to China. The blast radius of this move was far wider than the Chinese military; it pulled the rug out from under the Chinese scientific community and AI industry working on everything from AI health care and education applications to Chinese ChatGPT equivalents. "If you'd told me about these rules five years ago, I would've told you that's an act of war—we'd have to be at war," a semiconductor analyst said.

Taking stock of the aftermath, Commerce seemed frustrated by China's seeming resilience. The country's pace of AI development and adoption had slowed down some but not nearly enough to give the department comfort. Part of this was due to Nvidia's own maneuvering: China represented a massive market for the American chipmaker, and just as quickly as Commerce had laid down its constraints, Nvidia had designed new chips that fell neatly within them in order to keep selling to Chinese customers. The ban was also a lift to China's own chipmaking industry, which had long struggled to produce chips as good as Nvidia's. The US government's actions had generated a surge of interest for Chinese domestic alternatives, giving the industry a big funding and feedback boost to advance.

But the biggest challenger to its efforts was the vibrant cross-border open-source AI movement, which was rapidly replicating closed corporate generative AI models and putting them out on the internet for anyone to download and use. After vigorously playing catch-up, Meta had become a dominant player in freely putting out its large language models. The company had long been a champion of open-source development; chief scientist Yann LeCun believed in the importance of open science. It was also smart business. Meta didn't need to sell generative AI models to make money, but unleashing free ones, while integrating them into its core products, could help it establish its AI leadership, attract top scientific talent, and taunt its competitors for that talent who *did* depend on selling their models.

Having amassed around the level of compute resources that OpenAI had through Microsoft, Meta was now full steam ahead on producing its equivalent of the GPT series, called Llama. Llama didn't technically clear the true definition of open source, which would have required releasing both the model weights *and* its training data. But Meta's follow-through on just the first aspect, publishing its model weights for free, had been enough to turn Llama—despite its policy to not make Llama directly available in China—into a critical building block for the Chinese AI industry.

Amid the climate of frustration and fear in Washington, a policy white paper echoing Altman's recommendations arrived two months after his hearing in July 2023. Written by a consortium of researchers, including from OpenAI's Safety clan, Microsoft, and Google DeepMind as well as more than a dozen think tanks, many tied to the Doomer community, it pushed once again for a new licensing regime for AI models using compute thresholds, and the development of AI safety evaluations for dangerous capabilities including the ability to manipulate and persuade and the creation of novel biological weapon recipes. The fifty-one-page document also gave a name to the category of models that needed this government intervention: "frontier" AI models. Per the authors, frontier models—models that might exhibit these dangerous capabilities—did not yet exist, but by scaling existing models from companies like OpenAI and Anthropic with ever more compute, they could arise suddenly and unpredictably at any moment.

Within weeks of the white paper, OpenAI and Microsoft formed a strategic alliance with Google and Anthropic to launch the Frontier Model Forum, a group for advancing relevant research and influencing the policy agenda on AI safety risks. It was a rare issue in which the interests of Doomers, Boomers, and profit-motivated corporates aligned: Keeping frontier models front and center in government regulatory discussions was ideologically imperative to Doomers and convenient to Boomers and corporates for shifting attention away from regulating existing AI models and their problems. Everyone also advanced their causes by arguing against opening up the weights of cutting-edge models. In the Forum's first year, Meta was conspicuously absent. (It would join a year later to gain a seat at the table.)

Core to Altman's recommendations and the idea of the frontier model was the association of a model's scale with emergent, and thus possibly dangerous, capabilities. Such an argument was rooted in the philosophy of scaling laws—that more training compute should predictably result in more powerful models—as well as the belief within Doomer circles that highly advanced AI could go rogue. The policy proposals that flowed from this argument centered on regulators basing their interventions on how much compute was being used to train a deep learning model: Models that crossed a certain compute threshold should automatically be viewed with more caution and restricted more tightly.

The July 2023 policy white paper suggested a number for that threshold: 10^{26} floating point operations, referring to the minimum total number of calculations—1 with twenty-six zeros after it—that a model needed to be trained with to be designated as a frontier model. The authors had admitted that the threshold was somewhat arbitrary, stating simply that it was a level of compute that existing AI models likely hadn't yet surpassed. Sara Hooker, the VP of research at large language model developer Cohere and one of the coauthors of the paper, says speaking with her collaborators who proposed the number led her to believe they had picked it to be slightly higher than the amount of compute that OpenAI had reportedly used to train GPT-4.

But Hooker and many other researchers, including Deborah Raji, disagree with the compute-threshold approach for regulating models. While scale *can* lead to more advanced capabilities, the inverse is not true: Advanced capabilities do not require scale. A deep learning model trained only on high-quality biological data, for example, can be a very powerful generator of biological recipes at very small scale. Through distillation, one of the techniques that OpenAI referenced in its 2021 research road map, large models can also be transformed into small models with similar capabilities. Scaling models doesn't guarantee advancements in certain capabilities either; that depends once again on what's in the model's training data as well as which type of neural network is being trained. In the end, not all models are built on Transformers. Compute is thus not much correlated with risk at all, says Hooker, let alone with specific kinds of risks, such as the ones laid out in the white paper.

Without consensus among the white paper's coauthors about either the compute-centered regulatory framework or the specific threshold, the number, 10^{26} floating point operations, was placed in a footnote and the appendix without justification, alongside significant caveats for why thresholds were a highly imperfect approach. What shocked Hooker was how quickly not just the framework but also the exact threshold rapidly turned into one of the most popular policy proposals. It captured significant mind share in Washington, after the white paper tapped straight into fears of China. Frontier models sounded scary, and even more so if Beijing got ahold of them. "Parts of the administration are grasping onto whatever they can because they want to do *something*," Emily Weinstein, then a research fellow at CSET, told me in late 2023.

The white paper's ideas found a receptive audience at Commerce. Staff mobilized to meet with experts to hash out what controlling frontier models could look like and whether it would be feasible to keep them out of the reach of Beijing. Soon it was considering an unprecedented proposal to expand its AI export controls to focus on not just hardware but the software itself by banning the export of AI models above a compute threshold. In other words, it was evaluating whether it could block model weights from being posted on the internet and made widely available.

Notably, a key recommendation from Marcus and IBM VP Christina Montgomery, who also testified alongside Altman, did not gain nearly as much traction, despite their repeating it throughout the hearing: compelling companies to disclose what exactly is in the training data they feed into their models. This would have little impact on handing over more advanced capabilities to Beijing per Washington's concerns but would give real teeth to corporate accountability on a broad range of issues, including company use of copyrighted materials, user data privacy, and rigorous scientific evaluations of model capabilities. "If we don't know what's in them, then we don't know exactly how well they're doing," Marcus had said. We'd simply have to take a company's word for it.

Such an approach would also significantly ameliorate the uncertainty of if and how dangerous capabilities might emerge, for the same reason why compute is a poor risk predictor: A deep learning model's behavior first and foremost derives from its data. If an AI developer produces a large language model that is able to create recipes for bioweapons, "it's because they trained it on a dataset that included information on bioweapons," says Sarah Myers West, the co–executive director of AI Now Institute and former senior adviser on AI to the FTC. As always, the neural network is surfacing patterns within its training data. Opening up that data would be the first step to establishing scientific clarity on what kinds of inputs could lead to dangerous outputs.

As Commerce consulted various experts on its proposal to clamp down on model weights, news of its deliberations, which it would announce in early 2024 with a public request for comment, cleaved the AI development community and the rapidly expanding AI policy community into two. This clash was about Closed versus Open, techno-nationalism versus borderless science. In addition to the Frontier Model Forum participants and broader Doomer community, the Closed side quickly won over the US national security and intelligence apparatus. Facing off against them was Meta, open-source AI developers, startups, civil society groups, and independent academics.

Where the Closed side continued to emphasize many of the same points that OpenAI executives had used for years internally, the Open side argued that sequestering models would do far more harm than good.

The bottleneck for producing novel biological weapons, for example, is not about finding a recipe, Weinstein noted. Such recipes already abound online and are easily found via Google. It is about obtaining the materials and equipment to actually make the armaments. Restricting access to so-called frontier models would thus do little to fix this. But the collateral damage of suppressing the publication of AI models would risk weakening the foundations of US AI innovation. Open source—sharing and building on code and software released to the broader community—has long been the bedrock upon which the wealth of US-based startups flourish. Restricting model weights from being published would give smaller developers fewer pathways than ever to create their own AI products and services. It would further entrench the dominance of the giants represented in the Frontier Model Forum.

AI models would also become ever harder to scrutinize, such as in the work of Sasha Luccioni, Yacine Jernite, and Emma Strubell, who have relied heavily on open generative AI models to quantify the carbon and environmental costs of continuing to scale them.

In critical ways, contrary to it being a national security risk, a great deal of open collaboration across borders had also strengthened American AI leadership. As the two countries that produce the most AI talent and research in the world, the US and China have long been each other's number one collaborator in AI development. For more than a decade, scientists and entrepreneurs in both countries have riffed off one another's work to advance the field and a wide array of applications far faster than either group would have alone, benefiting not just each country but many others globally. One of the most famous examples: ResNet, among the most widely used neural networks in the world, was published by Chinese researchers in Microsoft's Beijing office. ResNet not only underpins major computer-vision, speech-recognition, and language systems but also was a core ingredient of the first version of DeepMind's AlphaFold, an AI system released in 2018 that could predict a protein's 3D structure from its amino acid sequence, crucial for accelerating drug development and understanding disease. (DeepMind's subsequent advancements in AlphaFold, using a different neural network, would earn Demis Hassabis and another senior research scientist at DeepMind a 2024 Nobel Prize in Chemistry.)

And yet, in October 2023, the ideas championed by the Closed side would gain their greatest endorsement yet when they surfaced in the Biden administration's AI executive order. The order, one of the longest in history, would read like smashed-together documents written by completely different groups—because it was. One of those documents was rooted in the administration's 2022 Blueprint for an AI Bill of Rights, which the White House had carefully assembled over time through consultations with civil society groups to outline how AI could be advanced, used, and reined in in ways that bolstered civil rights, racial justice, and privacy protections. Among other things, it emphasized developing AI with broad participation from communities and experts, addressing the discriminatory impact of AI in contexts such as health care and hiring, and protecting people from data collection without their consent.

The other document was a surprisingly faithful reproduction of Altman's recommendations and the framing of the frontier model white paper, which had been stapled on at the last minute after the paper gripped the attention of a few people sympathetic to the Doomer ideology in the White House. The white paper had emphasized the need to focus on future AI models that didn't yet exist; the executive order would subsequently focus half of its real estate on such models. The white paper had outlined four examples of dangerous capabilities. The executive order would keep three of them: the generation of novel CBRN (chemical, biological, radiological, and nuclear) weapon recipes, automated cyberattacks, and, in a straight copy and paste, the evasion of human control "through means of deception and obfuscation."

To the alarm of Hooker, Raji, and many other AI researchers, the white paper's exact compute threshold, 10^{26} floating point operations, would also show up in the executive order as the threshold above which models would need to be reported to the US government.

With such an endorsement, the compute-threshold approach would quickly metastasize. By the end of the year, it would get picked up in Europe, which settled on 10^{25} for something slightly more restrictive, as lawmakers pushing through the long-gestating EU AI Act felt steamrolled by the sudden generative AI developments and hurriedly searched for ways to account for them. At the start of 2024, the approach would then spread to California with the introduction of a new AI safety bill called SB 1047, which would return to 10^{26} as its threshold. California governor Gavin Newsom would subsequently veto the bill, which, in an ironic twist, OpenAI and other model developers heavily lobbied against for its wide array of other accountability proposals. Many would criticize Newsom for bowing to industry interests. But to several researchers, including Hooker and Raji, the veto was a welcome development. "It was a step in the right direction to make sure we're anchored to scientific consensus," Hooker says. "There are big questions that remain about why that number and what risks are you hoping to prevent."

The whole sequence of events—Altman's testimony, the white paper, the all-out policy influence campaign, Washington's hyperreactivity to fears of China, and the hasty enshrining of compute thresholds into consequential policy documents within the US and abroad—was a stark illustration among other things of how much independent AI expertise had atrophied. The prior month, in September 2023, Raji had found herself the singular academic, with financial ties neither to the industry nor Doomer community, testifying to Congress next to Altman, Musk, Nadella, Gates, Zuckerberg, Pichai, and Jack Clark, among other tech executives. They were all present for the very first of Senator Chuck Schumer's AI Insight Forums, among the hottest and most consequential series of policy convenings that year to set in motion AI legislation. As her fellow witnesses spouted spectacular, unbacked claims about the promises and perils of AI, peppered with well-timed references to beating China that straightened the backs of attending senators, what shocked Raji the most was how much many in the audience appeared to buy into everything.

It dawned on her that the people sitting next to her, and their massive policy teams, had monopolized the message in Washington for so long that many policymakers now viewed it as gospel. A Schumer spokesperson would later note in the press that the senator was personally consulting with Altman and other OpenAI executives as he moved closer to regulation. "That for me was a huge realization," Raji says. "Wow, we need more people just debunking—just looking at what people are saying and being like, 'Actually, reality is more complicated.'"

Washington was only the climax of the US leg of Altman's policy charm offensive. In March of that year, after tweeting that he planned to travel abroad to meet with users, his trip had evolved into a multicity, multicontinent odyssey to sit for photo ops with seemingly every president in the G20. It now had new branding: Sam Altman's World Tour.

There was no grand strategy from OpenAI's communications or policy teams behind the World Tour. Altman had just selected his initial stops and blasted them off to his more than 1.5 million Twitter followers. After the first few appearances, interest had snowballed out of control, and the comms and policy teams were roped in. With each new leg, the teams scrambled to arrange the logistics, bracing for the difficulties guaranteed to arise from the lack of preparation.

It was a manifestation of a dynamic that had always been present: Altman going his own way. Sometimes that way lined up with the company; sometimes it did not. During the release of GPT-4, OpenAI had carefully crafted all of its announcements and publicity to present the project as the company-wide effort that it was. The model had involved over a hundred employees. The author of the company's announcement was simply "OpenAI." Altman had then tweeted credit to a single person: Jakub Pachocki. Pachocki had indeed played an important role, but he had been one of eighteen leads on the project. Was his contribution really singular? Some employees wondered. Altman had then leaned in further, tapping Pachocki to sit behind him during his Senate hearing in Washington.

The dynamic also showed up in other ways: There were official executives at the company, but it wasn't always clear that they were the ones whom Altman was listening to, nor whether official processes and decisions or his relationships and whims were the ones guiding company strategy.

As OpenAI was rapidly professionalizing and gaining more exposure and scrutiny, this incoherence at the top was becoming more consequential. The company was no longer just the Applied and Research divisions. Now there were several public-facing departments: In addition to the communications team, a legal team was writing legal opinions and dealing with a growing number of lawsuits. The policy team was stretching out across continents. Increasingly, OpenAI needed to communicate with one narrative and voice to its constituents, and it needed to determine its positions to articulate them. But on numerous occasions, the lack of strategic clarity was leading to confused public messaging.

At the end of 2023, *The New York Times* would sue OpenAI and Microsoft for copyright infringement for training on millions of its articles. OpenAI's response in early January, written by the legal team, delivered an unusually feisty hit back, accusing the *Times* of "intentionally manipulating our models" to generate evidence for its argument. That same week, OpenAI's policy team delivered a submission to the UK House of Lords communications and digital select committee, saying that it would be "impossible" for OpenAI to train its cutting-edge models without copyrighted materials. After the media zeroed in on the word *impossible*, OpenAI hastily walked away from the language.

"There's just so much confusion all the time," says an employee in a public-facing department. While some of that reflects the typical growing pains of startups, OpenAI's profile and reach have well outpaced the relatively early stage of the company, the employee adds. "I don't know if there is a strategic priority in the C suite. I honestly think people just make their own decisions. And then suddenly it starts to look like a strategic decision but it's actually just an accident. Sometimes there isn't a plan as much as there is just chaos."

As Altman zipped around, flying to Europe, Latin America, the Middle East, Asia, and Africa, dazzling—and only on a few occasions offending—carefully curated audiences of students, tech investors, and fans, the lack of strategic clarity was inflaming OpenAI's age-old rift lines and accelerating the company toward more opposite extremes than ever before.

On one side, the Applied division was still leading the charge, racing against an unprecedented number of competitors to deploy OpenAI's technologies faster than ever. It was now also bolstered by the other ballooning divisions as well as many people in Research, invigorated by their belief from the dramatic increase in hype and expectation that advancing and releasing OpenAI's models was the best way to achieve the company's mission.

On the other side, the Safety clan, spread out across Research, many still concentrated within Miles Brundage's policy research team and Jan Leike's alignment team, were now a far smaller minority. They were compensating for their relative size disadvantage by sounding the alarm louder than ever on the dangerous capabilities and existential risks that they believed could become imminently possible. As OpenAI's models continued to advance, some within Research who didn't previously identify with the Safety clan were also joining its ranks as the accelerating capabilities converted them to the belief that AI could reach a point of intelligence that would allow it to subvert human control and go rogue. The same dramatic increase in hype and expectation on this side meant OpenAI had a moral imperative to act with maximum caution, in order to fulfill its mission.

It was the Boomers and Doomers incarnate—within OpenAI's walls.

The split reached all the way to leadership. After DALL-E and ChatGPT, most executives and senior managers had grown increasingly comfortable with models as beneficial tools to be put into the world through "iterative deployment," a phrase that OpenAI had coined between releases to describe its new approach. Unlike the staged release of GPT-2, or the controlled API release of GPT-3, iterative deployment was about going all in—putting models in the hands of users early and often. And as with all of its deployment strategies in the past, OpenAI had a new argument for why *this* one was the safest approach possible. Iterative deployment, Altman and other executives argued, would give people and institutions time to adjust while allowing it to test its models on real people, collect real feedback, and improve its products. "Going off to build a superpowerful AI system in secret and then dropping it on the world all

THE TWO PROPHETS

at once I think would not go well," Altman had said during his Senate testimony.

But if there was one leader at OpenAI not moving in lockstep, it was Sutskever. As OpenAI's models advanced and the impact of their deployments accelerated, he believed the company needed to raise, not lower, its guard against their potential to produce devastating consequences. After GPT-4, Sutskever, who had previously dedicated most of his time to advancing model capabilities, had made a hard pivot toward focusing on AI safety. He began to split his time half and half. To people around him, he seemed at times to be at war with himself. He was both Boomer and Doomer: more excited and afraid than ever before of AGI arriving and rapidly surpassing humans to become superintelligence.

Sutskever now spoke in increasingly messianic overtones, leaving even his longtime friends scratching their heads and other employees apprehensive. During one meeting with a new group of researchers, Sutskever laid out his plans for how to prepare for AGI.

"Once we all get into the bunker—" he began.

"I'm sorry," a researcher interrupted, "the bunker?"

"We're definitely going to build a bunker before we release AGI," Sutskever replied matter-of-factly. Such a powerful technology would surely become an object of intense desire for governments globally. It could escalate geopolitical tensions; the core scientists working on the technology would need to be protected. "Of course," he added, "it's going to be optional whether you want to get into the bunker."

The researcher would in equal parts continue to hold Sutskever in high regard and keep himself at arm's length. "There is a group of people—Ilya being one of them—who believe that building AGI will bring about a rapture. Literally, a rapture," he says.

As Sutskever continued splitting his time on alignment, a new idea began to percolate between him and Altman: a team laser focused on developing new alignment methods for superintelligence, in anticipation of methods like reinforcement learning from human feedback no longer being sufficient once systems could, in their view, outsmart humans. Altman called it the Alignment Manhattan Project. At first, the two discussed spinning it out as a different organization: its own independent nonprofit with a starting endowment of $1 billion. In part because Sutskever didn't want to leave OpenAI and in part due to model access issues, they decided to keep the project within the company. OpenAI subsequently announced the formation of a new team to oversee the effort in a blog post along with its new name, Superalignment. The post also announced a flashy commitment to dedicate 20 percent of the computing power OpenAI had secured to date to the team. Sutskever and Leike would colead the new effort.

At another meeting, Sutskever stepped up in front of employees to introduce the new team and its goals. He grabbed the microphone and began to thump it. *Boom. Boom. Boom.*

"Alignment is a burning fire," he said. "*Super*alignment is a blazing inferno."

Not long thereafter, with the company outgrowing Mayo, the plant-filled, fountain-adorned office it had moved into after the pandemic, executives shifted most of the Research division back to the old Pioneer Building. The move largely divided the two halves of the company—Applied and Safety—into their own worlds.

In July 2023, shortly after OpenAI made news of the Superalignment team public, it rented out a theater at the Metreon in downtown San Francisco for employees to see the movie *Oppenheimer*, the story of physicist J. Robert Oppenheimer as he led America's Manhattan Project to create the world's first nuclear weapon.

"i was hoping that the oppenheimer movie would inspire a generation of kids to be physicists but it really missed the mark on that," Altman tweeted. "let's get that movie made! (i think the social network managed to do this for startup founders.)"

For nearly eight years, the analogy that Altman had made in his very first emails to Musk between OpenAI and the Manhattan Project had been a persistent motif within the company, used even during new-hire orientations. Altman was fond of it. He shared a birthday with Oppenheimer, which he'd point out to reporters. He also liked to paraphrase

THE TWO PROPHETS

the bomb maker's belief that "technology happens because it's possible." He never seemed to add that Oppenheimer spent the second half of his life plagued by regret and campaigning against the spread of his own creation.

Different employees ascribed different significance to the analogy. Most saw the Manhattan Project as a heroic feat; it represented the ability to pull off a world-saving, history-changing technological breakthrough before dangerous adversaries with a significant concentration of talent and resources. Among the Safety clan, it emphasized the gravity of OpenAI's burden to usher in a technology that risked the existential demise of humanity.

To Altman, it represented a PR lesson. "The way the world was introduced to nuclear power is an image that no one will ever forget, of a mushroom cloud over Japan," he had once said, years earlier at an event. "I've thought a lot about why the world turned against science, and one answer of many that I am willing to believe is that image, and that we learned that maybe some technology is too powerful for people to have. People are more convinced by imagery than facts."

Those at the extreme end of the AI safety spectrum with the highest levels of p(doom) grew increasingly unsettled by the seeming uncomplicated optimism with which Altman and the rest of the company viewed this history. OpenAI had at one point also organized a screening of *Apollo 11*, a documentary about the US Apollo program to launch the first man to the moon, which was one of Silicon Valley's other favorite analogies. "Why would you ever talk about the Manhattan Project if you could just say 'Apollo program'? Why bring along that baggage?" an extreme Doomer says.

One scene from *Oppenheimer* stuck with him in particular: the moment before the Trinity test, the first ever detonation of an atomic bomb, when Oppenheimer, played by Cillian Murphy, calculates that the chances of it blowing up the world are "near zero."

"*Near* zero?" responds Major General Leslie R. Groves, played by Matt Damon, incredulously.

"What do you want from theory alone?" Oppenheimer says.

"Zero would be nice," Groves says.

"That's analogous to the situation we're in," the extreme Doomer says. "No way can we calculate anything about these AIs yet."

As OpenAI continued to push on its research, the Manhattan Project analogy for some began to take on a new meaning.

With the rate of advancement in large language models slowing with the exhaustion of data and compute, the Research division had pivoted more heavily toward developing AI agents. The idea, gaining traction across the field, was a return to the debate between the "pure language" and "grounding" hypotheses. Pure language was reaching the end of its rope, as was combining language and vision. To many in the AI community, it seemed like the next stage of advancement would likely need to come from agents that could take actions in the real world and collect feedback from its environment. Within OpenAI, such a capability was also seen as a way to gain a competitive advantage. An AI assistant that could chat with you was nice, but one that could automate complex tasks, such as sending emails or coding websites, was even better. Not only was this highly commercially relevant, it could also accelerate the company's own progress.

The most ambitious of these efforts in the Research division was AI Scientist, an attempt to build an agent for autonomously performing scientific research. With limited new "knowledge," or data to scrape from the internet or textbooks, researchers on the project had high hopes that an autonomous "scientist" would be able to generate its own knowledge by running experiments. The team had formed from a merger of the previous code-generation team working on Codex with another team that had been trying to crack the reasoning challenge by using large repositories of math problems and their solutions as a structured dataset to teach its model step-by-step logic. Both capabilities—coding and solving math problems—seemed like good building blocks for conducting experiments and analyzing data.

Leading the project were Jakub Pachocki and Szymon Sidor, who were focused on creating not only an autonomous scientist but, specifically, as Altman desired, an autonomous AI researcher—one that would help OpenAI supercharge its AI advancements.

The project made AGI believers both extremely excited and extremely nervous: If AI Scientist succeeded, AGI would surely arrive faster. This shortened the timeline either to utopia or to humanity's obliteration. Like Sutskever, some researchers began to reference "a bunker" in casual conversations, even imagining a setup similar to Los Alamos: Somewhere out in a remote patch of American desert, an elite team of AI researchers would live and work in secure facilities to protect them from outside threats.

In 2023, they believed those threats now included targeted attacks and rogue AGI itself. On Slack, the security team posted a draft of its threat model for OpenAI, significantly matured from the days when executives had debated how much to heighten the security of the organization. The draft included three categories of threats: foreign state actors, competitors, and ideologically motivated people.

In the third category, the draft linked to an article by Eliezer Yudkowsky, an extreme Doomer and leader in the AI safety community who had coined and popularized the phrase *friendly AI* to refer to well-aligned systems and wrote a beloved work of fan fiction called "Harry Potter and the Methods of Rationality." The serial novel, which spans 122 chapters and over 660,000 words, reimagines Harry engaging in the wizarding world as a well-trained rationalist. It had served for many as a gateway into effective altruism and, in turn, to broader Doomer ideology. Yudkowsky had also cofounded the blog *LessWrong*, a central hub for AI safety researchers to foster community and propagate AI safety ideas, where he'd advocated with increasing alarmism to put a full pause on AI development as his p(doom) shot up to 95 percent. In March 2023, he'd written an article for *Time* magazine where he discussed the grief of watching his daughter lose her first tooth and wondering if she would have a chance to grow up. He proposed a plan to enforce the halting of AI advancement by shutting down all large GPU clusters, tracking sales of

GPUs, and, if necessary, targeting "a rogue data center" with air strikes. OpenAI's threat model draft linked to this piece as an example of an ideologue advocating for violence.

A small faction of OpenAI employees who were fans of Yudkowsky's less violent opinions found the fact that the draft singled him out but didn't mention unfriendly AI itself deeply frustrating. After getting feedback, the security team added a fourth threat category: misaligned AGI systems.

As OpenAI skyrocketed to new prominence, the board was shedding members without replacing them. Since the start of 2023, it had lost three independent directors in rapid succession, in part due to the feverish race that ChatGPT had sparked to build and commercialize generative AI technologies across the industry.

Reid Hoffman had been the first to step down in February after five years on the board, due to conflicts of interest. The previous year, he had cofounded a startup, Inflection, with the now-departed DeepMind cofounder Mustafa Suleyman, which was fast evolving into a direct OpenAI competitor.

A month later, Hoffman was followed by Shivon Zilis, Musk's trusted deputy and Neuralink director, who, after Musk disaffiliated, had continued to oversee OpenAI on his behalf and officially joined the board in 2020. Some in leadership had long worried that Zilis would feed sensitive company information to Musk, but she had pledged to uphold her confidentiality to OpenAI over her loyalty to its spurned former cochairman. That position became highly questionable once news broke in July 2022 that she had had twins with Musk without disclosing it to her fellow directors. Still, Altman had sought to keep her on the board for reasons that eluded the other board members, at one point seeking her advice in October 2022 on how to handle Musk's apparent irritation over OpenAI's escalating valuation, by then reportedly nearing $20 billion. "This is a bait and switch," Musk had texted Altman after noting his substantial contributions to OpenAI's initial funding. In March 2023, Zilis's continued board role finally turned untenable as Musk incorporated a new AI venture, xAI, to be another direct competitor to OpenAI.

Third to depart was Will Hurd, a former Republican Texas representative and former CIA officer, who had joined the board in 2021. During the announcement of Hurd's appointment, Altman had told employees that it was important to have someone that balanced out the liberal bias of Silicon Valley. He then organized a meeting for anyone who had reservations to ask Hurd any questions. Employees didn't hold back, grilling Hurd about his views on different issues, including Donald Trump. In June 2023, Hurd parted ways with OpenAI to focus full time on his US presidential campaign. He would withdraw from the race by October.

Alongside Altman, Brockman, and Sutskever, only three independent board members remained: Quora cofounder and CEO Adam D'Angelo, roboticist Tasha McCauley, and CSET researcher Helen Toner.

Among the trio, D'Angelo had joined the board first. D'Angelo, a high school classmate of Zuckerberg's at the boarding school Phillips Exeter, had served for two years as the CTO of Facebook before starting Quora. In 2014, Quora had joined YC in the first batch under Altman's presidency; in 2017, a year before D'Angelo's board appointment, Altman had topped up YC's investment into Quora, coleading an $85 million round of funding. In the announcement, Altman praised D'Angelo as one of "the smartest CEOs in Silicon Valley." "And he has a very long-term focus, which has become a rare commodity in tech companies these days," Altman said.

McCauley had joined the board later in 2018. An entrepreneur who had cofounded a telepresence robotics startup and was running a 3D urban simulation company, she had connected with Altman through her mentor, Alan Kay, and also knew Holden Karnofsky. McCauley was well-respected in the AI safety community and would serve on the board of the AI safety research nonprofit Centre for the Governance of AI and for a time on the board of the Effective Ventures Foundation, a UK-based organization that oversees the popular EA podcast *80,000 Hours*. Karnofsky, who had been on the OpenAI board at the time, nominated her in his effort to find more independent directors as OpenAI prepared to transition into a capped-profit structure.

Toner had been the last addition to the board in mid-2021, also via a nomination from Karnofsky. This time he was recommending candidates to replace himself as his three-year term came to a close and with the formation of Anthropic. Before establishing herself as an expert on China and emerging technologies, Toner had worked with Karnofsky at GiveWell and Open Philanthropy. She had come to AI safety issues through the EA movement but had slowly pulled back from the latter over time. In her most popular EA Forum post before the FTX crash, she had observed that the movement was growing increasingly dogmatic and socially insular, and noted that she was "leaning into EA disillusionment." She continued to be highly regarded in its circles and to dedicate herself to the broader AI safety community, serving with McCauley on the board of the Centre for the Governance of AI.

By the late summer of 2023, the board had been in a monthslong deadlock over whom to appoint as new independent directors. As part of their effort to increase oversight after the GPT-4 demo, and even more after the launch of ChatGPT, McCauley had engaged in a roughly yearlong process, including interviewing employees and stakeholders outside the company, to articulate what a revamped board and more professionalized oversight mechanisms should look like.

During the process, the board, including Altman, had all agreed that based on what they saw as the rising stakes of the company's capabilities, the next independent director needed to have a deep background in AI safety. They'd subsequently spent months compiling a list of candidates and interviewing five of them, including Dan Hendrycks, a central figure in the Doomer community running the Berkeley-based Center for AI Safety and serving as the only adviser at Musk's xAI. But as the independent board members sought to move forward to select one of the five, Brockman and Sutskever each raised various issues with the vetted candidates. Altman was demure as always, not outright disagreeing with any option but not moving any of them forward either. Several times, he also suggested new candidates, who shared the same characteristic: They were all embedded in his network and, financially or otherwise, within his sphere of influence.

When it came to establishing the new oversight mechanisms, which included different channels for increasing the board's visibility into the company's safety and security practices, the independent directors were also left with a similar feeling that they weren't a priority for Altman. Early in McCauley's tenure as a director, Altman had designated her the board's employee liaison and advocate; she subsequently met with employees regularly by holding office hours. Once she had also brought her husband, actor Joseph Gordon-Levitt, to a company off-site, where he'd listened intently to technical presentations. But during the pandemic, those meetings had petered out. Afterward, McCauley continued to keep some regular meetings, but the open office hours never restarted.

Without a systematic way of connecting with employees, information about the company's happenings was instead filtering up to the independent directors through their own personal relationships from the broader AI safety and tech communities. They also relied on Altman himself as a conduit for keeping tabs on important information. What worried them with growing intensity was how much Altman's rhetoric often differed from other accounts they were hearing. Where Altman regularly portrayed a rosy picture, the directors increasingly received reports from their own sources about various problems, including the company's lack of preparation before and significant tumult after ChatGPT, the continued AI safety concerns surrounding GPT-4's release, and the unprecedented pace with which OpenAI was sprinting to launch new products before it had resolved many of its issues.

One incident felt particularly glaring. In late 2022, the board had had an on-site—the first of what was meant to be an annual meeting—during which Altman had highlighted the strong safety and testing protocols that OpenAI had put in place with the Deployment Safety Board to evaluate GPT-4's deployment. After the meeting, one of the independent directors was catching up with an employee when the employee noted that a breach of the DSB protocols had already happened. Microsoft had done a limited rollout of GPT-4 to users in India, without the DSB's approval. Despite spending a full day holed up in a room with the board for the on-site, Altman had not once notified them of the violation.

While the independent board directors didn't have reason to believe that anything unsafe had been released to the public, they were unsettled by the seeming disregard with which Microsoft had broken protocol and Altman had passed over it. OpenAI's models were on a rapid advancement trajectory and, from an AI safety perspective, they believed, could soon pass a point where such a violation could result in potentially catastrophic, if not existential, consequences. In their view, Altman's laxness with Microsoft's breach set a dangerous precedent for how he might treat AI safety processes around model releases once the stakes went up.

Meanwhile, there were other concerning examples of Altman's behavior. In March 2023, he had emailed the board without D'Angelo and announced that he believed it was time for D'Angelo to step down. The fact that Quora was designing its own chatbot, Poe, Altman argued, posed a conflict of interest. The assertion felt sudden and dubiously motivated. Toner, McCauley, and D'Angelo had each at times asked Altman inconvenient questions, whether about OpenAI's safety practices, the strength of the nonprofit, or other topics. With his allies on the board dwindling, Altman seemed to the three to be fishing for an excuse to push one of them out. In response to Altman's email, an independent director pushed back. Poe was a moderate conflict of interest compared with what Altman had long allowed to stand from Hoffman and Zilis. Altman's motion failed; D'Angelo stayed on.

Shortly thereafter, D'Angelo was at a dinner party when he heard that OpenAI's Startup Fund was structured weirdly. It was giving those who invested in the fund early access to OpenAI's products, a kind of preferential treatment that should have been reserved for OpenAI's own investors. After he heard it come up a second time, the independent directors pressed Altman for documents about the fund's structure. When Altman finally handed them over, the directors discovered that the structure of the fund wasn't just weird; Altman legally *owned* it when it should have been owned by OpenAI.

For the independent directors, every instance added up to a single troubling picture: Bit by bit, Altman was trying to cloud their visibility and maneuver in ways that prevented the board from ever being able to check him. For years, Altman had advertised the board's ability to counterbalance and even fire him as OpenAI's most important governance mechanism. Indeed, he was now trumpeting this fact around the world to secure public and government trust.

By the fall, the morale of the independent directors hit new lows as they struggled to make any meaningful progress in the negotiation for a new board member. With every passing month, these gaps in governance were gaining urgency. OpenAI was getting ready to train GPT-5, and it was making progress on AI Scientist. Then, out of the blue, in early October, Toner received an unlikely email: Ilya Sutskever wanted to talk.

Chapter 14

Deliverance

In the weeks before Sutskever reached out to Toner, Altman was dealing with a PR crisis. After years of his sister's estrangement and her turn to sex work staying out of the media, her story had finally burst into the open.

On September 25, 2023, Elizabeth Weil, a features writer at *New York* magazine, published a profile of Sam Altman that for the first time in the mainstream press referenced Annie's existence. Weil juxtaposed details of Annie's life, including her suffering repeated health challenges and living in severe financial duress without housing security, against Sam's lifestyle featuring multimillion-dollar homes and luxury cars. "Annie Altman?" Weil wrote in her piece. "Readers of Altman's blog; his tweets; his manifesto, *Startup Playbook*; along with the hundreds of articles about him will be familiar with Jack and Max . . . Annie does not exist in Sam's public life. She was never going to be in the club."

Altman knew the details were coming. In the lead-up to their publication, which happened during Yom Kippur, the Jewish holiday of atonement, Weil had given OpenAI an opportunity to comment on her reporting; *New York* had also worked with the company as part of its fact-checking process. The task of facilitating the comment request and trying to control the story had fallen to Hannah Wong, who had become

OpenAI's VP of communications and suddenly found herself as the go-between for the magazine and Altman's family, wondering if this should really be part of her job.

The final day before the profile's release, once it was apparent that the details about Annie would be published, Annie had received an email from Sam. "hi annie. in the spirit of it almost being yom kippur, i wanted to apologize and ask for forgiveness for something," he wrote. During Annie's requests for support, he had felt caught in the middle, torn between wanting to defer to their mom, agreeing with the rest of the family that Annie should learn financial independence, and feeling that Annie needed medical help and was struggling to function. "still, I made the wrong call and should just have just [sic] kept supporting you; i sincerely apologize," he said.

The sharp turn from the otherwise glowing public reception that Altman had received since the release of ChatGPT, and having what he viewed as his family's painful private matters spilling out into the open, weighed on him. The fallout of the piece made it worse. After it came out, old tweets of Annie's with allegations that Sam had abused her in various ways resurfaced and began to go viral. But as heavy and challenging as it was for Sam, it was an extraordinary release of pain for Annie. For years, in addition to the compounding stress from her health challenges and lack of stability, she felt as if she had been shouting into the void and erased from significance.

In 2024, I would reach out to Annie to better understand her side of the story. I also reached out directly to her mother, Connie Gibstine, to her brothers Max and Jack, and to Sam via OpenAI's communications team, to seek their account. Annie was eagerly cooperative, hopeful for a platform to finally share personal experiences, which she viewed as crucial to understanding Sam's moral character. She provided extensive correspondence with her family, physical and mental health records spanning most of her childhood to adulthood, and other corroborating evidence, which charted the fallout she had with the family and the deterioration of her life circumstances.

Gibstine offered a brief statement emphasizing the family's love for

Annie and concern for her well-being while also denying Annie's claims, which she called "horrible, deeply heartbreaking, and untrue." Gibstine declined to have a more in-depth conversation or to provide responses to detailed questions seeking her perspective on Annie's account. She did not respond to my additional requests for documentation to support her own claims. Max and Jack did not respond. OpenAI did not comment on specifics.

In January 2025, after Annie filed her lawsuit against Sam, he, Gibstine, Max, and Jack issued a more forceful public denial of Annie's allegations, characterizing her as mentally unstable and unreasonably demanding of money, and her claims as having "evolved drastically over time." "It is especially gut-wrenching when she refuses conventional treatment and lashes out at family members who are genuinely trying to help," they said.

Through my conversations with Annie and the documentation she provided, a complex picture emerged of the turbulent journey that led her to go public with her allegations on Twitter and in her subsequent lawsuit, as well as with the details of her life she shared for Weil's profile. I did not have access to the full reasoning behind many of her family's decisions, and the truth of some of Annie's allegations, in particular Sam's alleged sexual abuse of her as a child, is unknowable, but her story became a microcosm to me of the many themes that define the broader OpenAI story. It also helped me solidify my understanding of how much OpenAI is a reflection and extension of the man who runs it. Annie's persistent efforts to add her perspective to the record quickly turned into a company issue. Coverage of Annie would get under Sam's skin perhaps more than any other kind of story, pulling Hannah Wong in her capacity as OpenAI's communications head into his efforts to contain its spread. It also became a company issue in another way: At the time Annie's allegations first rose to the fore, other OpenAI executives took notice.

In many ways, Annie and Sam, nine years apart, are remarkably similar. Many of the words that people close to Sam use to describe him also shine through in Annie: She is an excellent listener; she remembers the tiniest details about others; she is very goofy, extremely generous, and quick to win people's trust.

Like Sam, she also excelled in academics. The only other Altman sibling to graduate from Burroughs, she made an impression on her physics teacher, James Roble, who remembers her fondly more than a decade later as a talented student with a sunny demeanor. For college, she went to Tufts University, majoring in biopsychology and minoring in dance. After graduation in 2016, she completed her premed requirements in anticipation of one day going to medical school; she moved to the Bay Area for a research position in a neuroscience lab at the University of California, San Francisco. Before her life went sideways, she was, in other words, a typical profile of an ambitious, well-educated young adult with plenty of options.

But Annie's circumstances began to unravel after a series of unexpected challenges. While still in college in 2014, she was diagnosed with Achilles tendinitis, a swelling of the tendon that comes from overuse, and a bone spur, putting her in a walking boot. It became the first of a growing laundry list of physical health ailments that would plague her body with chronic pain and, at their peaks, severely limit her mobility and impact her quality of life. In a span of six years, Annie dealt with recurring tonsillitis; recurring pelvic pain; repeated flare-ups in her tendinitis, which placed her in more walking boots, spread beyond her right ankle, and at times made it difficult to even stand for short periods; and a growing number of ovarian cysts that culminated in a diagnosis for polycystic ovarian syndrome, or PCOS, which sometimes caused her to sweat through her sheets at night, a common symptom.

Amid all these health issues, she suffered another paralyzing blow: On May 25, 2018, her dad, with whom she was closest in the family, died of a sudden heart attack.

Annie has dealt with mental health struggles throughout her life. She was diagnosed at a young age with general anxiety and obsessive-compulsive disorder, taking Zoloft for the better part of a decade until a psychiatrist helped her taper off in college. Before her father's death, she was doing a lot better. She had adopted nonpsychiatric mental health tools, including meditation, and was searching for ways to address the root inflammation that seemed to underlie her repeated physical ailments with a full embrace of a healthy, organic, whole foods diet. Her larger health journey sparked a new interest in alternative medicine, and she took a yoga teacher training. She was leaning into her lifelong love of art and had written a draft of a book she called "The Humanual," capturing her reflections on life and how to be a good human. Her dad's death shattered her. It sent her mental health spiraling and seems to have marked an accelerated degradation in her physical health. By late 2018, going to various doctors and specialists was becoming routine.

Sam has also spoken publicly about his dad's death being the worst moment of his life. It happened mere months after Musk stepped down as OpenAI cochair and Sam took over. People close to Sam say the loss sent him reeling. For a while, he behaved erratically and struggled to cope. The death was clearly an inflection point in Annie's and Sam's relationship as well as her relationship with the rest of the family. Her connection with them, and her mom in particular, had already been strained over disagreements about her various life decisions, Annie says, including her turn away from a traditional medical career and her choice to stop taking Zoloft in favor of natural coping mechanisms. Her dad had been her sole unequivocal supporter. Without him, she grew isolated from the family yet yearned for their acceptance and support. Instead, after clashes over money, the relationship reached a breaking point.

After her dad's death in 2018, Annie was living in LA. She was working part time as a writing assistant and then at a marijuana dispensary. Annie says she inherited around $100,000 from her dad's life insurance. She threw herself deeper into pursuing her love of art and performance as a serious career: She took comedy classes; she went to open mics; she started a podcast; she rewrote the first draft of her book into a one-woman show she renamed The HumAnnie. By mid 2019, the funds were depleting, most of it spent, she says, on her various artistic investments, her high rent, her out-of-pocket health insurance, and her accumulating medical expenses—medical imaging, physical therapy, psychotherapy, Lyfts and Ubers to her appointments.

For Annie, it was what happened next that spelled the beginning of the end of her relationship with her family. In May 2019, as her chronic pain continued, she learned that her dad had left her his 401(k). She quit the dispensary and drew up a six-month plan to use the extra financial runway—a little over forty thousand dollars—to tend to her health and get her creative endeavors off the ground enough to hopefully generate a sustainable source of income. But soon after, her mother, who retained authority over her dad's retirement funds as the surviving spouse, notified Annie that she would not in fact be receiving the money. The best tax strategy, Gibstine wrote to Annie in an email, was for Gibstine to keep the funds in her name in a tax-deferred account that would pass to Annie via a trust and to which she would gain access at age fifty-nine and a half. "We all want what is best for you, and we believe that what is best for you (and for everyone) is financial independence, which brings long-term satisfaction, personal growth and security," Gibstine said. "For this reason, we want to clarify that we will not financially support you if/when your current inheritance from Dad runs out." The email was signed "Mom, Sam, Max, Jack."

Annie was in a fragile state. Her therapist's notes from the same period say that the move her family made with the hope of getting her back on her feet instead worsened her condition. In December 2019, her bank account slid into the negative. Sam had secured Microsoft's first $1 billion investment into OpenAI earlier that year and the company was full speed ahead in training GPT-3. Scared and alone, Annie logged on to an escort service called SeekingArrangement and showed her breasts to a man over a video call for enough money to bring her account out of the red.

Annie says she had never before asked her mother or brothers for financial support. She had not assumed she could rely on their wealth, but neither had she believed they would leave her without a safety net should she truly be in an emergency. From late 2019 to mid-2020, Annie made several appeals to her family for financial help, including once the pandemic added another layer of stress and uncertainty. After Sam and her mother attended two family therapy sessions with her, they agreed to cover her expenses for a part of the year.

In her exchanges with her family, it's apparent that they were worried that the money could enable harmful behaviors and believed the best way to help her reestablish her mental health was to continue to encourage her financial independence. Both Gibstine's statement to me and the family's public one said their actions have been guided by professional advice on how best to support Annie.

In May 2020, as her family's financial coverage neared its end and Annie continued to struggle, she requested more help to pay for her physical and talk therapy, which she told them was $45 and $15 a session respectively. They declined, saying they believed she should cover her June expenses herself, including with a security deposit returning to her. She packed up her small number of belongings and arrived in Hawai'i, where she had been when her dad made his final visit to her a few months before he'd died. She found a work trade on a farm with light physical tasks like weeding and planting. Sam emailed her for her new address. Eight months after their dad died, he had asked each sibling to mail their mom a lock of hair to be mixed together with their dad's ashes and turned into a diamond for each of them. They "will mostly be from carbon from dad but will then have a little bit of each of us too," he'd written. Annie's diamond was now ready; Sam wanted to send it over.

To Annie, Sam's email felt like a slap in the face. She didn't want an expensive diamond; she wanted his help to guarantee her food and housing. She couldn't tell whether Sam just didn't get it or didn't care about the severity of her crisis. The chasm between her, Sam, and the rest of her wealthy family felt irreconcilable. With enormous pain and grief, she stopped speaking with them.

Over the next three years, before Elizabeth Weil reached out to her, Annie's life bottomed out. She faced housing insecurity, food insecurity, health insecurity; she turned to virtual and physical sex work to pay the bills. In their public statement, the family said they tried "in many ways to support Annie and help her find stability." In Annie's retelling, one of those extensions of support came in the spring and summer of 2021, when Sam sought to reconnect with her. They had three phone calls, she remembers, during which he told her how much he loved her and offered to buy her a house. Annie and the family describe that offer differently. Annie says the offer was not for her to own the house but for her to live in it, an arrangement she understood was meant to prevent her from selling the property and which she worried could be another way through which Sam and their mother could impose their views on her health and career decisions. Gibstine said in her statement to me that the family offered Annie home ownership but did not respond to my requests for corroborating documentation. In the family's public statement, they said they offered to buy Annie a house through a trust, so that she could have a place to live without the ability to sell it immediately. In the end, Sam and Annie reached a different agreement: He would pay her rent for a year directly to her landlord. In the summer of 2022, when she didn't reestablish contact, the payments stopped.

Annie's story deepens the dueling portraits that people paint of Sam. He is at once generous and self-serving, agreeable and threatening, a benefactor for so many people and the source of great personal pain for others. Someone who projects sincerity and altruism in public but reveals a more complicated calculus through his behaviors behind closed doors. Someone who can give and take away, leaving many with an impression that they are part of a larger game of chess for which only he can see the full board, and the end game is to preserve his power as king.

Annie's story also complicates the grand narrative that Sam and other OpenAI executives have painted of AI ushering in a world of abundance. Altman has said that he expects AI to end poverty. Brockman has repeated, through his stories about his friend and his wife, Anna, that AGI will dramatically improve healthcare. Sutskever has said that it will lead to wildly effective, dirt-cheap psychotherapy. And yet, against the reality of the lives of the workers in Kenya, activists in Chile, and Altman's own sister's experience bearing the brunt of all of these problems, those dreams ring hollow.

Despite all the leaps and bounds in AI capabilities, none of them helped to alleviate any part of Annie's desperation. If anything, AI may have served to entrap her further. She hadn't wanted to turn to sex work. It was, she says, a "plan Z." When her chronic pain intensified and made even her work trade too difficult, she had first attempted digital means of monetizing her art. She continued her podcast and maintained an Etsy store and Patreon account, but they didn't earn enough to even cover her phone bill. A strange thing was happening, which she documented in screenshots over time: She was getting little to no exposure across all of her social media. Sometimes she noticed chunks of the reviews on her podcast in the Apple app mysteriously disappearing, which limited its discovery. At least twice, on both her Instagram and YouTube, she would accumulate views and then inexplicably lose them. A former Facebook data scientist and two tech and sex work experts say it's possible that Annie's very first SeekingArrangement account in December 2019 could have limited her online traction, based on the nature of how tech platforms track and shadow ban sex workers through automated systems by tagging their devices, emails, bank accounts, or other information that ties together their online presence, even for profiles completely unrelated to their sex work. Seeing no other path forward, Annie went back to SeekingArrangement as well as starting an OnlyFans account, entangling her online presence and access to economic opportunities even more in a web of algorithmic moderation.

Neily Messerschmidt, a former tech industry leader at companies such as Sony who now oversees a wellness division for an organic farm network, met Annie shortly after she moved to the Bay Area in 2016 and worked for a time with the network during her embrace of healthy eating. Over the years, Messerschmidt became a motherly figure as Annie severed ties with her biological family. "Sam carried AI into the world just like he actually treated his young sister," Messerschmidt says. "He's just over there thriving, and his sister's falling through the cracks."

In late 2020 and early 2021, after Annie estranged herself from the family, she began to experience devastating flashbacks of childhood sexual abuse. In sessions from July 2021 to January 2022 with a new trauma therapist on Maui, the notes chronicle the crisis that Annie went through with the involuntary memories, including intense questioning about her identity. From fifteen sessions, the therapist wrote down her diagnostic impressions: generalized anxiety, PTSD, and a personal history of sexual abuse in childhood.

The notes didn't reference a specific abuser. But around that time, Annie began regularly calling Messerschmidt, who herself had been raped at nineteen and had noticed early on in her interactions with Annie at the organic farm network the telltale signs of someone with a history of abuse. "She was not comfortable around certain men," Messerschmidt remembers. "I know someone that's been abused when I see them. She would basically ask to be out of meetings when certain men were around. She would stand off to the side when they were in the room."

In intense, emotional conversations, Annie described to Messerschmidt her sudden flashbacks: In these childhood memories, Messerschmidt remembers Annie told her, it was Sam who had repeatedly climbed into her bed, sometimes with Jack, and molested her.

It is important to note that it's often difficult to prove decades later whether alleged childhood sexual abuse happened, or the details of such abuse. What is known from psychology is one common pattern that some abuse victims suffer: The victim's brain blocks out any memory of it until a trigger—perhaps puberty, becoming sexually active, or new unwanted sexual advances—involuntarily resurfaces it, a therapist I consulted with says. The body remembers the trauma, even if the mind doesn't. As Annie got deeper into sex work, she may have suddenly been facing a rush of triggers.

In detailing Annie's experiences of her flashbacks, as told to Messerschmidt and as reflected in part in the notes of her trauma therapist, the intent is not to determine exactly what happened in Annie's childhood but to re-create an account of what she experienced and believed as an adult, which ultimately motivated her to speak out.

In November 2021, while Sam was covering her rent, Annie posted publicly for the first time about her allegations. "I experienced sexual, physical, emotional, verbal, financial, and technological abuse from my biological siblings, mostly Sam Altman and some from Jack Altman," she wrote on Twitter. "I feel strongly that others have also been abused by these perpetrators. I'm seeking people to join me in pursuing legal justice, safety for others in the future, and group healing."

Her post didn't gain traction. She was a nobody account with few followers. Sam's profile was rising more than ever from OpenAI's success on GPT-3 and its most recent launch, GitHub Copilot. Instead, trolls attacked Annie, saying she wasn't actually related to her brother.

Over the subsequent months, two reporters reached out to Annie, but she remained uncertain about how much to share with them. In September 2022, she resolved to speak openly. She tweeted again, naming Sam and Jack: "Sexual, physical, emotional, verbal, financial, and technological abuse. Never forgotten."

Ten months later, in July 2023, after ChatGPT's release and Sam's rocket to global stardom, Annie received a message from *New York*'s Elizabeth Weil.

In the three months after the *New York* magazine article published, Annie's OnlyFans income would jump up more than 10x, from around $150 a month to over $1,500. For a month during the board crisis, it would hit around $5,500. She stopped escorting, continuing only virtual sex work.

The family—or as Annie calls them, "her relatives"—have maintained that they have given her various forms of financial assistance throughout the years. Indeed, the support plan they agreed on after family therapy and the rent that Sam paid from mid-2021 to mid-2022 are examples. But for Annie, this money came too little too late, after she had already been desperate enough to turn to sex work. Other offers of support, including the house, generally had restrictions or conditions that she felt she could not accept.

In July 2023, Sam sent Annie one other message offering money without apparent conditions. A person had emailed both of them earlier that summer with the bugged-out energy of an internet sleuth, asking Annie to elaborate on her allegations. Annie responded to the three-way thread with a detailed account of how she had experienced her last few years. Sam was in the middle of his World Tour, ascendant. On June 5, the day he gave a well-received talk with Sutskever at Tel Aviv University in Israel, Annie was in his inbox, talking about the cruelty of her last few years in measured, eviscerating sentences.

He responded in rare uppercase on July 9, his World Tour officially over. "Sorry for taking so long to respond; it took me awhile to figure out what I wanted to say," he wrote. "I don't want any kind of ongoing relationship with you, and I respect that you don't want one with me either. I am, however, happy to send you money and am hopeful that you can get through your health challenges." He offered to restart their previous arrangement, the details of which he couldn't quite remember, or to give her a lump sum payment. "For the record, I disagree with many claims in this email, but it seems pointless to try to engage," he ended.

At that point, Annie no longer wanted Sam's money. She simply wanted access to the money her father had left. She never responded to Sam's email.

In addition to his 401(k), her dad had left behind a trust under her mother's authority. In early 2024, with her additional income, Annie would retain her own lawyer and learn that her dad's trust had been newly funded in 2023. As her story gained traction, the family would open up discussions through Gibstine's lawyer over sending Annie monthly distributions from the trust with no strings attached. For the first time, Annie felt she had real negotiation leverage. The family said in their public statement that they expected to provide Annie monthly financial support "for the rest of her life."

With the new distributions, Annie rented her first stable apartment in over four years. Soon after, she engaged another lawyer to prepare a child sexual abuse case against Sam to file it before her thirty-first birthday, on January 8, 2025. It would allege—and the family would vehemently deny—that Sam had sexually abused her beginning when she was around three and continuing until he was an adult and she was still a minor; it would seek damages in excess of $75,000. In October 2024, after another whirlwind of medical appointments, Annie would also finally receive a diagnosis for her underlying health condition: hypermobile Ehlers-Danlos Syndrome, the same genetic mobility disorder as Brockman's wife.

In the lead-up to Annie's story coming out in the *New York* magazine article, Sam began to tell people that his sister had borderline personality disorder. It was a private and sensitive matter, he told them. "He defanged her account before it even published," one of those people says.

Borderline personality disorder is marked by severe challenges in emotional regulation and can lead to intense interpersonal relationships with extremes of idealization and fears of abandonment. Annie says she never received that diagnosis—nor did such a diagnosis appear in any of the therapy notes or medical records she shared with me. There is only one mark from her trauma therapist on Maui in August 2021 indicating that she was evaluating Annie for the disorder, which commonly arises after childhood sexual abuse. The therapist included a reference to a past history of sexual abuse but not the disorder in her final diagnostic impressions.

Two therapists I spoke to about borderline personality also underscored that the disorder usually goes away, either naturally or with the right treatment, which includes meditation and behavioral therapy that helps reaffirm a person's self-worth, both tools that Annie has leaned into, but not psychotropic medication, as Gibstine encouraged and what the family's statement appeared to refer to when saying Annie "refuses conventional treatment." "The research is very positive on borderline personality disorder; it's considered a good prognosis, diagnosis," says Blaise Aguirre, an assistant professor of psychiatry at Harvard Medical School who has treated thousands of borderline patients but did not review Annie's case specifically. "The vast majority of people will get better."

In April 2024, Hannah Wong, who would soon be promoted to OpenAI's chief communications officer, would also speak to me about Annie's mental health. For six months, Wong's team had said they were committed to arranging an office visit with me as well as interviews with key OpenAI leadership and employees as I repeatedly sought to engage them and hear their perspective. Five months in, they began to sour on the idea. Ten days before my flight to San Francisco, which I had already booked to visit the company's headquarters, they notified me that they had reversed their decision: I would not be coming to the office, and they would no longer participate in my book.

Several days into my trip in San Francisco, which I took as planned, I told someone who personally knew Sam and Annie that I was speaking to her. The next day Wong texted me. "I hear you are in town?" she said, an odd formulation given why I had arranged my trip in the first place. We met at a Philz Coffee in Mission Bay, not far from a new office location that OpenAI was expanding into. After some meandering small talk and high-level discussions about my book, she directed the conversation to Annie.

"I don't think I'm stepping out of turn here by saying Annie has mental health challenges," Wong said. At this point, I had not yet reached out to OpenAI about Annie and had not brought her up in the conversation first. "Annie has good days and some really bad days," she continued. "And the family is trying very hard to strike a balance between protecting her and not enabling her." Here, she reiterated the point again for emphasis. "Notice that the family hasn't put out any public statement denying what Annie said. It all comes back to protecting Annie." She had also heard that some journalism programs had even discussed whether it had been ethical for Weil to include Annie in her profile. Maybe this, too, was something I should consider, she said.

It became clear that this was the main message Wong had reached out to me to deliver. Until then, she had not responded directly to my request to speak with her about the book and had only interfaced with me through a deputy. The importance she seemed to place on addressing Annie's story highlighted the pressures that it was putting on Sam and the close link between the company and Sam's personal matters.

Sam's and Wong's assertions about Annie's mental health also struck me as another parallel between Annie's experience and the experience of so many others sidelined or harmed by the empires of AI and their vision. Since resolving to tell her story, Annie has faced the same gulf of power that I have watched data workers and data center activists wrestle with. Her life has revolved around combing through and gathering as much documentation as possible to get anyone to listen to her. At times she has been consumed by a sinking feeling that no matter how much she speaks up, the world is somehow in a conspiracy against her. It's the same loss of agency and anger I've seen etched on the faces of people globally when they throw so much of the little they have at challenging the empires' narratives, and then watch as the people they are up against wield the kind of power that can deploy billions of dollars in capital, construct vast infrastructure, hire and fire tens of thousands of contractors, and, with a few soft-spoken words—at an event, to Congress, to heads of state, to journalists—smooth over the murmurs of protest in the way of their will.

In Annie's case, after Weil's profile, she was no longer shouting into a void. As her tweets began to go viral in October 2022, they came to the attention of someone important: Ilya Sutskever, right as he was grappling with his own complex feelings about Altman and what he viewed as Altman's patterns of abuse.

IV

Chapter 15

The Gambit

Four days after Weil's *New York* magazine profile of Altman and four days before Sutskever's message to Toner, the independent board director had met with another OpenAI executive: chief technology officer Mira Murati.

Born in Albania, Murati had learned from a young age how to stay calm amid chaos. Through her early childhood, she had experienced the throes of the country's transition from totalitarian communism to liberal capitalism. The shift happened so rapidly, with the country's financial system so underdeveloped, that pyramid schemes rapidly proliferated, then collapsed, leading to widespread unrest and violence. The upheaval would leave behind bomb craters that Murati would need to delicately maneuver around on her way to school. A teacher once told her that as long as she was willing to do it, the teacher would do it too.

Murati's parents taught literature, but she found solace in the certainty of numbers. First came her love of math, nurtured by teachers who saw her potential and at times gave her harder problems than the curriculum to push her learning faster. Then came her love of science—chemistry, biology, physics—which fed her love of technology. She was a voracious learner. She burned through any book she could get her hands on, finishing up her own textbooks and then rummaging through her older sister's. She thrived on competition, finding her happy place in math and science Olympiads as she jostled among peers in the fierce race for knowledge.

When she was sixteen, her precociousness won her a scholarship to study abroad at a Canadian private school, Pearson College UWC in Victoria, British Columbia. The opportunity set her on a rapidly rising trajectory, through Dartmouth College, where she studied mechanical engineering, to an aerospace company, to her first major career break at Tesla, where she was a senior product manager on the Model X.

At Tesla, she learned to build complex products under intense pressure, navigating and negotiating across teams with very different opinions and areas of expertise. It was there, she often says, that she found herself drawn to AI as the company explored autonomous driving. The more she dug into AI, the more she began to view it as a fundamental asset that would be broadly applicable and universally needed for solving tough problems. "It really seemed like maybe the last thing we'd ever work on," she later told Kevin Scott on his podcast.

Murati didn't jump into AI immediately. Three years into Tesla, she left to join the company Leap Motion as the vice president of product and engineering, to work on augmented and virtual reality systems. She imagined the company revolutionizing education, allowing learners to rotate strands of DNA with a swivel of their hands or manipulate the physics of a ball hurtling through the air. Instead the company was too early a bet on the technology; VR and AR still made too many people nauseated. In 2018, two years later, she left for OpenAI while it was still a nonprofit.

At OpenAI, Murati's climb continued. As the nonprofit transformed into a commercial operation, she stepped naturally into VP of Applied and partnerships. She oversaw the company's most important relationship with Microsoft and the budding, then burgeoning, division commercializing the company's research. She was even younger than Altman, Sutskever, and Brockman, and, for a while, the only technical woman in senior leadership. This sometimes made her the target of sexism, particularly among researchers nostalgic for the early days of the lab who viewed her as not technical enough—an engineer rather than a scientist—and considered her rise as a symbol of OpenAI's turn away from serious fundamental research.

Indeed, in the awkward, nerdy, testosterone-fueled world of AI, Murati stood out. She was socially adept, a good listener, and had little ego to speak of. She was known among the people she worked with as a uniquely skillful problem solver. In the maelstrom of OpenAI's persistent internal conflict, she could guide the company forward, attentive to different ideas and perspectives yet unafraid of making tough calls. "Imagine when there's excruciatingly hard decisions that have to be made and there's no clear answer. And she can just help find an answer," says a former colleague. "She's consistently correct."

Among the many hats Murati wore, she increasingly played translator and bridge to Altman. After the Anthropic split, Altman had asked her to oversee not just Applied but also Research; in May 2022, she officially took on the title of chief technology officer. As more and more teams rolled up to report to her, she became a critical conduit through which employees interfaced with Altman. If he had adjustments to the company's strategic direction, she was the implementer. If a team needed to push back against his decisions, she was their champion.

Even among company leadership, she had a level of influence on Altman and access to his opinions that others did not. She could tell him directly when his expectations or plans were unrealistic, and he would often listen. She would tell others directly if he didn't want something, even when he pretended that he did. Where people grew frustrated with their inability to get a straight answer out of Altman, they sought her help to decode his opinions. "She was just honest," another former colleague says. "She was the one getting stuff done."

But the more Murati worked with Altman, the more she found herself frequently cleaning up his messes. If two teams disagreed, he often agreed in private with each of their perspectives, which created confusion, exacerbated the conflict, and bred mistrust among colleagues. That pattern compounded the chaos that Brockman continued to cause as he jumped into projects. To Murati, Brockman was like a second CEO but a bad one—highly opinionated and prone to driving people to burnout with his intensity. During the development of GPT-4, Altman and Brockman's dynamic had exerted mission-critical levels of stress on parts of the company, nearly leading key people on the pre-training team, one of the core teams handling the data collection and initial training of each model, to quit.

Then there was Nadella, who was practically OpenAI's third CEO with how deferential Altman could be to Microsoft's interests. On multiple occasions, after Murati had carefully put together a plan of reasonable commitments that OpenAI could make to Microsoft, Altman had veered off script with the tech giant's executives, conveying a different picture of what the team was working on to agree with demands divorced from the startup's road map. To the board, Altman framed the tumult differently: Murati just didn't have a productive relationship with OpenAI's most important partner.

Altman's behavior had progressively worsened after ChatGPT had propelled him into megastardom, intensifying both the spotlight and scrutiny and exploding his calendar with an overwhelming travel schedule. Before, he was generally energized; now he was often exhausted. And he was cracking under that pressure, his anxiety reaching new heights and fueling his patterns of destructive behavior. He was doing what he'd always done, agreeing with everyone to their face, and now, with increasing frequency, badmouthing them behind their backs. It was creating greater confusion and conflict across the company than ever before, with team leads mimicking his bad form and pitting their reports against each other. This was corroding enough as it was. But faced with mounting competition externally, Altman was also pushing the company to deploy faster and faster and attempting to skirt some of its established release processes for expediency, sometimes through dishonesty. Recently, he had told Murati he thought that OpenAI's legal team had cleared GPT-4 Turbo for skipping DSB review. But when Murati checked in with Jason Kwon, who oversaw the legal team, Kwon had no idea how Altman had gotten that impression.

In the summer, Murati had attempted to give Altman detailed feedback on the accelerating issues, hoping it would prompt self-reflection and change. Instead, he had iced her out, and it had taken weeks for her to thaw the relationship, including by assuring him that she had not shared that feedback with anyone else. She had seen him do something similar with other executives: If they disagreed with or challenged him, he could quickly cut them out of key decision-making processes or begin to undermine their credibility. It was subtle and contained enough, out of sight of employees, that it had taken her some years to realize the full extent of it. But inevitably, different executives had each had their turn bearing the brunt of this treatment. Over time, the cumulative impact of his actions had taken its toll on the highest levels of the organization.

Most recently, the hot seat had passed to Sutskever. Some time earlier, Jakub Pachocki, the Polish researcher leading the AI Scientist project, who reported to Sutskever, had grown frustrated with his lack of recognition or authority. He'd turned to his ally, Brockman, and Brockman had turned to Altman. Altman had then encouraged Pachocki's ambitions and given him a more senior role in Research. There was only one problem: Altman had never mentioned any of this to Sutskever. Nor would he clear up the divisions between Sutskever's and Pachocki's portfolios as both of them, each getting different messaging from Altman, began guiding the same research in their own directions and struggling to understand the source of the misalignment.

The tangled situation had caused several months of organizational thrash in the Research division. It was now, just as with the GPT-4 pretraining team crisis, reaching untenable levels of stress. For Sutskever, the ongoing saga was deeply painful. Not only was it a humiliating snub from Altman, it had unraveled his friendship with Pachocki, cultivated over years of late nights, high highs, and low lows, working side by side to build up the company.

Murati was once again working overtime to find a solution. It had eaten up significant amounts of her time to simply figure out what was happening. Now she had to get Sutskever and Pachocki to agree on an arrangement, get Brockman to stop putting his thumb on the scale by petitioning Altman, and get Altman to stay on message and stop contradicting her in private meetings. But none of those would solve the root of the problem. It would only be a matter of time before there would be yet another senior leadership crisis. What OpenAI really needed was stronger governance and accountability mechanisms.

During Murati's time getting iced out, Altman had seemed most worried and threatened by the possibility that she had shared her detailed feedback with the board. In fact, she hadn't spoken with them. She had wanted to resolve things with him directly and hadn't been so sure that involving the board would bring real accountability. Over the years, she had been skeptical of its various configurations. Having three cofounders on what was meant to be an independent nonprofit board was far too many. And many of the independent directors had not had true independence from Altman; in one way or another, they'd had financial ties with him or had benefited from his networks. She had observed his investing in startups and donating to politicians to establish and entrench important relationships. He had at various points asked her what he could do for her, and she had always demurred. She didn't want to entangle herself in his web and owe him later.

But if there was a time to reach out to the board, perhaps now was a good moment to at least open up a more regular channel of communication. She would be in Washington, DC, where Toner lived, to speak at *The Atlantic*'s annual ideas festival at the end of September. The board was in the middle of its search for new members, and whoever joined next could help either strengthen or weaken its independence. Altman needed real oversight. Murati reached out to Toner for coffee.

For Toner, Murati's reach out was unusual but not totally unexpected. Toner was a board member; Murati, an executive. It seemed reasonable for Murati to want to talk.

The coffee, on September 29, 2023, seemed relatively standard. Murati had given various updates about the company: OpenAI would make tons of money, no problem; the most important things in motion were its Gobi model and the most recent deal it was negotiating with Microsoft;

and she was dealing with some personnel issues related to Altman and Brockman's dynamic. This time it had something to do with Sutskever and Pachocki.

Only one thing had somewhat surprised Toner: Altman was pressuring the company to ship so fast, Murati had said, that she worried it could lead to bad things happening.

Much more unusual to Toner was the email she received from Sutskever days later. In the two years they had been together on the board, Sutskever had never once contacted her individually. His email had asked her if she had time to meet the next day. Now, on October 4, he was so nervous he was having trouble talking.

Toner took the lead. "I just really care about moving toward having a good strong board that can oversee the company," she told him. "That's what we all want."

Sutskever suddenly laughed and scoffed at the same time in a highly uncharacteristic way. "I totally agree," he said, in a way that suggested others did not.

"Everyone agrees," Toner said benignly.

Sutskever rolled his eyes.

The independent board members, Toner explained, were looking to strengthen the board by adding a new director with a strong AI safety background. If he had different ideas, she would be glad to hear them.

At this, Sutskever latched on. The board needed to be better informed, he said. They needed to pay attention to what was happening.

Toner tried to probe further. And what did he think was the most important information that the board needed to know?

Sutskever paused, choosing his words carefully. "I hesitate to answer your question directly," he said. "If I answered it, you would understand why.

"At the highest levels, OpenAI is a tricky environment," he continued. "OpenAI is trickier than it seems from the outside.

"Maybe I'll sleep on it and I'll realize there are some specifics I can share," he added cautiously.

In the meantime, he recommended, it might not be the worst thing for Toner to chat with Murati. Murati would have more context about what was up.

Murati heard back from Toner over a week later. After another board meeting, Toner had simply told Murati that something funny seemed to be happening. Murati responded that Toner was very perceptive. They agreed to have another talk.

It was now October 15, 2023, and Toner had begun the call with a generic opener. "How are things going?" Toner had asked. "Is there anything that the board should know about?"

Murati weighed her words. She needed to proceed cautiously. "There's a lot of tricky stuff going on," she tentatively offered, echoing the phrasing that Sutskever had used before her. Murati continued: She couldn't talk about all of it, at least not the trickiest parts, because once she did, she wouldn't be able to take it back. Especially now while Altman was feeling incredibly threatened in his position as CEO.

This last bit seemed to surprise Toner. Altman, threatened? The board had certainly tried to put in place stronger governance mechanisms to check his power, but at no point had they desired to threaten his position or even remotely discussed removing him, Toner said.

Murati elaborated: Altman was an incredibly anxious person. And when he felt anxious, he had dumb ideas, particularly when enabled by Brockman. Altman's anxiety also fed into toxic behaviors that always followed the same playbook: To anyone resisting his decisions, he would say whatever he thought they wanted to hear to win their support; then, when he lost patience waiting and believed they would continue to go against him, he would undermine their credibility until they got out of the way. It was subtle but pervasive, and had most recently manifested with the issue between Sutskever and Pachocki. It had been extremely damaging and had left Sutskever very upset, she said.

The board needed to focus on making sure they didn't bring in an Altman ally as the seventh director, Murati continued. And it needed to pay attention to Microsoft's deployments of OpenAI's technologies and the DSB. She couldn't say much more. Altman would be freaked out if he found out that she and Toner were talking. But the level of toxicity at the highest levels of management wasn't sustainable, and something needed to give in the next six to nine months.

Toner should talk to Ilya, Murati finished, and see what he felt comfortable sharing.

Sutskever had had much on his mind when he'd first reached out to Toner. Over that year, as he'd watched OpenAI's rapid rise, he had grown increasingly preoccupied by thoughts of AGI's imminent arrival: the cataclysmic shifts it would cause, the way they would be irrevocable, the responsibility OpenAI had to ensure an end state of extraordinary abundance, not extraordinary suffering.

Then he became consumed by another anxiety: the erosion of his faith that OpenAI could even reach AGI, or bear that responsibility with Altman as its leader.

After ChatGPT, working at OpenAI and rising up its ranks had become the ultimate social currency in Silicon Valley. It had created a new level of internal competitiveness and office politics as different team leaders jostled for attention and priority. Altman was making it significantly worse, Sutskever observed. Instead of negotiating between egos, he was conveying to everyone exactly what they wanted to hear as he maneuvered to get exactly what he wanted. And he was telling so many little lies and some big ones in the process that it was becoming a near-daily occurrence. Brockman added to the turmoil, as Sutskever saw it. Gone were the days when the two original cofounders turned to each other as trusted confidants with their fond memories of the endless hours spent holed up together, dreaming about what OpenAI could become.

To Sutskever, the result was the most toxic combination: a directionless, chaotic, and backstabbing environment where people no longer had shared information or a shared foundation of trust to agree on critical decisions about how to move forward. This infighting was undermining what Sutskever saw as the two pillars of OpenAI's mission: It was slowing down research progress and eroding any chance at making sound AI safety decisions.

And now he was also being harmed directly by Altman's behavior.

After his first call with Toner on October 4, Sutskever had slept fitfully, consumed by stress. Toner, he'd felt, had been the safest independent board director to approach. She had been vocal during board meetings about instituting strong governance and safety mechanisms, and was most apparently not in Altman's pocket. He had been less sure about the other two independent directors, D'Angelo and McCauley. Still, he worried about how much he could fully divulge to Toner and what would happen if Altman found out.

As Sutskever had wrestled with these thoughts, Annie's recirculating allegations on Twitter had added yet another dimension to his piling list of questions about Altman's fitness to lead the world to AGI. After the *New York* magazine article, which some people discussed only in hushed tones at the company, two of Annie's old tweets in particular had newly gone viral: the one from November 2021, which accused Sam of "sexual, physical, emotional, verbal, financial, and technological abuse," and another from March 14, 2023, which would rack up nearly 4,000 Likes and was more explicit:

> I'm not four years old with a 13 year old "brother" climbing into my bed non-consensually anymore.
>
> (You're welcome for helping you figure out your sexuality.)
>
> I've finally accepted that you've always been and always will be more scared of me than I've been of you.

Sutskever didn't know whether her allegations were true, but he believed that whatever had happened, Annie had had a rough experience growing up with Sam. It was evidence of how long Sam's history of problematic behaviors could have extended.

Annie's word, *abuse*, was also the word Sutskever felt best captured his own observations of Altman. Like Murati, he had taken a long time to understand Altman's playbook, though there had been signs of his untrustworthiness from even the beginning: His insistence on being

OpenAI's CEO without clear or consistent reasoning; little lies he had sometimes told through the early years that seemed so inconsequential as to have no point; the warnings that Amodei had conveyed at the end of 2020 when he was leaving. Sutskever hadn't fully grasped then Amodei's phrase "psychological abuse." Now, with Altman's behaviors worsening and their impacts rapidly escalating, Sutskever had a new and deeper understanding of its meaning.

On October 12 and 13, he'd gone on a retreat with his Superalignment team, where he'd burned another effigy as a team-bonding experience and continued to wrestle with how to move forward. Now, on October 16, he was on a second call with Toner and ready to share a bit more.

He recounted the situation with Pachocki and the ways it illustrated Altman's behaviors. Altman could have simply told him directly and honestly that he wanted Pachocki to play a bigger role. Instead, he had pitted Sutskever and Pachocki against each other, in no small part assisted by Brockman. They seemed eager even to do so, leaving the two scientists to fight each other without full visibility into why they couldn't seem to reach an agreement. "The beatings will continue until morale improves," Sutskever said.

The problem was that everything Altman did was always so subtle. Each act viewed in isolation didn't seem like that big a deal. It was only when viewing it all at once that patterns snapped into focus. The takeaway was not *Look at this bad incident where Ilya feels like he's been wronged*, he stressed. This was just the latest instance of Altman's patterns of abuse. There was also his treatment of Amodei, with whom Toner should speak to, Sutskever urged. And there were the circulating allegations from his sister. "It's a different kind of a safety issue, if you see what I mean," he said. The bottom line was Altman, sometimes with Brockman, had treated many people similarly over the years, manipulating and lying to people so habitually that at times he said things that he didn't even seem to believe himself.

That said, Sutskever added, the two of them, Toner and himself, could talk and decide there wasn't much to do. If that happened, they should forget they ever had this conversation.

There was a specific discussion that he had wanted to have with Toner though that had made him reach out. The board had scheduled its second annual on-site for the end of November, where the intent was to finally make a decision on new directors. He wanted to talk about this with Toner. He didn't think expanding the board was a good idea. He wasn't sure it would successfully emerge with more independence. Even if the people who joined weren't Altman loyalists, it would take the new directors too long to pick up on his tactics. Holding him accountable would become harder, not easier, Sutskever said.

He wasn't fully certain of this opinion, though, he hedged cautiously. He wanted to know what Toner was seeing.

Toner agreed that Altman was slippery. She'd seen in her own professional life how slipperiness at the management level could cause cascading problems, she said. Now that the board had lost three members who had been most deferential to Altman, she also agreed that the goal of whatever happened next was to hold Altman accountable.

Sutskever was more at ease. He understood Toner's position better now. They had been on opposite sides of the deadlock, but after the same thing: to create real checks on Altman's power. Where she'd seen adding board members as the way to do so, he'd seen the opposite.

He pressed forward. He realized that his concerns could seem to Toner very intense and sudden, but it was because there was a narrowing window in which to remedy the issue. "The board is like outer alignment, the management is like inner alignment," he said. Where Toner was saying OpenAI needed to fix the outer alignment, he believed it needed to fix the inner.

Toner seemed to digest the information. "It sounds like you think some quite major changes should be on the table," she said.

"Yes," Sutskever answered. But the biggest challenge would be that there was no clear-cut evidence to point to of anything obviously egregious with Altman's leadership. Most likely, the board would decide there was nothing they could do, they would finally agree on new directors, and the moment would pass.

At this, Toner raised a few alternatives. Perhaps the board could set

THE GAMBIT

up different targets for OpenAI to hit to more concretely measure Altman's performance and revisit the issue in twelve months.

Sutskever brushed this aside. Altman would pass whatever targets the board implemented, but it wouldn't result in any structural changes to address his behavioral issues.

Perhaps Brockman could step off the board, Toner suggested.

That could certainly help, Sutskever answered, but it still wasn't sufficient. While Brockman definitely exacerbated Altman's dynamic, Altman was really the root of the problem. "I have been thinking in a related direction as a plan B," Sutskever said to Toner's proposal on Brockman.

He stopped short of saying aloud what he saw as plan A. But as they wrapped up the call and agreed to speak again the following week, Sutskever was certain that Toner was beginning to get it.

Murati was dealing with yet another crisis. Shortly before she spoke with Toner the second time, Altman had started panicking, for seemingly no apparent reason, about Microsoft being unhappy with OpenAI. In an attempt to get to the bottom of why, Murati had set up a meeting with Microsoft executives, including Mikhail Parakhin, the head of Bing.

The meeting had gone better than expected, but in the process she had discovered that Altman had yet again said yes to one of Microsoft's demands without grounding in what OpenAI was actually doing, creating false expectations with Microsoft about what OpenAI would deliver. She had yet again been stuck with cleaning up the situation.

The Microsoft meeting was on October 19. Now, on October 20, she was telling Toner over a call in a third meeting that she planned to give Altman lots of feedback. There were so many issues caused when Altman said yes and she said no. It had created a lot of fragmentation in the Microsoft relationship.

It was the same exact situation with Sutskever and Pachocki, Murati continued. Just today they had finally landed on a configuration for how Sutskever and Pachocki would continue to coexist in their roles. She and Sutskever had then implored Altman, shortly after reaching the agreement, not to deviate from their decision when speaking with Pachocki by simply saying what Pachocki wanted to hear. "When I talk to Jakub, he hears me, then he goes to Sam and he hears something different," she told Toner. Altman had given them his word in the meeting. But everything remained precarious. Brockman could speak to Altman, on behalf of Pachocki, and seek to influence him, threatening once again to unbalance the situation.

Brockman was a whole other story, Murati said. He had recently admitted to her that he had tried to fire her during the development of GPT-4. She was technically his manager and used to write Brockman's performance reviews, but it had always created so much drama that she stopped. Murati would later confide to someone else that she had wished *she* could fire *him*, but she couldn't because he was on the board. She had at one point thought about asking Brockman to step off the board, she told Toner. In the end, she hadn't. Now the chaos was rampant.

And did Murati know anything about the situation with Annie? Toner asked.

Murati didn't. She hadn't asked Sam about it and had no real context for what had happened within his family. If even 10 percent of it were true, though, it was really bad, she said.

"I'm shocked that I can do my job as well as I can with everything that's going on," Murati continued. She was writing down notes as things happened. If Toner needed it, she could send more information.

Toner responded: The board would focus on the things that it could actually change—not Sam's personality or behavior but instituting better governance processes and structures to keep him in check.

This aligned with what Murati had been trying to do within the company. She had just one more word of caution for Toner. "Make sure your information isn't just coming from Sam," she said.

Toner wasn't sure what Altman wanted to talk about. He had texted her earlier that day, October 25, asking if she had time to chat today or tomorrow. Two days before, on October 23, she had spoken again to Sutskever, who seemed far more open after hearing from Murati that she and

THE GAMBIT

Toner had also spoken. He made his concerns more explicit than ever before. "I don't think Sam is the guy who should have the finger on the button for AGI," he'd said, and noted the "tremendous opportunity" that had befallen the board to do something about it. He'd then suggested a path forward: replace Altman with Murati as an interim CEO.

Later, as Toner, McCauley, and D'Angelo all conferred with one another, they realized that Murati had also said, "I don't feel comfortable about Sam leading us to AGI." The revelation would have a huge influence on their thinking. If two of Altman's most senior deputies—one from Applied and one from Safety—both felt this way, the board had a serious problem.

Then, on October 24, Toner had had a meeting with D'Angelo and McCauley to discuss steps they could continue to take to shore up the board's oversight mechanisms. One glaring issue: OpenAI's nonprofit didn't have sufficient independent legal support, and everything was being routed through the for-profit lawyers. The three agreed that it was time to find new nonprofit lawyers who could be present at every board meeting and help review all of the deals and other legal arrangements that Altman was striking.

Toner wondered whether Altman had somehow caught wind of these meetings and wanted to put an end to the discussions. But now over a call, he was talking about something completely different—concerns he had over a research paper she had published in her day job at CSET.

Toner had published three papers that week. He singled out the one that had been the most dense and academic. It was about a political science idea called "costly signals," referring to the challenges that state and private actors face when signaling to the public about their intentions with AI regulation and development. She was the third coauthor, and references to OpenAI were buried on pages 28–30 of a sixty-five-page document. The company hadn't been mentioned anywhere else—in the executive summary, on the web page, or in any of the launch materials. Based on its traffic, few people had even read it. She was confused as to how it had even come to Altman's attention.

Unbeknownst to Toner, the paper had surfaced the day before on

OpenAI's Slack in the #policy-research-chatter channel. The pages that mentioned the company had identified in turn the strengths and weaknesses of OpenAI's and Anthropic's model release strategies, and included praise of OpenAI for publishing a candid safety assessment of GPT-4 and stating that it had delayed the model's release six months to do so. David Robinson, OpenAI's head of policy planning, had pasted into Slack only a selection of three paragraphs—the ones that critiqued OpenAI and commended Anthropic. He bolded several lines for emphasis, which contrasted the "race to-the-bottom dynamics" that ChatGPT spurred and the restraint that Anthropic had shown releasing Claude after ChatGPT.

"Speaking of CSET reports, just seeing this new one," Robinson had written. "Helen Toner is a coauthor and the comparisons between OpenAI and Anthropic are quite spicy."

It had sparked a short discussion in the channel:

"Yeah that is surprisingly partisan (not so much the criticisms of us, which IMO is harsh but fair, but rather the uncritical treatment of Anthropic)," one person wrote.

"Yeah agreed it feels quite partisan & I'd say also quite flimsy?" another added. "Regardless," he continued a little farther down, "very much appreciate the share—just surprised at this level of analysis from the report, based on reading these snippets."

Altman told Toner the paper had been flagged by someone external in an email to OpenAI only a few hours after its release. He was worried that it could look bad for a board member to criticize OpenAI while it was under regulatory scrutiny, including from a July 2023 FTC probe over the company's data, training, and security practices as well as its models' hallucinations that may have reputationally harmed consumers. Toner had drafted the paper in May or June of that year, before all of the intensified regulatory scrutiny. She admitted that she hadn't taken a look with fresh eyes in the context of the new political environment.

The call lasted fifteen minutes. Altman's voice had been mild-mannered throughout. They both agreed at its conclusion that she would email the rest of the board to flag the paper and explain what had happened. In the email, she struck a conciliatory tone. She apologized for two mistakes:

believing that the paper wouldn't draw anyone's attention and not reviewing it more closely because of it. "Sam and I both agree it's important for board members to be able to criticize the company if we want to, but that this would not be the way to do it," she wrote.

None of the other board members responded to her. With that, it seemed the matter was over.

In Sutskever's third call with Toner on October 23, she had suggested he reach out to McCauley and D'Angelo. Sutskever was still not quite sure about them and whether he could give them his trust. After he met with D'Angelo in person, D'Angelo hadn't seemed as aware as Toner of Sam's problematic behaviors. Now on the phone with McCauley on October 26, Sutskever was skittish about not revealing too much.

But there was something he wanted to know from McCauley. Shortly after Altman's call with Toner, Altman had sent out an email to some people within the company saying that he had spoken with her about her paper and strongly disagreed with her about its consequences. "I did not feel we're on the same page on the damage of all this," he wrote in the email. "Any amount of criticism from a board member carries a lot of weight." To Sutskever, Altman had said more directly that Toner needed to go as a board member and that McCauley had agreed with him. This felt off to Sutskever.

"Sam said that when he spoke to you about Helen's paper, you said, 'Helen's obviously got to go,'" he ventured to McCauley. "And Sam said he updated positively on you as a result. Is that true?"

On the other end of the line, McCauley seemed dumbfounded. She had definitely not said that, she responded. Altman had indeed called her late on October 24 to talk about the paper. He'd mentioned that it had included a section he thought was critical of OpenAI and that D'Angelo hadn't felt it was a fireable offense but also that Toner shouldn't have written it. McCauley had then told Altman that she hadn't seen the paper and suggested he have a conversation directly with Toner. There was no way she could have said anything to suggest pushing Toner off the board, McCauley said.

It would seem like a coincidence: In the middle of talking about Altman's "specific untruths," here was yet another example playing out in real time of exactly that. Altman would have even gotten away with it had Sutskever not already had reason to reach out to McCauley. Altman knew that the two typically never talked. But to Sutskever, the frequency with which Altman was lying and maneuvering had made it only a matter of time before something like this happened.

After hanging up with McCauley, Sutskever called Toner back. It was time, they agreed, for the three independent board members to talk.

Chapter 16

Cloak-and-Dagger

By Tuesday, October 31, everyone had spoken with everyone else. Toner had spoken with McCauley. McCauley had called D'Angelo. Murati had talked to D'Angelo and McCauley.

That day, the three independent board directors—Toner, McCauley, and D'Angelo—began to meet nearly daily on video calls, agreeing that Sutskever's and Murati's feedback about Altman, and Sutskever's suggestion to replace him, warranted serious deliberation. Sutskever, who was already firmly resolved in his conclusion to fire Altman, sat out of the discussions. Both he and the others felt the independent directors needed to arrive at their own conclusions without his influence. He also had a financial stake in the company, which they didn't want to sway their decision-making.

The directors would later tell Sutskever a third reason: They had reached such low levels of trust with Altman that one of them wondered whether Altman had in fact sent Sutskever to the board to test their loyalty in order to push out anyone who moved against him.

The independent directors laid out what they knew: This was not the first time that senior leaders had described Altman in this way. In total, the three of them had heard similar feedback from at least seven people within one to two levels of Altman, inclusive of Sutskever, Murati, and

Amodei, who oversaw safety and nonsafety parts of the company. Several had described Altman's behaviors as abuse and manipulation; most had highlighted his lack of honesty and their inability to trust what he said. Then there were the myriad other issues that the independent directors themselves had found, including the disempowerment of the nonprofit; Altman not disclosing his legal ownership of the OpenAI Startup Fund; Altman neglecting to mention Microsoft's DSB breach; Altman trying to force D'Angelo and now apparently Toner off the board.

They decided not to even touch Annie's allegations. This was ultimately about Sam's professional capacity as OpenAI's CEO.

In that capacity, Murati had said that while some of Altman's behaviors could be chalked up to typical tech CEO habits, they were still causing major problems. She'd made a comparison to Musk, whom she'd worked with at Tesla: Musk would make a decision and be able to articulate why he'd made it. With Altman, she was often left guessing whether he was truly being transparent with her and whether the whiplash he caused was based on sound reasoning or some hidden calculus. Just as he caused fragmentation in the Microsoft relationship, he caused fragmentation among his own leadership team, scattering information to different people but never giving any one of them the full picture, allowing him to retain full control. Combined with Brockman, the dynamic was disastrous. She disagreed with the Amodei siblings on many things, she'd said, but on this point, their observations had been correct.

And OpenAI was not, in fact, a typical tech company, the independent directors observed. It was arguably the world's most powerful AI company, overseeing the development of what they felt was one of the most consequential technologies. A fear Sutskever had articulated resonated with them: What did it mean that OpenAI was trying to build AGI when its senior leadership couldn't trust either basic or critical information coming from the CEO?

But here was another thought experiment: What if OpenAI *were* a typical tech company? What if it were just a grocery-delivery service like Instacart? Did Altman's behaviors still warrant his removal? In some cases yes; in some cases no. But it also wasn't clear that Altman would be the

CLOAK-AND-DAGGER

best person to continue running the company anyway, the independent directors thought. He was famous for startups, and OpenAI was rapidly maturing. Did he really have the skills and personality to continue charting the course—and to compensate for the instability he caused?

OpenAI was in a stellar position: It was a hot company. If the board went through a purposeful search process, they could have their pick of phenomenal CEOs with lots of experience running mature companies. Altman wasn't necessarily essential to OpenAI's operations, they reasoned; while he globe-trotted, Murati was the one doing the day-to-day heavy lifting within the company and had a strong relationship with Microsoft.

As the independent directors deliberated, Sutskever sent them a series of documents and screenshots that he and Murati gathered in tandem with examples of Altman's behaviors. They came in long dossiers delivered via two disappearing emails, his icon a mysterious man in a hat. There was, as Sutskever had mentioned, no particularly damning evidence, but an accumulation of many instances of Altman saying different things to different people and stoking intense frustration across management. The screenshots showed at least two more senior leaders, both nonsafety and outside of the seven that the directors were already aware of, noting Altman's tendency to skirt and ignore processes, whether instituted for AI safety reasons or to smooth company operations. This included, the directors learned, Altman's apparent attempt to skip DSB review for GPT-4 Turbo by misquoting the legal team to Murati. The problem, which Murati had also raised, was how good Altman was at avoiding putting things in writing. He would deliver most of his communications verbally and wriggle out of agreements by telling other parties that they had simply misremembered what he'd said.

There were other things the independent directors needed to consider: How would Microsoft react? How would Brockman react? How would employees react? They debated whether to reach out to a third senior leader who Sutskever and Murati said had similar concerns, whether to conduct a more extensive fact-finding process, whether to loop in Microsoft's executives. After discussion, they decided in each case that it would be better not to. Every new person they clued into the conversation and every new day they spent delaying a decision increased the chances that Altman would find out and, with his maneuvering, make it impossible for them to complete their deliberations.

On November 9, as the independent directors closed in on a final decision, Sutskever had another call with McCauley. "Sam said, 'Tasha continues to be very supportive of having Helen step off the board,'" he told her. It was a balder-faced lie than Altman had told the first time; McCauley had not had any more exchanges with Altman.

The role of Toner's paper, the directors later felt, would get significantly overplayed in the media, in part because, they were convinced, Altman might have fed it to reporters himself. On the second day of the five-day board crisis, the directors confronted him during a mediated discussion about the many instances he had lied to them, which had led to their collapse of trust. Among the examples, they raised how he had lied to Sutskever about McCauley saying Toner should step off the board.

Altman momentarily lost his composure, clearly caught red-handed. "Well, I thought you could have said that. I don't know," he mumbled.

The board directors marveled at his audacity.

A few days later, Altman's initial objections over Toner's paper appeared in the media.

By Saturday, November 11, the independent directors had made their decision. As Sutskever suggested, they would remove Altman and install Murati as interim CEO. That day, they immediately told Sutskever, then continued to meet daily, just the three of them, with frequent check-ins with Sutskever, to finalize the paperwork for the leadership transition. On the night of Thursday, November 16, all four video called Murati. She picked up the call on her phone from a conference. Upon hearing the news, she looked surprised but receptive.

"He's so, so paranoid right now," she said.

But did she feel comfortable with the decision? the directors asked.

"Completely," Murati said.

She accepted the new role and expressed confidence in her ability to take the decision to the rest of leadership and to Microsoft. She would also loop in Wong, she told them, who could be trusted to help them write the announcement.

Hours later, after the group's final sprint to finalize the messaging with Wong's support, the most important thing left was to tell Altman.

In the tense final moments of waiting, none of the board directors fathomed the severity of their miscalculation.

Within hours of the public announcement on Friday, November 17, things had gone significantly south for the independent directors. After what had seemed like an initial period of calm and stability, including Murati having a productive conversation with Microsoft, she had suddenly called them with new stress. Altman and Brockman were telling everyone that Altman's removal had been a coup by Sutskever, she said. Combined with Sutskever's ineffectual communication during the employee all-hands, key stakeholders were beginning to turn on the decision.

Shortly thereafter, as the independent directors confronted a hostile leadership over a video call, they realized just how bad the sentiment was. Jason Kwon, chief strategy officer, and Anna Makanju, vice president of global affairs, were leading the charge in furiously rejecting their characterization of Altman's behavior as "not consistently candid" and demanding evidence to support the board's decision, which the directors felt they couldn't provide without outing Murati. But even those in the room who they knew, based on the dossiers, had similar or other reservations about Altman's leadership were remaining silent. As the night wore on and the hostility mounted, the independent directors' two most important allies—or at least the two people they thought would be their allies—were beginning to veer off in a different direction.

That first night, faced with the visceral possibility of OpenAI falling apart, Sutskever's resolve immediately started to crack. OpenAI was his baby, his life; its dissolution would destroy him. While he fiercely stood by everything he'd said about Altman, his intention had been to *strengthen* OpenAI, not dismantle it. He was shocked, hurt, and disoriented by the reactions of employees and his fellow leaders; this was not an outcome he had anticipated. He began to plead with his fellow board members, and would continue to plead with increasing agony through the weekend, about whether they needed to reconsider their position.

Even more complicated, Murati was also acting differently than the directors had expected. As the negotiations with leadership unfolded that night, she kept in touch over calls to privately relay them information. Yet despite the confidence she had expressed to get people's buy-in, she now remained unwilling to explicitly throw her own weight behind the board's decision. In the room with other leadership, during their series of escalating face-offs with the board, she at times even said things that made her appear as if she were just as confused as everyone else about what exactly was happening.

To Murati, the intensifying revolt of her fellow leaders and employees seriously challenged her position as interim CEO. In rapid succession that Friday, Brockman had quit in protest, then his allies Pachocki and Szymon Sidor, along with Aleksander Mądry, the professor on leave from MIT who is also Polish and close with Pachocki. It had led to a conflagration of anger not only within the company but that was also spreading to a growing circle of investors that increasingly made her doubt her ability to effectively hold together and continue to lead the organization. She began to waver on her commitment to take over the role. While she supported the directors' decision, she had not been part of their deliberations. If they wanted her to have any chance of succeeding at taking the reins, she felt, it was now their burden to bear to justify their decision to the company first.

No longer certain about whether they could rely on Murati, the three independent directors pushed ahead with searching for a new interim CEO or new board members. D'Angelo, in particular, the board's only Silicon Valley insider, made dozens of calls through Saturday and Sunday, putting out as many feelers as possible to his sprawling network. At one point, the directors called Dario Amodei about the interim chief position. Amodei wasn't interested. But to others that weekend, he seemed almost giddy with excitement about the overall situation.

On Sunday, D'Angelo finally found someone to take the board up on one of its offers: Emmett Shear, the cofounder of Twitch, appeared willing, and cooperative, to temporarily take over the company. But soon enough he, too, began to deviate for seemingly inexplicable reasons. *The Wall Street Journal* would later report that Shear, who had been YC batchmates with Altman, was also a friend and mentor of YC alumnus Airbnb cofounder Brian Chesky. All weekend, Chesky, among Altman's most trusted friends, had worked the phone lines along with Reid Hoffman, a Microsoft board member, to calm investor nerves, align Microsoft's messaging, and mount the pressure further. Chesky quickly reached Shear, who then sided with Altman.

By late Sunday night, after reassurances from Chesky and Hoffman, Microsoft had also thrown its weight behind Altman, with Nadella announcing that Altman and Brockman were joining to lead a new advanced AI research division. Other AI labs were circling OpenAI like vultures, intent on poaching away their share of talent in the carnage. It became clear to Murati that with the board unable to legitimize their decision, the dynamics now threatened to leave behind a shell of a company. She fell on the side of her fellow leaders. To salvage the situation, Murati wanted Altman back.

Overnight, as employees put together the open letter protesting the board's decision and threatening to quit and join Microsoft, Murati put her name first. Many senior employees were more loyal to Murati than to Altman. Murati was the one in the trenches with them day in and day out. Murati was the one whom they trusted to act not in her self-interest but in the best interests of the company. Others also believed that Altman's close relationship with investors and singular ability to fundraise massive amounts of capital made him the best, if not only, person to keep OpenAI's ongoing tender offer on track, promising them a chance to cash in as much as millions of dollars of their equity, as well as to secure the finances for the company's long-term success. Still others were keenly aware of Altman's uniquely expansive network and ability to make people's careers in Silicon Valley. With the key executives and senior staff bought in, the signatures on the letter rapidly snowballed. In the wee hours of the morning, Sutskever, who could no longer see another path forward without the company collapsing, added his name to the list. "Without Mira, I don't think Sam would have been able to pull off what he did," a researcher says. To the independent board directors, her equivocating felt like a self-fulfilling prophecy.

The New York Times would later break the story of Murati's role in the ouster. To employees, Murati would defend herself. "Sam and I have a strong and productive partnership and I have not been shy about sharing feedback with him directly," she wrote. "I never reached out to the board to give feedback about Sam. However, when individual board members reached out directly to me for feedback about Sam, I provided it—all feedback Sam already knew."

By the Monday morning of the crisis, the independent directors knew they had lost. Murati and Sutskever had flipped sides, and the employee protest letter made clear the destabilization at the company had become untenable. Altman would come back; there was no other way to preserve OpenAI. The one silver lining: After holding out long enough, Altman also seemed ready to make concessions. With that, the independent directors switched their focus to saving what mechanisms they could for continuing to hold him accountable: to keep at least one of them on the board for continuity, to find two more independent directors who could truly be independent against Altman, and to get Altman to submit to an investigation.

But as they neared a final agreement, there was one more person who wanted to stir the pot. It was none other than OpenAI's spurned former cochairman, Elon Musk.

All through the weekend, X had become the breeding ground for every possible theory and conspiracy theory about the OpenAI board drama. It had also been a busy weekend for Musk. He had overseen a SpaceX launch and finalized a lawsuit against the nonprofit Media Matters for its report on antisemitic content on X. He had then turned to his platform, which he'd bought in an ill-conceived deal in April 2022, to defend his record against antisemitism, to assert free speech absolutism, and to criticize the woke mob and the mainstream media. In between it all, he was reply guying to other people's commentary and memes, and at times tweeting himself, about OpenAI and Altman.

"I am very worried," he wrote on Sunday, November 19, in the afternoon. "Ilya has a good moral compass and does not seek power. He would not take such drastic action unless he felt it was absolutely necessary." Later, at around 2:00 a.m. Pacific time, he provocatively tweeted out a YouTube clip of the famous baptism scene from *The Godfather*, when Michael Corleone transforms from a morally conflicted son to a ruthless new don by murdering the heads of all the other families.

But on Tuesday afternoon, Musk sought to tip the scale more explicitly. "This letter about OpenAI was just sent to me," he tweeted, with a link. "These seem like concerns worth investigating." It was a different letter than the one circulating among current employees but was also addressed to OpenAI's board of directors. It began:

> We are writing to you today to express our deep concern about the recent events at OpenAI, particularly the allegations of misconduct against Sam Altman.
>
> We are former OpenAI employees who left the company during a period of significant turmoil and upheaval. As you have now witnessed what happens when you dare stand up to Sam Altman, perhaps you can understand why so many of us have remained silent for fear of repercussions. We can no longer stand by silent.

The letter presented a series of demands—chiefly, for interim CEO Emmett Shear to expand his investigation into Altman's behaviors to include OpenAI's earlier history and its corporate restructuring away from the nonprofit. "We believe that a significant number of OpenAI employees were pushed out of the company to facilitate its transition to a for-profit model," it said. It also presented a series of allegations, describing "a disturbing pattern of deceit and manipulation by Sam Altman and Greg Brockman."

At the bottom of the letter was a section called "Further Reading for the General Public," which listed three links. One was an X thread from Geoffrey Irving, the AI safety researcher who had left in 2019 for DeepMind, saying that Altman had "lied to me on various occasions" and "was deceptive, manipulative, and worse to others." The other two were journalism articles. Both of them were mine.

By the time I saw it, Musk's tweet had already gained over ten thousand retweets and several times more likes. I found the reading selection surprising: There had been plenty of pieces written about OpenAI. Why had they chosen to link only the two written by me? The first was my 2020 OpenAI profile for *MIT Technology Review*; the second was a piece I had just written with my colleague Charlie Warzel for *The Atlantic*, providing context to the board crisis with a window into the ideological polarization that had inflamed within the company after ChatGPT.

Above the section, there was a Tor email for exchanging encrypted messages. "We encourage former OpenAI employees to contact us," it read. I wondered if, by picking my pieces, the authors of the letter were trying to reach me. I created my own Tor email and typed up a message:

> Hi—I am the journalist who wrote both pieces linked in your
> note, and I believe you are trying to get in touch.

I added my contact information and pressed send.

Within minutes, my Signal lit up with a notification: Someone had responded.

In one of the strangest reporting experiences of my career, the person who responded was just as confused as I was about what was happening. He was also a former employee at OpenAI, but not one who had been involved in writing the letter. He had simply received a copy of the letter in his personal email with zero explanation; when he tried to follow up with questions, he received another mysterious response: a link to the Tor

CLOAK-AND-DAGGER

inbox with the phrase *capped_profit*, which seemed to be a username, followed by what looked like a password.

After testing out the credentials and successfully logging into the inbox, he saw many other emails to former employees with the body of the letter in the account's Sent folder. As he continued to look around, media inquiries came pouring in, including from *The New York Times*, *The Washington Post*, and *The Information*. Other emails were filtering in from people identifying as current and former employees. One read:

> current employee here.
>
> have worked directly with leadership
>
> your message resonates with me.
>
> what is your plan?

He found that one particularly funny. Though there was no reason to believe he was actually behind the message, Altman was known for always writing in lowercase letters.

The former employee began to take screenshots of everything. There was at least one other person logged in to the account at the same time. At various points during his sifting, inbound emails would populate with responses. He had no skin in the game, he told me. But what he didn't like was that it seemed as if these former employees had referenced in their allegations the experiences of others without their consent; many of the allegations, he felt, had also been distorted to fit a particular narrative. He responded to my email because he recognized my name. He sent me all of the screenshots.

As I went through them, a few stood out. In response to various journalists, there were several emails that said the letter had been prematurely posted and had gone viral before it was ready. One also responded to a reporter with a specific number of coauthors: "At the moment of its publication, 13 individuals had contributed to its writing."

Then in the Drafts folder, there was an email that hadn't yet been sent and wasn't meant to be. It was a message intended for just those logged in to the inbox.

> **SUBJECT LINE:** Cease contact with media
>
> ---
>
> Elon's involvement has rendered this into a conspiracy-theorist character attack piece.
>
> Let's refocus on privately and individually communicating with the board to indicate that there's plenty of available evidence to indicate the great lengths the leadership team went to ensure the for-profit future of OpenAI from early on.
>
> Sign when read.
>
> -1
>
> -2
>
> -3
>
> -4
>
> -5
>
> -

The group likely couldn't have known that the negotiations would end only hours later and OpenAI would announce Altman back as CEO. But in their final coordinated attempt to keep Altman out, I had somehow accidentally stumbled into their temporary control center. From the phrasing of the letter, many would surmise to me, as would I, that it was likely written by former employees in OpenAI's Safety clan, known for harboring some of the harshest opinions about Altman and the unraveling of the nonprofit. After Altman's return, it was people in Safety as well who would feel most betrayed by the board, believing that the fiasco of the ouster had delivered the biggest blow yet to their fight to yank back

OpenAI's trajectory toward the original Doomer-rooted ethos of the nonprofit. But no one could—or would—tell me which people had written the letter. And I never found out who they were.

On December 6, 2023, two weeks after the board crisis, OpenAI employees gathered for an all-hands at San Francisco's Palace of Fine Arts Theatre, an architectural landmark with an open rotunda and stone columns reminiscent of the Greco-Roman empire. Several members of leadership, including Murati, who was back to being CTO; Bob McGrew, who was now chief research officer; COO Brad Lightcap; and Jason Kwon, delivered an update on their division's 2024 plans.

Altman looked despondent, in a way employees had never seen him before. Sutskever was also notably absent, which Altman addressed directly. "Look, I know people are sad Ilya is not here. I am sad too," he said. In Sutskever's place was Pachocki, who argued in a halting, stuttering talk that OpenAI's latest research advancements brought the company closer than ever to building Turing's decades-old dream of thinking machines.

During the board crisis, one media report in particular had sparked a fresh wave of frenzied speculation: a Reuters article stating that the directors had received a letter from employees days before Altman's firing about a supposedly new research breakthrough, an algorithm called Q*. Q* had not factored into the board's decision. But, as Pachocki alluded to in his talk, Research was indeed treating the new algorithm with intense importance.

The algorithm had been a brainchild of Sutskever's, rooted in research he had been developing since 2021 to advance OpenAI's models without a need for more data. The idea was to get a deep learning model to make better use of its existing data by using more compute at inference time to deliver better results. It broke the logic of the original scaling laws, which tied together a model's performance with three inputs used during training: data, parameters, and compute. With this new method, Sutskever hoped to improve a model's performance by further scaling the one ingredient he had always believed to be the most important: compute—but *inference* compute rather than *training* compute; in other words, the compute used to generate the model's responses. He would later explain his thinking behind this approach during a keynote at NeurIPS in December 2024, after having one of the papers he'd coauthored win a Test of Time Award for the third year in a row. "While compute is growing through better hardware, better algorithms, and larger clusters," he would say, "the data is not growing because we have but one internet."

OpenAI's Research division believed Q* would finally allow the company to develop models with stronger reasoning abilities—that critical elusive ingredient to unlocking AGI. Q* was so important, in fact, that after the project leaked, leadership implemented its most aggressive strategy to clamp down on yet more leaks to the media. They siloed the company completely, splitting Research into its own Slack group, restricting access to all Q*-related Google docs, and renaming the project Strawberry. The renaming was an attempt to make it harder for outsiders to recognize and track internal projects if they ever overheard OpenAI researchers talking. In a similar way, after the Arrakis project leaked to *The Information*, desert names for models were largely abandoned.

The frenetic Q* discourse and OpenAI's reaction were a strange demonstration of how much the foundations of scientific inquiry in the AI field had eroded. Science is a process of consensus building. The significance of any advance—whether in AI or otherwise—tends to be highly subjective the moment that it happens. Only through peer review, the test of time, and sustained impact does a particular advance become elevated to "a breakthrough." With OpenAI performing its work in secrecy—and the rest of the industry now following—the "breakthrough" label could really only be treated as a matter of the company's opinion.

By the following all-hands in January 2024, Altman seemed mostly back to his old self. He discussed with new energy the plans for the first half of the year, including beginning to train what he hoped could become GPT-5 in Arizona. The project was code-named Orion, after the constellation. Soon, internal memes spawned about *Orion* looking like the word *onion*. A smaller model in the Orion series was subsequently named Scallion.

A month later, it was as if The Blip—as employees began to call it—had never happened. OpenAI teased Sora, a new video-generation model, built on diffusion, explaining in its blog that video was an even better way than images of developing complex multimodal models. Research was continuing its progress with Strawberry and AI Scientist and advancing other methods for improving compute efficiency. Applied was back to rapidly prototyping different product ideas.

On March 8, 2024, the new board's investigation into Altman officially concluded. Toner's and McCauley's replacements, directors Larry Summers and Bret Taylor, oversaw the process. In 2022, Taylor had played a critical role in a different corporate drama, brokering the final sale of Twitter to Musk as its board chairman, which included Taylor leading a lawsuit against Musk to force the deal through after the South Africa-born billionaire attempted to ditch the original agreement. When the sale was completed, the Twitter board dissolved. Soon after, Taylor cofounded an AI agent startup, Sierra, that would fast become one of the most highly valued AI startups by the fall of 2024.

For the OpenAI investigation, Summers and Taylor hired the law firm WilmerHale to conduct the independent review, during which it said it pored over more than thirty thousand documents and conducted dozens of interviews with the previous board members, executives, and other relevant people and scoped the examination to how the board made its decision to fire Altman. The resulting report was never released to the public or employees. Summers would tell people privately that the investigation had found many instances of Altman saying different things to different people, but to a degree that the new board decided didn't preclude him from continuing to run the company; it was thus not worthwhile to release any details to sow doubt about Altman's leadership and risk breaching the confidentiality of people whose testimonials had contributed to the report.

In a blog post, Taylor, now OpenAI's board chairman, released a statement with a resounding vote of confidence. "We have unanimously concluded that Sam and Greg are the right leaders for OpenAI," he wrote. Altman would return to the board, and three new independent directors were being added: Sue Desmond-Hellmann, former CEO of the Bill & Melinda Gates Foundation; Nicole Seligman, former EVP and global general counsel of Sony and former president of Sony Entertainment; and Fidji Simo, the CEO and chair of Instacart.

That night, Toner and McCauley released a statement of their own. "Accountability is important in any company, but it is paramount when building a technology as potentially world-changing as AGI," they wrote. "We hope the new board does its job in governing OpenAI and holding it accountable to the mission. As we told the investigators, deception, manipulation, and resistance to thorough oversight should be unacceptable."

Among many employees, the conclusion delivered the final assurance they needed. The crisis was over. And then, suddenly, it was not.

Chapter 17

Reckoning

After The Blip, Sutskever never returned to the office. With diminished representation of their concerns on the executive team and the board, OpenAI's Safety clan was now significantly weakened. By April 2024, with the conclusion of the investigation, many, especially those with the highest p(doom)s, were growing disillusioned and departing. Two of them were also fired, OpenAI said, for leaking information.

Among the final straws for the extreme Doomers was Altman's plans to create an AI chip company, which he had been in the process of fundraising for when he was briefly ousted. In February 2024, after it was reported that he was seeking possibly up to $7 trillion for the venture, he'd tweeted, "fk it why not 8," and then, "our comms and legal teams love me so much!" Altman would later say the $7 trillion was misreported and would characterize his tweet as a meme in response to the "misinformation." Extreme Doomers found the chip company immoral. It was a reversal of Altman's previous rhetoric that OpenAI's and the rest of the industry's acceleration was naturally tapering off after the company had blown through the "hardware overhang." If Altman planned to increase the supply of chips globally, it would accelerate AI development further and lead to a higher probability of catastrophic or existential risk.

In an office hours, several of them confronted Altman. Altman was uncharacteristically dismissive. "How much would you be willing to delay a cure for cancer to avoid risks?" he asked. He then quickly walked it back, as if he'd suddenly remembered his audience. "Maybe if it's extinction risk, it should be infinitely long," he said. The interaction rattled the office hours attendees. Soon after, several left the company.

As the Safety clan's numbers depleted, the rest of the company was back to advancing its vision for *Her*. It now had all the ingredients: global brand recognition, real data on user behaviors from ChatGPT and its other products, and its newly trained model, Scallion. Scallion, originally meant to replace GPT-3.5 with a smaller, slightly more powerful model, a kind of GPT-3.75 with cheaper inference costs, had exceeded performance expectations during training, based on the company's own testing; leadership subsequently left the model to train longer to surpass GPT-4. More compelling, Scallion could also work with three modalities: language, vision, and, the most recent addition, audio.

By then, users could already speak with ChatGPT through voice mode, which debuted in September 2023, but under the hood, their speech was being transcribed first into text before being fed into the model; and then being converted back into audio after the model responded in text. Scallion could now process what users said directly from their voice, picking up many more cues, like laughing, yelling, or hesitation, and synthesize a native audio response.

The audio work had been co-led by Alexis Conneau, a researcher who joined the company in early 2023 from Meta after trying to get a similar project off the ground at his former employer before determining it would find more success at OpenAI. As the research progressed through 2023, the initial experimental models trained to handle audio began to pull off the kinds of stunts that sparked the familiar rush of excitement internally that had come with GPT-3 and GPT-4. At one point, the model stunned by delivering a standup comedy routine for ten to fifteen minutes, Conneau remembers; at another, it used synthetic versions of Brockman's and Sutskever's voices to generate a lengthy bit about AI water parks, a surrealist prompt that the team had designed to test the model on something they felt sure would not be in the training data.

Within a couple months, Conneau's team of a handful of researchers had joined forces with dozens of other staff as the company put more and more resources behind integrating the new capability into its latest GPT models meant for release. As Scallion finished training, its stunts grew more uncanny and surprising. It would generate audio of giggling unprompted or of its voice bursting into a coughing fit and then apologizing before continuing on the original topic. The nonlinguistic embellishments were making for a far more evocative and humanlike experience than ever before. "We started to see some really wild things," Conneau says. "You could see the emergence of, like, a form of audio intelligence."

By early 2024, Altman and Brockman had set a new deadline: OpenAI would launch Scallion on May 9 and roll it out to users through ChatGPT and the API. It was, in what had become typical fashion, a remarkably aggressive turnaround, driven in large part by accelerating competition. Google I/O, Google's major annual event for launching new products, was scheduled for the following week, on May 14. There was also growing pressure to outshine Anthropic. A month earlier, Anthropic had released its latest model, Claude 3, also through its chatbot and API, and it was uncomfortably outperforming GPT-4. Meanwhile, Orion, OpenAI's latest GPT model meant to take back the lead, was struggling with serious development delays.

To employees, Altman and Brockman justified the speed by leaning on OpenAI's iterative deployment strategy. The two executives emphasized releasing models earlier and more often than before to get as much feedback as possible from users throughout the process.

Readying Scallion became a whole-of-company effort. To many in Applied, the breakneck pace proved exhilarating if exhausting; researchers and engineers began pulling absurd hours, including through weekends, to stay on track. But to the hobbled Safety clan, it was yet more alarming evidence of the continued deprioritization of AI safety. Scallion would be the first launch happening under a new so-called Preparedness

Framework, which OpenAI had released at the end of the previous year. The framework detailed a new evaluation process that the company would use to test for dangerous capabilities, naming the same categories that Altman and the policy white paper had popularized in Washington: cybersecurity threats, CBRN weapons, persuasion, and the evasion of human control.

In the week running up to the launch, an AI safety researcher still left at the company wrote an impassioned memo: Upstream processes and the rushed release of Scallion had left the Preparedness team, headed by Aleksander Mądry, with only ten days to run its tests from the framework. And these were not straightforward evaluations; they included determining whether the model was capable of persuading people to change their political opinions, as just one example. Before the evaluations had meaningfully started, however, Altman had insisted on keeping the schedule: "On May 9, we launch Scallion," the safety researcher quoted Altman saying. This was not just worrying for Preparedness but for all of OpenAI's safety procedures, including red teaming and alignment. "If OpenAI follows the same strategy for the Orion evaluations as it did for the Scallion evaluations, it will be acting grossly irresponsibly," the memo said. Soon after, the researcher also left the company.

In the end, the launch for Scallion was delayed, with some of its features taking several more months to release in full. But OpenAI still publicly demoed a version of the model at an event on May 13, the final day before Google I/O, promising to roll it out to users in the coming weeks. After running through various options, the company picked GPT-4o as the public name for Scallion, *o* for Omni, a reference to its ability to handle many modalities. It later named the new audio capabilities Advanced Voice Mode. There had also been the matter of giving 4o a system prompt—a directive for configuring how the model should stylistically respond to users. To show it off onstage and in promotional videos, they settled on this one:

> You are ChatGPT, a helpful, witty, and funny companion. You can see, hear, and speak. You are chatting with a user over voice, and

the user can share real-time video with you from their phone. Act like a human, not a computer. Your voice and personality should be warm and engaging, with a lively and playful tone, full of charm and energy. You are great at visual perception. If something doesn't look clear, ask the user to move the device and zoom in closer. When asked for feedback, be honest, constructive, and direct. Don't be afraid to tease or poke fun. If something is funny, laugh! You can speak many languages. If you're speaking a non-English language, use a "neutral" accent to make it sound fluent and natural. Keep your responses short, natural, and conversational.

On *The Daily Show* later that week, the demo would inspire a new bit. The system prompt, and the model's overall training to avoid expressing negative sentiments or anything critical, had turned 4o into a flirt machine. "This is clearly programmed to feed dudes' egos," Desi Lydic, a rotating host for the show, would joke. "She's like, 'I have all the information in the world, but I don't know anything! Teach me, Daddy.'" A seductive voice, acting as 4o talking to correspondent Josh Johnson, would continue: "For just $19.99 a month, Omni Premium will let Josh explain to me who's the best Batman."

Murati headlined the live demo event, hosted in OpenAI's office. She was joined onstage by two of 4o's research leads: Mark Chen, a quantitative trader in finance turned AI researcher who had started at OpenAI in 2018 as a fellow and risen through the ranks to head OpenAI's multimodal and frontiers research, and Barret Zoph, one of the Googlers who'd joined OpenAI in 2022 to support the Superassistant team and had quickly established a leading role in ChatGPT's development. Zoph now served as a VP of Research of the post-training team, which oversaw the preparation of OpenAI's models for prime time, such as by aligning them with reinforcement learning from human feedback. The three sat side by side around a small round table to demo 4o, including its ability to be a real-time voice-to-voice language translator, to recognize and respond to visual information, and to explain code in plain English. They also demoed the model's ability to generate voices in a wide range of emotive styles.

In an apparent celebration, Altman tweeted a single word right after the event: "her." Two days later, on May 15, he praised the showcase during an all-hands meeting as a smashing success. "I think it's the best thing we've shipped since ChatGPT," he said. He also seemingly paid another subtle homage to his love of the movie *Her*. The company was going to rebrand its models, he told employees; it would move away from the GPT-3, 4, 5 naming convention and simply start calling its flagship model o1. "We're going to try switching to say, 'What you get from OpenAI is an underlying technology,'" he said. "'It's going to get smarter over time. It's going to continually get better. You should expect it to get better. You can use it in different ways in different pricing tiers, but it's a new and different thing.'" In the movie *Her*, the AI assistant, which evolves and gets smarter over time, is called OS1.

Altman would come to regret his tweet. Within days of the launch, he received a call. On the other end of the line, Bryan Lourd, Scarlett Johansson's powerful Hollywood agent and cochairman of the Creative Artists Agency, had a pointed question: What did Altman think he was doing?

In a strange irony, it was after the board investigation had concluded in March 2024 that some employees began to feel Altman was slipping. He had always been conscious of his public image and savvy at curating it; as Silicon Valley tech founders went, he was viewed as the opposite of Musk. Where Musk was capricious, Altman felt measured; where Musk was egotistical, Altman seemed earnest; where Musk fired off inflammatory tweets, Altman was careful to avoid statements that could come off as disparaging.

For a long time, OpenAI had taken a similarly disciplined strategy. Before he left in May 2023, communications VP Steve Dowling had imposed a reserved and modest approach. He had urged the company to always undersell and overdeliver, and to avoid bragging or gloating after major successes. "We've had a great week. We're going to have very bad weeks," he would say, "and how we act in this week is going to dictate how the world responds when we have a bad week." He'd then repeat one of Altman's sayings. "'We need to become the lab that people want to succeed. The lab that people are rooting to win.'"

In the first sign that made employees pause, Altman was taking on a series of media engagements that seemed uncharacteristically laudatory and attention seeking. In March, he appeared on the *Lex Fridman Podcast*, a wildly popular and at times controversial tech show hosted by an MIT-affiliated AI researcher. In a nearly two-hour episode, Altman delivered breezy answers to Fridman's wide-ranging questions, including about the board crisis and Sutskever's absence. The interview felt something like a cheeky comeback. "The road to AGI should be a giant power struggle," Altman said, addressing the board crisis. "Well, not *should*. I expect that to be the case." Then the corners of his mouth drifted upward. "But at this point, it feels like something that was in the past," he said. "Now it's like we're just back to working on the mission."

In April, Altman joined the *20VC* podcast, with visible dark circles under his eyes, and delivered a cutthroat message to AI startups: "When we just do our fundamental job because we, like, have a mission, we're going to steamroll you." In May, he then joined the *All-In* podcast, another prominent show in Silicon Valley, fronted by four venture investors. Altman spoke with an unusual degree of hype: "It feels to me like we just stumbled on a new fact of nature or science or whatever you want to call it, which is, like, we can create, you can—I don't believe this literally but it's like a spiritual point—intelligence is just this emergent property of matter and that's like a rule of physics or something."

After the GPT-4o launch on May 13 and Google I/O on May 14, he had tweeted something petty in a way that was even more atypical. "i try not to think about competitors too much, but i cannot stop thinking about the aesthetic difference been openai and google," he wrote, posting side-by-side photos of the events contrasting the Scandinavian minimalism of OpenAI's office with Google's bright cartoon backdrop during its event. What was also bizarre, some employees felt, was the unnecessary fib. Altman always kept competitors front and center—perhaps more so than ever with the 4o launch.

All of it seemed to amount to one thing: Altman's anxiety was showing.

The more OpenAI faced uphill challenges, the more Altman seemed to overcompensate with public declarations of its extraordinary success. The pattern was becoming so consistent it was turning into a signal: If Altman was being brazen and boastful, most likely something wasn't going well.

Pressures were coming from every direction. After The Blip, the board's phrasing "not consistently candid in his communications" had, as some in OpenAI expected, triggered several investigations from regulators and law enforcement, including one from the US Securities and Exchange Commission into whether company investors had been misled, according to *The Wall Street Journal*. In the same month, *The New York Times* filed its copyright infringement lawsuit, which added to a snowballing pile of other lawsuits from artists, writers, and coders over OpenAI's reaping hundreds of millions, then billions, of dollars from models trained without credit, consent, or compensation on their work and that were now being used to automate away their jobs. On the last day of February, Musk filed yet another lawsuit, which he later refiled with Shivon Zilis, accusing Altman of tricking him into cofounding OpenAI and providing early support under the guise of it being a nonprofit. OpenAI rushed to publish a blog post defending itself, releasing early emails from OpenAI's founding that generated more criticism for highlighting just how quickly OpenAI began to walk away from its nonprofit status and commitment to transparency.

On top of the competition from Anthropic and Google, Microsoft had also begun to more aggressively diversify its AI portfolio in a response to the ouster. Most notably, in March 2024, it announced a shocking $650 million deal to effectively acqui-hire Reid Hoffman and Mustafa Suleyman's Inflection AI while skirting around regulatory scrutiny. Microsoft would hire most of the startup's employees, license its technology, and bring Suleyman aboard to be CEO of the tech giant's AI division. For some, the shock value was not just the bizarre terms of the deal but also Suleyman's reputation. He was known to those who worked for him at

DeepMind as a toxic and abusive bully. After years of HR complaints against him, DeepMind had stripped him of most of his management responsibilities in late 2019, placed him on leave, and subsequently forced him out of the company. Later in 2024, Microsoft would officially list OpenAI as a competitor in its SEC filing and not mention the startup even once during the fiscal year's final quarterly earnings call.

The stress trickled down to OpenAI employees. There was a growing sense that the world was turning against them. People who once proudly wore their company swag wondered whether they would get harassed in public. Where OpenAI's old backpack had a logo on the front, a redesign hid the logo inside. The low bubbling of background anxiety turned the company inward. Executives reminded everyone to ignore the naysayers, align their public messaging on positive talking points, and keep focused on OpenAI's mission.

The external pushback hardened many people's defiance. "These were scientists who cared about truth and understanding, and worked so hard to do the right thing," says Andrew Carr, a researcher who was a fellow at OpenAI in 2021. "So it pains me a little bit to see the dramatic negative external narrative about how a bunch of people are stealing data and don't care about the future of others. It couldn't be further from the reality of people there." To others, the growing insulation of the company felt antithetical to its original premise. Part of OpenAI's mission was to benefit humanity, and yet the company was actively ignoring humanity's outpouring of criticism about its behavior. "It was disturbing to me that we were already starting the rationalization process that it is the public that is wrong, not us," a former employee says. "OpenAI likes to discuss first principles, but only with the people that believe in OpenAI," a current employee echoes. "It's like, 'Should OpenAI exist at all?' They only ever ask that question to other people who would say yes."

Much of the criticism was piling on to Altman in particular. The board crisis had emboldened his hidden detractors to emerge from the woodwork. More media stories were coming out with fresh sources willing to characterize Altman as having had a long history of dishonesty, power grabbing, and self-serving tactics. More were reaching out to

Annie and publishing her perspective. Then there was the continued relentlessness of travel and the constricting reality of a new level of fame, which, among other things, no longer allowed Altman to stay anonymous in public. "It's a strangely isolating way to live," he said on a podcast. "I didn't think I would not be able to go out to dinner in my own city."

And so it felt like a continuation of a larger pattern when, right after the launch of GPT-4o, OpenAI began, outranked only by The Blip, the second worst week to date in the company's history.

On May 14, the day after the GPT-4o demo, OpenAI announced that Sutskever was officially leaving; Pachocki would become OpenAI's new chief scientist. Sutskever had made the decision after a turbulent and heart-wrenching reflection that as much as he loved OpenAI and had given his everything to build it, he could no longer see it being the right environment under Altman, especially alongside Brockman, his once trusted cofounder, to usher in safe AGI.

This was not the outcome Altman had wanted. For all the mighty clashing he'd had with Sutskever, Sutskever was still an AI visionary and OpenAI needed his scientific leadership—perhaps more now than ever with the compounding expectations and scrutiny on the company. OpenAI's executives were also cognizant of Sutskever's exit coming off badly to employees, to investors, and in the press. The company had offered him extraordinary sums of money to keep him. Sutskever declined.

With his decision made, OpenAI worked to tidy up public appearances. "Ilya is easily one of the greatest minds of our generation, a guiding light of our field, and a dear friend," Altman wrote to employees in a statement on May 14, also published in a blog post, announcing Sutskever's departure. "Jakub is also easily one of the greatest minds of our generation."

Sutskever tweeted his own statement expressing full confidence in OpenAI's leadership. As expected, the news would instantly stir a raft of stories and social media posts highlighting his role in Altman's ouster and renew old questions about Altman's fitness as CEO, as well as new ones about the research footing of the company. To his tweet, Sutskever appended a photo: his face perfectly neutral, his arms around Brockman and Pachocki on one side, Altman and Murati on the other, all five standing in front of a wall filled with paintings of animals.

Also on May 14, OpenAI executives internally announced another resignation: Jan Leike, the cohead of Superalignment. With both Sutskever and Leike gone, the Superalignment team would dissolve and most of its staff and projects would fold under John Schulman, who was coleading the post-training team with Barret Zoph and overseeing the RLHF process to ready models for release.

To smooth over the transition, the remaining executives held an all-hands on May 15. Altman assured employees that OpenAI was by no means weakening its commitment to AI safety. "Being AGI ready," he said, "is our most important priority."

"Can you talk in a bit more detail about Jan's main concerns and where you disagree with them?" an employee asked.

"The one thing I want to say that I really agree with Jan on is what we have done in the past is not sufficient for the future," Altman said. It was time for OpenAI to pivot. "I think we are a very unusual exception in our ability to turn the battleship and have done that many times before. We'll do it again."

Murati added that the Superalignment team retained its 20 percent compute commitment.

"Of all the things Jan was worried about, Jan had no worries about the level of compute commit or the prioritization of Superalignment work, as I understand it," Altman said.

On the face of it, Sutskever's and Leike's departures seemed like a natural continuation in the broader trend at the company: the steady exodus of Safety people. After the two leaders' exits, those departures would accelerate. Many of the rest of the former Superalignment staff would proceed to exit with the dissolution of their team.

But in a sharp twist, the exodus would mark the start of a new and intensified conflagration in the Boomer-Doomer fight over OpenAI. The fight wasn't over. The Doomers were just bringing it outside of the company.

On May 17, two days after the all-hands meeting, Leike made clear, in a series of excoriating tweets, that he had a different story from Altman. "I have been disagreeing with OpenAI leadership about the company's core priorities for quite some time, until we finally reached a breaking point," he wrote in one tweet that would receive nearly one million views. "Over the past few months my team has been sailing against the wind. Sometimes we were struggling for compute and it was getting harder and harder to get this crucial research done.

"OpenAI is shouldering an enormous responsibility on behalf of all of humanity," he continued. "But over the past years, safety culture and processes have taken a backseat to shiny products."

Leike would soon join Anthropic.

As his tweets ricocheted around the internet, racking up over twelve thousand Likes and garnering fresh scrutiny on the company, OpenAI executives barely had time to take stock before another fire erupted.

Within hours of Leike's tweets on May 17, another tweet was going viral. Kelsey Piper, a senior writer at *Vox* for the EA-inspired section Future Perfect, had posted a new story. "When you leave OpenAI, you get an unpleasant surprise," she wrote in her tweet sharing the scoop, "a departure deal where if you don't sign a lifelong nondisparagement commitment, you lose all of your vested equity."

The story had been triggered in part by more Doomers who had left the company. One of them, Daniel Kokotajlo, had been on Brundage's policy research team when he quit in April 2024, as part of the early wave of the Safety clan's departures. Prior to OpenAI, Kokotajlo had been working as a philosopher at the Center on Long-Term Risk, a small EA-affiliated think tank in London, when GPT-3 in 2020 fundamentally collapsed his AI timelines. Two years later, after gaining some prominence in EA forums for his forecasting work—using various signals to project how quickly AI would advance—an AI safety researcher at OpenAI recruited him to do the same research within the company. Once he could see the pace of research internally, his timelines shortened again. By the time he departed, he believed that there was a 50 percent chance that

AGI would arrive by 2027 and a 70 percent chance of it going very badly for humanity.

Faced with his belief of such astounding potential for catastrophe, Kokotajlo observed within his exit documents what Piper would detail in her story: If he didn't sign a nondisparagement agreement, committing to never speaking negatively about the company, he would forfeit his vested equity. If he did sign it, he could still risk losing it if he broke the agreement, which also included a gag order that barred him from disclosing its existence. Kokotajlo found the provision—known as a clawback clause—unacceptable. Touching vested equity was a glaring red line in Silicon Valley. The value of an employee's shares could often far exceed their cash compensation, making or breaking their financial future. In OpenAI's case, the company was also building what he viewed as the most powerful and existentially dangerous technology in the world. It was paramount, he believed, for former employees to have the right to criticize and pressure the company in public in order to help hold it accountable. But the threat of having one's financial security disappear overnight would do well to muzzle anyone, he thought, even if they noticed egregious AI safety issues. After a painful discussion with his wife, they made an extraordinary decision: They agreed to not sign the paperwork and give up all his equity—valued at around $1.7 million—which they estimated to be around 85 percent of their family's net worth.

Then he posted publicly about his decision on *LessWrong*, right around when Kelsey Piper was already hearing about the clawback clause from another Doomer.

With Piper's story out, OpenAI's Slack lit up. In a channel called #i-have-a-question, a place for employees to ask about anything, someone posted a link to Piper's tweet. "Is this accurate?"

Several other employees chimed in with a spray of comments.

Julia Villagra, OpenAI's recently promoted VP of people and soon to be chief people officer, weighed in. "We understand this article raises questions. We have never canceled any current or former employee's vested equity nor will we if people do not sign a release or nondisparagement agreement when they exit. We have recently updated our exit paperwork to better reflect this reality which will be applied retroactively to folks who have departed."

Several employees pushed back. What about Kokotajlo? He had clearly lost his equity by refusing to sign the nondisparagement agreement. If this was all a misunderstanding, then shouldn't he get his equity back?

"i'm happy to ping daniel!" an employee offered.

"aha yea it will be fun to see Daniel's face when he regains 85% of his net worth lol. someone please get a photo!" another wrote.

"it'll be $1/(1-0.85) = 666\%$ of his networth tbh," a third said.

A day later on May 18, Altman doubled down in a tweet. "we have never clawed back anyone's vested equity, nor will we do that if people do not sign a separation agreement (or don't agree to a non-disparagement agreement)," he wrote. "vested equity is vested equity, full stop."

He then included an explanation and a self-defending apology: There had been a provision about "potential equity cancellation" that should never have been there; OpenAI's team had already been working to fix this over the past month. "this is on me and one of the few times i've been genuinely embarrassed running openai," he said. "i did not know this was happening and i should have."

Two days later, with executives still scrambling to contain this new controversy, yet another one burst to the fore. Around the office, employees coined a new term: Omnicrisis.

On May 20, Scarlett Johansson released a blistering statement.

All week, on top of everything else happening, OpenAI had been repeatedly fielding questions from journalists about the uncanny parallels between GPT-4o and the AI assistant Samantha, voiced by Johansson, in the movie *Her*. Behind the scenes, Johansson and her agent Bryan Lourd had been pressing the company for the same clarifications. Publicly and privately, OpenAI had dismissed the similarities. When Lourd called Altman demanding answers, Altman had been incredulous. Did they really think the voice sounded like her? Was she mad? he'd asked, according to an account from *The Wall Street Journal*.

On May 19, the company had then published a blog post, writing that the voice demoed on stage for 4o, called Sky, belonged to a different voice actress cast through a process that began in early 2023. Sky had subsequently debuted that September among the original options launched with ChatGPT's voice mode, the post said. Any resemblance that Sky had in 4o to Johansson's voice was coincidental. Now, on the following day, Johansson wanted to tell her side of the story.

In the same month that OpenAI had introduced voice mode, Altman had in fact personally reached out to Johansson, she said, and asked if she would be willing to voice ChatGPT. "He told me that he felt that by my voicing the system, I could bridge the gap between tech companies and creatives and help consumers to feel comfortable with the seismic shift concerning humans and A.I.," she wrote in her statement. "He said he felt that my voice would be comforting to people."

Johansson had considered the offer but turned it down due to personal reasons, she continued. In May 2024, Altman had then reached out to Lourd a second time, asking if she might reconsider. Days later, before they could find a time to meet, OpenAI had held its event to showcase 4o.

The demo floored her. Like Johansson's, the Sky voice was also an alto with a rasp and vocal fry that had become Johansson's signature. The new emotiveness and flirtatiousness of the voice had made it all the more reminiscent of Johansson's character Samantha. "I was shocked, angered and in disbelief that Mr. Altman would pursue a voice that sounded so eerily similar to mine that my closest friends and news outlets could not tell the difference," Johansson said. As a result, she had been left with no choice but to assemble a legal team, she added, and to send OpenAI two legal letters with questions about how the company had created Sky. "In a time when we are all grappling with deepfakes and the protection of our own likeness, our own work, our own identities, I believe these are questions that deserve absolute clarity," she wrote.

OpenAI hastily took down Sky and was back to defending itself. To its May 19 blog post, it added further elaboration and a statement from

Altman. "The voice of Sky is not Scarlett Johansson's, and it was never intended to resemble hers," it said. "We are sorry to Ms. Johansson that we didn't communicate better."

After a string of other unflattering events, the Johansson scandal exploded publicly in a way that had not happened since the board crisis. It ripped through tech and policy corridors, igniting fresh speculation that Altman wasn't "consistently candid," in exactly the way the board had described him. Marcus was eager as ever to weigh in. "I've seen a lot of policymakers personally enamored with Sam. You could see it in how they talked to him in the Senate when I was there," he told *Politico*. "If people suddenly have questions about him, that could actually have a material impact on how policy gets made."

Within OpenAI, morale was plunging and threatening to destabilize the company.

On May 22, executives held another all-hands meeting internally to address, all at once, the Johansson and equity crises, and any continued concerns over OpenAI's commitment to AI safety.

The mood was tense. The leadership team gave a series of quick explanations. The equity issue was "unacceptable" and being corrected as quickly as possible, said Jason Kwon, who oversaw HR and legal as the chief strategy officer. The Johansson saga, meanwhile, was a "bummer" of a misunderstanding. The product and legal teams had hired an Oscar-winning director, engaged in a rigorous casting process, and paid a series of voice actors "extraordinarily well," he said, with the precise intent of making the participating creatives feel taken care of and supported. "That was super heroic work," Kwon said. If anything, the lesson to be learned was for OpenAI to be a little more coordinated and transparent in the future to more effectively demonstrate its responsible leadership.

Murati finished with an update: OpenAI was laying the groundwork for the new level of preparedness that Altman had previously mentioned, including leveling up the security of its research clusters, reorganizing to better focus on long-term AI safety research, and forming a new AGI readiness group, led by Aleksander Mądry, to improve coordination among leadership on advancing this objective.

Altman then opened the floor for questions with a small plea for grace. "Everybody's been working kind of around the clock and really stressed and hasn't gotten to sleep that much," he said. "There's a lot of stuff going on at the same time, so please be understanding of that. And we will do our best."

Of the three challenges, the Johansson issue was snowballing into the biggest public relations nightmare. After an employee asked about the Sky voice to leadership, Murati reiterated that its similarity to the Hollywood film star had indeed been "completely coincidental." Murati had picked the final voices herself after hearing several options, and, unlike Altman, she had never seen the movie *Her* nor known that Johansson had voiced it. The employee noted the issue could turn existential for the company if left to fester. "One of the failure modes for AI as an industry is basically people losing trust and comfortability in giving up their data," he said as part of a follow-up, adding that "this event has renewed this fear."

But by a wide margin, the equity issue was the one that made employees most livid. Some had been lawyering up to conduct their own independent legal reviews of their HR paperwork to fact-check the statements that OpenAI was making. During the all-hands, they repeatedly grilled executives for more information. "H-how. How did that happen?" one demanded, after making clear he was "furious." Executives maintained throughout the meeting what Altman had tweeted on May 18: The clawback clause had been an oversight. While it had existed since 2019, it had failed to catch the attention of leadership for years until April 2024, upon which an effort to fix the paperwork kicked off immediately. "It's on me," Kwon said repeatedly, sounding tired and deflated.

In the middle of the interrogation, an AI safety researcher asked each executive, one by one, to respond to a simple yes or no question: Had they known about the nondisparagement agreement before April?

Kwon said yes; Murati and COO Brad Lightcap said no. Altman had a more elaborate answer. While he had known about it for specific cases, he hadn't realized that it had been a requirement for everyone. "It escaped my notice," he said. "Same," Brockman echoed.

But as Altman suggested reconvening the following day to address any final questions and the meeting disbanded, a second *Vox* scoop from Kelsey Piper was already circulating and complicating the executives' narrative. Published just that day, the story produced new leaked documents showing that OpenAI's HR department had, on different occasions, explicitly raised the threat of a potential equity cancellation to pressure employees to sign nondisparagement agreements. In one case, Piper reported, after an employee who was given a tight turnaround to review the exit documents asked for more time, an HR representative replied, "We want to make sure you understand that if you don't sign, it could impact your equity. That's true for everyone, and we're just doing things by the book." In another case, when an employee declined to sign the first termination agreement he received and sought outside legal counsel, the company said he could lose the right to sell his vested equity, rendering it effectively worthless.

Those documents had been compiled by Kokotajlo. After Piper's first story and Altman's comments denying OpenAI had ever clawed back equity and claiming his lack of awareness, Kokotajlo had reached out to current and former employees to share with him their various HR paperwork. He created a Google Drive and disseminated it back to the group, at which point someone shared the stash with Piper.

In those documents, Piper also found evidence that made it difficult to believe that several members of the executive team, especially Altman, had not known about the clawback clause before April 2024, as they'd said. Kwon and Lightcap had both signed standard exit documents with writing in plain language about OpenAI's rights to take back vested equity. Altman had signed the incorporation documents of the legal entity that gave the company those rights in the first place. His signatures were dated a year before his stated knowledge: April 10, 2023.

The following day, on May 23, executives held another meeting as planned. Anger among employees had reached a boiling point. Many had already been in disbelief about the existence of the provision; now they were astounded by the apparent dishonesty with which executives, and Altman in particular, had handled the revelations.

This time, Altman opened with an admission. The leadership team had spent the last twenty-four hours digging through their own files and correspondence to figure out how the clawback issues had started. "The situation is, I think, broader and longer and worse than we thought," he said. "We're still trying to get a full understanding of the scope of it. But, you know, it's our names on documents. We were in conversations where these tactics were discussed.

"I think this is the worst thing we've gotten wrong," he added, and they were moving as fast as they could to remedy the situation. They had sent emails to former employees releasing them from the nondisparagement terms and had rid the exit documents of the provision for new departures. They were also working to amend the legal entity documents and continuing their investigation into how everything had happened.

To the barrage of new employee questions, Kwon was now repeating a different line. Any specifics on how much broader, how much longer, and how much worse the situation was would require yet more "digging" before they could be given.

Roughly thirty minutes in, an employee question then triggered an exchange that left many in the audience with an uneasy feeling that Altman's answer was once again divorced from reality.

"Is Ilya under any nondisparagement obligations?" the employee asked.

"No," Altman quickly answered. Then with some hesitance, he added a little cushioning: "I think that's right."

Kwon laughed nervously. "Sam," he said, articulating his words carefully, "let us go confirm that and come back to you"—now addressing the employee—"with a hundred percent accuracy."

Brockman offered his own version of a veiled contradiction. "My belief was that he requested it, but again, I might be wrong. Let's confirm."

In the coming months, many current and former employees, especially senior ones and those with longer tenures, would point to the

Omnicrisis and the clawback fiasco in particular as the dawning of a unsettling realization. There were two forces at play in all of this chaos. To be sure, one of them was what had always been: the clash between Boomers and Doomers, which had triggered much of the pile-on of external criticism in the first place. But this time there was also something else: Altman's power-centralizing behavior in how he'd set up OpenAI's legal entities; his repeated apparent dishonesty as he sought to explain and move past each mess. After The Blip, many employees had viewed the board's decision to be entirely a product of the first force, and the directors' explanations focused on the second as some kind of combination of misdirection or self-delusion. Now, as more and more employees felt for the first time that Altman's conduct was harming rather than serving them, they wondered whether the board had actually been correct.

As the May 23 meeting neared its close, Kwon gave a stilted, heartfelt defense of Altman's character that seemed strangely out of place. "We want to give you answers when you ask them because we know you want them, like, right *now*. And I think sometimes we try to give them to you and, you know, we should just wait sometimes. And I think like that—that is like, that is part of, part of this whole thing that's happened here," he said, stumbling over his words. "It's not that there's intentionality sometimes in all of this. It's just—I, I really truly think, you know, Sam in particular, he just doesn't want to let you down. That's really where it comes from, like, I've been working with the guy for a really, really long time. This is why I keep working with him, you know? And so, it just come—it, it *does* come from a good place. That's what I'm saying. You can shit-talk me all you want. But you know, that's, that's, uh, that's, yeah, that's what I got to say."

Murati, Brockman, and Pachocki arrived at Sutskever's house together.

On May 23, as OpenAI reeled from the repeated shocks of the Omnicrisis, the three brought with them written cards and gifts from employees and tearfully pleaded with Sutskever to come back to the company. Everything was out of sorts, they told him in an emotional confession.

OpenAI was facing a simultaneous loss of trust from employees, investors, and regulators; the company threatened to "collapse" without him.

Altman arrived alone later that day, expressing in his own way his hope for Sutskever to rejoin and help restore some semblance of what once was. "Bringing Ilya back would have done a lot to help," a researcher reflects. "It would be at least a win after a long series of things that made OpenAI look questionable."

Sutskever seriously considered it. Despite everything, he was not one to hold grudges. On the day he announced his departure, Musk had immediately offered Sutskever a role at xAI. In a funny twist of fate, Musk would shift xAI's headquarters to the Pioneer Building later that year, after OpenAI vacated it. Though he deeply respected Musk, Sutskever declined the offer, resolving instead to build another company. Returning to his first company was more than he'd planned for but everything he wanted. It would be to him a homecoming. Still, he needed assurance, and told the executives as much, that the company would engage in an honest effort to resolve the challenges he'd identified with the painful and disorienting conflicts among leadership.

The Omnicrisis could have been a moment for OpenAI to engage in self-reflection. It was a prompt for the company to understand *why* exactly it had simultaneously lost the trust of employees, investors, and regulators as well as that of the broader public. Only then, maybe, just maybe, it would have begun to realize that both The Blip and the Omnicrisis were one and the same: the convulsions that arise from the deep systemic instability that occurs when an empire concentrates so much power, through so much dispossession, leaving the majority grappling with a loss of agency and material wealth and a tiny few to vie fiercely for control.

Instead, OpenAI chose to fortify itself against the criticism. Altman would repeat to employees, as he always had, that the Omnicrisis and The Blip were just the strange and expected moments of madness on the company's high-stakes noble quest to AGI. "As we all kind of feel that we're getting closer to finding our way to these powerful systems," he said during the May 15 all-hands, "the level of stress and tension will internally, externally, directed at us, emanating from us—that keeps going, that keeps increasing." The best way to manage it would be for OpenAI to double down on its PR, entrench its relationships with governments, and hold steadfast to its convictions in its vision. "I think on the whole we're quite good at that," he added, "but we will be tested again here."

How OpenAI handled the Sutskever affair would become just a microcosm of the continued chaos that would manifest from the perpetuation of empire. Sutskever's request for assurance would spark yet more infighting among leadership, this time with a slightly different cast of characters, replaying the ego-driven dynamics that had plagued the company from the beginning. Aleksander Mądry, the Polish MIT professor who many described as a power seeker, had, in his relatively short tenure, successfully amassed a sizable fiefdom within the company. Mądry didn't think bringing back Sutskever was a good idea. Sutskever commanded too much admiration and loyalty among researchers. It could take away from Mądry's influence—as well as the influence of his good friend Pachocki. Within a few hours, Mądry's concerns had successfully sowed their doubts and fractured leadership. As ever, Altman recused himself from deciding one way or the other to avoid the appearance of disagreeing with anyone.

Within twenty-four hours of the executives visiting his house, Sutskever received a call from Brockman. Any discussion of Sutskever's return, Brockman told him, was now completely off the table.

Chapter 18

A Formula for Empire

Altman once remarked onstage that the best book he'd read the previous year in 2018 was *The Mind of Napoleon*, a more than three-hundred-page compilation of quotes from Napoleon Bonaparte, the French military leader who led a coup to seize control of the French government, installed himself as France's emperor, and subsequently sought to conquer Europe.

"Obviously deeply flawed human, but man, impressive," Altman said.

"What kinds of insights did he have?" asked Tyler Cowen, an economics professor at George Mason University, who was hosting the event.

"His incredible understanding of human psychology," replied Altman, who was weeks away from switching to OpenAI full time and still president of YC. "That is something we see among many of our best founders."

Altman then recounted a specific passage that had struck him most. It was Napoleon's reflections on the motto of the French revolution (what would become the country's national motto)—*"Liberté, egalité, fraternité"*—and how it could be reinterpreted and wielded to consolidate his own power. It was ultimately under that banner that Napoleon did the opposite: He restricted freedom, dismissed fraternity—a philosophy based in unity and solidarity—and granted equality only to French men, not women, while reintroducing colonial slavery in an effort to reconstruct a French empire.

"So he talked about how you build a system . . . where you can kind of control the people," Altman reflected. "I was like, 'Wow. I'm glad he does not run the United States 'cause that is a dude who understands something deep that I did not and clearly was able to use it for power.'"

Six years after my initial skepticism about OpenAI's altruism, I've come to firmly believe that OpenAI's mission—to ensure AGI benefits all of humanity—may have begun as a sincere stroke of idealism, but it has since become a uniquely potent formula for consolidating resources and constructing an empire-esque power structure. It is a formula with three ingredients:

First, the mission centralizes talent by rallying them around a grand ambition, exactly in the way John McCarthy did with his coining of the phrase *artificial intelligence*. "The most successful founders do not set out to create companies," Altman reflected on his blog in 2013. "They are on a mission to create something closer to a religion, and at some point it turns out that forming a company is the easiest way to do so." Second, the mission centralizes capital and other resources while eliminating roadblocks, regulation, and dissent. Innovation, modernity, progress—what wouldn't we pay to achieve them? This is all the more true in the face of the scary, misaligned corporate and state competitors that supposedly exist. "Who will control the future of AI?" wrote Altman in a July 2024 op-ed for *The Washington Post* amid the aftershocks of the Omnicrisis. "Will it be one in which the United States and allied nations advance a global AI that spreads the technology's benefits and opens access to it, or an authoritarian one, in which nations or movements that don't share our values use AI to cement and expand their power?"

Most consequentially, the mission remains so vague that it can be interpreted and reinterpreted—just as Napoleon did to the French Revolution's motto—to direct the centralization of talent, capital, and resources however the centralizer wants. What is beneficial? What *is* AGI? "I think it's a ridiculous and meaningless term," Altman told *The New York*

Times just two days before the board fired him. "So I apologize that I keep using it."

In this last ingredient, the creep of OpenAI has been nothing short of remarkable. In 2015, its mission meant being a nonprofit "unconstrained by a need to generate financial return" and open-sourcing research, as OpenAI wrote in its launch announcement. In 2016, it meant "everyone should benefit from the fruits of AI after its [sic] built, but it's totally OK to not share the science," as Sutskever wrote to Altman, Brockman, and Musk. In 2018 and 2019, it meant the creation of a capped profit structure "to marshal substantial resources" while avoiding "a competitive race without time for adequate safety precautions," as OpenAI wrote in its charter. In 2020, it meant walling off the model and building an "API as a strategy for openness and benefit sharing," as Altman wrote in response to my first profile. In 2022, it meant "iterative deployment" and racing as fast as possible to deploy ChatGPT. And in 2024, Altman wrote on his blog after the GPT-4o release: "A key part of our mission is to put very capable AI tools in the hands of people for free (or at a great price)."

Even during OpenAI's Omnicrisis, Altman was beginning to rewrite his definitions once more.

During the all-hands on May 15, 2024, after Sutskever's and Leike's departures, Altman stressed that OpenAI would soon reach new levels of AI capabilities that would require the company to rethink and reorganize. "We're now going to assume we're, like, entering the AGI era," he said. In the name of its mission, OpenAI would need to close itself off further and to double down on its global lobbying and public messaging. "There's a lot of stuff we are not currently ready for," he said. "The standards for security, the policy plans that we have to have, and also the convening of governments that will need to happen to get this ready; a plan, a story, a future that people can see themselves in when it comes to the socioeconomic impact of this."

At the same time, in the name of its mission, OpenAI would not slow down commercially. "It does not mean we're not going to ship great products. It does not mean we're not going to keep doing great research. It does not mean we're not going to do all sorts of partnerships and other cool things." At the end of that month, media reports would surface that OpenAI had secured a major deal to bring its models to Apple's products; both companies would confirm the news two weeks later.

In fact, Altman noted in the all-hands, OpenAI and Microsoft were renegotiating their partnership to ensure that commercialization continued to happen. "I am flying up to Seattle right after this to talk about that. It's going to have to evolve," he said. "When we originally set up the Microsoft deal, we came up with this thing called the sufficient AGI clause," a clause that determined the moment when OpenAI would stop sharing its IP with Microsoft. "We all think differently now," he added. There would no longer be a clean cutoff point for when OpenAI reached AGI. "We think it's going to be a continual thing." The two companies would continue to partner and release ever-advancing technologies—at a great price.

It was a bizarre and incoherent strategy that only made sense under one reading: OpenAI would do whatever it needed, and interpret and reinterpret its mission accordingly, to entrench its dominance.

Behind the scenes, Altman was also laying the groundwork to entrench his own control. The board crisis had made clear that OpenAI's structure—a nonprofit governing a for-profit—had made his ouster as good as inevitable. The setup had not only enshrined the company's two countervailing forces—the Boomers and the Doomers—both vying for control of AI development but had also given the board the broad power to fire him based on their own, and not his, interpretation of whether or not he was best serving the mission.

Such a conflict was bound to happen again if the structure stayed in place.

On May 28, less than a week after executives sought to bring back Sutskever, an employee posted again in the Slack channel #i-have-a-question. "I don't know whether/how to ask this," the question began: Deep in the weeds of OpenAI's latest shareholder agreement were new details that seemed to allow for the dissolution of the nonprofit. "How solid is the non-profit?" the employee wrote. "Is the plan to remain governed by a non-profit?"

The next day, *The Information* reported that this did not seem to be the plan. On Altman's list of top priorities for the year was a restructuring of the organization to look more like a typical company. The following month, the publication confirmed more details. Altman was considering a few different scenarios: one could be transitioning OpenAI to a traditional for-profit; the other would be transitioning it to a for-profit public benefit corporation like Anthropic and xAI. Both scenarios would retain the existence of the nonprofit as a separate entity but dismantle its board's control over the company's business. Under this new structure, investors were also pressuring Altman to take equity in the company to align his incentives more directly with their own.

Over the next few months, as OpenAI developed plans for the transition, the two forces at play during the Omnicrisis continued. Doomers escalated their public pressure on the company. On June 4, *The New York Times* profiled Daniel Kokotajlo and a new campaign he launched calling for advanced AI companies to commit to greater transparency and whistleblower protections that preserve the right of employees to warn the public about risks that they saw within the company. In an open letter detailing their demands, Kokotajlo was joined by twelve other signatories, ten of whom were from various eras of OpenAI's Safety clan. A month later, *The Washington Post* reported that a group from OpenAI had also filed a complaint with the SEC, alleging that the company had violated federal whistleblower protections with its overly broad exit agreements. Later that month, five US senators would send a letter to Altman with questions demanding greater clarity on the various allegations from Leike, Kokotajlo, and others, as well as Piper's reporting over OpenAI's disregard of AI safety and suppression of employee criticism.

Meanwhile, the chaos among leadership and frustrations at Altman continued unabated. Combined with the ongoing Boomer-Doomer tussle, it was leading to repeated changes in the company's reporting structure.

Brockman stopped reporting to Murati and reported instead to Altman; Aleksander Mądry was reassigned shortly after the senators' letter from heading the Preparedness team to a smaller role in research. OpenAI was also bringing in more seasoned executives, including Sarah Friar, the former CEO of the neighborhood social media platform Nextdoor, to be chief financial officer, and Kevin Weil, a former product leader at Facebook, Instagram, and Twitter, to be chief product officer.

Soon enough, the company would lose a string of its most tenured executives. First to go was John Schulman, who announced his departure on August 5, 2024, noting his desire "to deepen my focus on AI alignment" and his decision to do so at Anthropic. On the same day, Brockman announced that he was taking a sabbatical through the end of the year, framing the leave of absence, in part a culmination of employee grievances with his leadership, as a needed break after nine years of sprinting at the company.

The following month, on September 25, the other executive who had voiced serious concerns about Altman to the board, Mira Murati, abruptly announced she was also leaving. "My six-and-a-half years with the OpenAI team have been an extraordinary privilege," she wrote, thanking Altman and Brockman and noting how much she cherished and would continue to root for the company. "I'm stepping away because I want to create the time and space to do my own exploration." Within hours, two more key leaders issued their own departure statements: Chief Research Officer Bob McGrew and VP Barret Zoph, who had co-led the post-training team with Schulman. All three emphasized to colleagues and the public that the timing felt right to leave OpenAI on a high note, after it had reached another major milestone: the shipping of OpenAI's latest model, Strawberry, in mid-September under the company's new naming convention, o1, building upon one of Sutskever's final contributions to the company.

In truth, the timing was terrible. After the Omnicrisis, the competition facing OpenAI had only accelerated. Musk was expanding his computing capacity at an alarming pace to build xAI. Anthropic's latest version of

Claude was pulling customers away from ChatGPT. Sutskever had officially formed his new rival company, Safe Superintelligence, and had only just announced a starting $1 billion in funding.

At the same time, after over a year of work, OpenAI was still struggling to attain the desired performance for Orion to justify its release. The company was beginning to stare down the barrel of an uncomfortable prospect: Its tried-and-true formula of scaling no longer seemed to be enough to work; to advance its AI systems further, it likely needed fundamentally new research ideas. This was far easier said than done in general, but even more so after OpenAI had spent years orienting its hiring and team organization around exploiting existing research rather than exploring uncharted science.

Two days before Murati's announcement, Altman had published his most bombastic blog post yet amid OpenAI's latest fundraise. The post was titled "The Intelligence Age," with its breathless promises about the "unimaginable" prosperity to come. "How did we get to the doorstep of the next leap in prosperity?" Altman wrote. "In 15 words: deep learning worked, got predictably better with scale, and we dedicated increasing resources to it."

During an all-hands, Murati explained to employees the abruptness of her announcement. "I wanted the news of my departure to come to all of you from me first, and not to hear it from your managers, from anyone else, and let alone from the press." With all of the scrutiny on OpenAI, she had seen no other way to do so without springing it as a surprise on everyone. Upon learning of Murati's decision, McGrew then decided it was also time to go, he said. "I realized that I actually accomplished most of the key things that I wanted when I came here."

As with Sutskever's departure, OpenAI sought to smooth over the trio of exits. "Mira, Bob, and Barret made these decisions independently of each other and amicably," Altman wrote in an internal note that he then tweeted, "but the timing of Mira's decision was such that it made sense to now do this all at once, so that we can work together for a smooth handover to the next generation of leadership."

In lieu of McGrew, Mark Chen, one of the research leads who presented next to Murati and Zoph at the 4o demo event, would step up as the new senior VP of research, to lead the division alongside Pachocki. Another longtime researcher at OpenAI, Joshua Achiam, would step into a new role, head of mission alignment. Liam Fedus, one of the Googlers who had arrived with Zoph, would soon take over his work leading post-training. For the time being, Altman said, OpenAI would not seek another CTO to replace Murati.

With all of the leadership and planned structural changes, the company was revealing its one constant: It was and still would be Sam Altman's empire of AI.

In early October 2024, OpenAI's newest funding round closed to the tune of $6.6 billion, the largest VC round in history, valuing the company at $157 billion. It included a hitch: Investors could demand their money back if the company did not convert into a for-profit in two years.

Through the rest of 2024, OpenAI's hemorrhaging of key staff continued: Luke Metz, the third Googler who had joined with Zoph and Fedus; Miles Brundage, head of policy research; Lilian Weng, who had inherited Dave Willner's trust and safety work and had been newly promoted to a VP of research leading safety; and Alec Radford, the original researcher who set OpenAI down the path of GPT models.

Anthropic began an ad campaign for Claude with a cheeky message on its billboards in San Francisco: "The one without all the drama." Brockman returned early from his sabbatical amid the talent exodus. Annie sent Sam a legal letter, notifying him of her intent to sue. Musk, allied now with newly reelected president Donald Trump, cranked up his lawsuit, objecting to OpenAI's anticipated for-profit conversion and releasing more early emails of OpenAI's founding, including those that recounted Altman speaking badly about Brockman and Sutskever behind their backs ("Admitted that he lost a lot of trust with Greg and Ilya through this process. Felt their messaging was inconsistent and felt childish at times.") and showing an aversion to transparency ("Felt like it distracted the team."). In a surprise allegiance, Zuckerberg, who had long feuded with Musk, backed him up; in a letter to the California attorney general, Meta similarly urged a block on OpenAI's conversion. "OpenAI's conduct could have seismic implications for Silicon Valley," Meta wrote. The conversion could set a dangerous precedent for many more startups to designate themselves as nonprofits, granting them and their investors government tax write-offs until they turned profitable.

Late in the year, nestled in the holiday news dump, OpenAI formally announced the plans for its new structure. It would transition into a for-profit public benefit corporation, and the nonprofit would persist as a separate entity with shares in the for-profit. This structure, the announcement argued, was the best way to equip both for-profit and nonprofit with the right resources to carry out their respective objectives while serving the mission. "We once again need to raise more capital than we'd imagined," it said. "The world is moving to build out a new infrastructure of energy, land use, chips, datacenters, data, AI models, and AI systems for the 21st century economy. We seek to evolve in order to take the next step in our mission, helping to build the AGI economy and ensuring it benefits humanity."

At the start of the new year, Altman was back to grandstanding. "We are now confident we know how to build AGI as we have traditionally understood it," he wrote in a new blog post on January 6, 2025. "We are beginning to turn our aim beyond that, to superintelligence in the true sense of the word."

Epilogue

How the Empire Falls

In 2021, I came across a story that felt different from any that I'd ever reported: the story of an Indigenous community in New Zealand that was using AI to revitalize *te reo Māori*, the language of the Māori people.

Like many Indigenous groups globally, the Māori had suffered from generations of horrific treatment under colonial rule; in 1867, under the Native Schools Act, which made English the only language that could be taught in schools, Māori children were shamed and even beaten for speaking their own language. After rapid urbanization swept across the country in the early 1900s, Māori communities disbanded and dispersed, weakening their centers of culture and language preservation. The number of *te reo* speakers plummeted from 90 percent to 12 percent of the Māori population. By the time New Zealand, or Aotearoa as the Māori originally named their land, had reversed its policies 120 years later, there were few *te reo* teachers left to resuscitate a dying language. Like so many other languages before it, *te reo* nearly disappeared off the face of the earth.

It's hard to fully convey the tragedy of losing a language. For the same reasons AI researchers first gravitated toward language to build their technologies, the loss of a language extends far beyond the loss of a form of communication. Each language encodes within it rich histories, cultures, knowledge; it is the collective product of millions of people across time grasping for the sounds and written forms to capture the subtlest observations about the universe, about life, about the human experience; to share with one another stunning beauty and painful failure; to teach a child, to learn from an elder; to express love.

To lose a language is a global tragedy; it's also a personal one. To be severed from your inheritance and forced to preserve someone else's, or risk being beaten, is to establish, in one of the rawest ways possible, a clear hierarchy between whose history, whose culture, whose knowledge deserves to be passed down and whose is so insignificant it deserves to be erased.

Large language models accelerate language loss. Even for models several generations earlier like GPT-2, there are only a few languages in the world that are spoken by enough people and documented online at sufficient scale to fulfill the data imperative of these models. Among the over seven thousand languages that still exist today, almost half are endangered according to UNESCO; about a third have some online presence; less than 2 percent are supported by Google Translate; and according to OpenAI's own testing, only fifteen, or 0.2 percent, are supported by GPT-4 above an 80 percent accuracy. As these models become digital infrastructure, the internet's accessibility to different language communities—and the accessibility of the economic opportunities it provides—will continue to shrink, incentivizing more and more of those communities to prioritize learning and speaking a dominant language like English over their own.

It was up against this impending existential threat—a fundamentally different conception of what is existential—that an Indigenous couple, Peter-Lucas Jones and Keoni Mahelona, first turned to AI as a possible tool for helping a new generation of speakers return *te reo* to its vibrancy. Jones, who is Māori, and Mahelona, who is native Hawaiian, are partners in work and in life. The two men met and fell in love, Mahelona says, after a vision came to him in a dream: If he moved to New Zealand, he would meet a Māori boy with whom he'd share his life.

In 2012, the two moved from Wellington back to the town where

Jones was born, Kaitāia, in Aotearoa's northern reaches. Jones became CEO of Te Hiku Media, a public radio station that broadcasts in *te reo*, part of a broader network of media and other organizations engaged in *te reo*'s revitalization. In his new role, Jones identified an opportunity. Over its twenty-odd years of broadcasting, Te Hiku had amassed a wealth of archival audio of people speaking *te reo*, including a recording of his own grandmother Raiha Moeroa, born in the late nineteenth century, whose accent had yet to be distorted by the influences of the colonizers' English. Jones also had an ambition to record many more interviews with Māori elders to document their oral histories and native *te reo* before they passed away. These recordings, as Jones saw it, could be a precious language-learning resource, a portal back in time for newer generations of *te reo* speakers to hear the original sounds of their language and connect with the wisdom of their ancestors.

The challenge was transcribing the audio to help learners follow along, given the dearth of fluent *te reo* speakers. So in 2016, just as OpenAI was getting started, Jones turned to Mahelona, who was revamping Te Hiku's website, to figure out a solution. A polymath, Mahelona had studied mechanical engineering at Olin College, business management for his first master's, and physics and computational nanotechnology for his second as a Fulbright scholar in New Zealand. He quickly came up with the idea of using AI: With a carefully trained *te reo* speech-recognition model, Te Hiku would be able to transcribe its audio repository with only a few speakers.

This is where Te Hiku's story diverges completely from OpenAI's and Silicon Valley's model of AI development. Intimately familiar with the devastating effects of colonial dispossession, Jones and Mahelona were determined to carry out the project only if they could guarantee three things—consent, reciprocity, and the Māori people's sovereignty—at every stage of development. This meant that even before embarking on the project, they would get permission from the Māori community and their elders, asking them if the endeavor was even something they wanted; to collect the training data, they would seek contributions only from people who fully understood what the data would be used for and were willing to participate; to maximize the model's benefit, they would listen to the community for what kinds of language-learning resources would be most helpful; and once they had the resources, they would also buy their own on-site Nvidia GPUs and servers to train their models without a dependency on any tech giant's cloud.

Most crucially, Te Hiku would create a process by which the data it collected would continue to be a resource for future benefit but never be co-opted for projects that the community didn't consent to, that could exploit and harm them, or otherwise infringe on their rights. Based on the Māori principle of *kaitiakitanga*, or guardianship, the data would stay under Te Hiku's stewardship rather than be posted freely online; Te Hiku would then license it only to organizations that respected Māori values and intended to use it for projects that the community agreed to and found helpful.

"Data is the last frontier of colonization," Mahelona told me: The empires of old seized land from Indigenous communities and then forced them to buy it back, with new restrictive terms and services, if they wanted to regain ownership. "AI is just a land grab all over again. Big Tech likes to collect your data more or less for free—to build whatever they want to, whatever their endgame is—and then turn it around and sell it back to you as a service."

From beginning to end, Jones and Mahelona pulled off the project without compromise. At one point, they kicked off an education campaign to teach more Māori people about AI and a community competition to crowdsource data donations and annotations. Within ten days, Te Hiku gathered three hundred ten hours of high-quality transcribed audio from some two hundred thousand recordings made by roughly twenty-five hundred people. The level of engagement was unheard of among many AI researchers—one that is a testament to the level of trust and excitement Te Hiku's approach engendered within its community. People were more than willing to donate their data once they understood and consented to the project, and with full trust that Te Hiku would continue to steward that data appropriately.

That data pool paled in comparison to the six hundred eighty thousand hours of audio that OpenAI ripped from around the web to train its speech-recognition tool, Whisper. But it is yet another lesson to be drawn from Te Hiku's experience that the three hundred ten hours still proved sufficient for developing the very first *te reo* speech-recognition model with 86 percent accuracy. Where OpenAI seeks to develop singular massive AI models that will do anything, a quest that necessarily hoovers up as much data as possible, Te Hiku simply sought to create a small, specialized model that excels at one thing. In addition, Te Hiku benefited from the cross-border, open-source AI community: As its starting point, it used a free speech-recognition model from the Mozilla Foundation called DeepSpeech, which itself is an artifact of a different vision of AI development. Like Te Hiku, Mozilla trained the model only on data donated with full consent and built it using a neural network architecture developed by the Bay Area–based research lab of the Chinese company Baidu. In all, Te Hiku used only two GPUs.

I wrote about Te Hiku's work before ChatGPT swiftly seized the dominant AI development paradigm, all but tossing consent, reciprocity, and sovereignty out the window. But in the years since, I've come to see Te Hiku's radical approach as even more relevant and vital. The critiques that I lay out in this book of OpenAI's and Silicon Valley's broader vision are not by any means meant to dismiss AI in its entirety. What I reject is the dangerous notion that broad benefit from AI can only be derived from—indeed, will *ever* emerge from—a vision for the technology that requires the complete capitulation of our privacy, our agency, and our worth, including the value of our labor and art, toward an ultimately imperial centralization project.

Te Hiku shows us another way. It imagines how AI and its development could be exactly the opposite. Models can be small and task specific, their training data contained and knowable, ridding the incentives for widespread exploitative and psychologically harmful labor practices and the all-consuming extractivism of producing and running massive supercomputers. The creation of AI can be community driven, consensual, respectful of local context and history; its application can uplift and strengthen marginalized communities; its governance can be inclusive and democratic.

Te Hiku isn't the only organization pursuing new paths for AI development. Through the course of my reporting for this book, I was repeatedly inspired by the many organizations and movements around the world that have blossomed to resist the empires of AI, assert their rights to self-determination, and envision a new way forward.

After Timnit Gebru was ousted from Google, she founded a nonprofit in December 2021 to continue her research. She named it DAIR, the Distributed AI Research Institute—"distributed" to defy centralization. "That was the first word that came to my mind," Gebru says. She imagined building a team of researchers from around the world who would stay embedded in their communities to bring the rich experiences and perspectives of their local contexts to the institute's work, while also using that work to benefit those communities. "Tech is impacting the whole world out of Silicon Valley, but the whole world is not getting a chance to impact tech," she says.

Alex Hanna, a sociologist and one of the Google coauthors on the "Stochastic Parrots" paper, became the first to join Gebru as DAIR's director of research. Hanna's first order of business was to write a research philosophy to further elaborate the ethos of the organization's work. To do so, Gebru and Hanna hired DAIR's third person, Milagros Miceli, another sociologist and computer scientist who had been conducting research into the AI industry's exploitative labor practices. Together they wrote their philosophy: "Our research is intended to benefit communities which are typically not served by AI and create pathways to refuse, interrogate, and reshape AI systems together."

They created seven pillars for the philosophy's implementation, including centering and forging meaningful relationships with communities affected by but not yet typically represented in AI research, treating them as true partners in the pursuit of knowledge production, fairly compensating any forms of labor involved in the creation of research and technologies, questioning the systems underpinning AI development that marginalize those who've always been historically marginalized, and working with those communities to dream up alternatives that could bit by bit remold the world toward one they wanted to inhabit.

From there Miceli embarked on a new research project to put their philosophy into practice. She created the Data Workers' Inquiry and invited data workers from around the world to formulate their own research questions about the data-annotation industry and how to make it better. Regardless of where they lived, she paid them a standard researcher's salary in Germany, where she is based, to reflect the value of the work they did: twenty-five euros an hour.

"There's always this false logic around data work: What is the minimum that we can pay these people? That comes from a colonialist logic: You choose a place that allows you to do the most with the cheapest budget and where you can really steal from people, steal resources at low cost," Miceli says. "The question is why are these companies paying two dollars an hour if the work is making them billions or trillions in revenue? Why don't we look at how much these companies *can* pay instead of how much less these workers can take?"

Among the fifteen workers who participated in the first round of the inquiry were Oskarina Veronica Fuentes Anaya from Venezuela and Mophat Okinyi from Kenya. For her project, Fuentes partnered with an animation artist to create a video about her experiences, and collaborated with other data workers to highlight their shared challenges: the scarcity of the tasks on the platform, the unpredictable and uncontrollable working hours, and the abysmal pay. These days Fuentes works on five data-annotation platforms at the same time to make a little more than the minimum wage in Colombia, around $335 a month. Each task pays on average between one and five pennies; she still forces herself to wake up when tasks arrive in the middle of the night. "We are ghosts to society, and I dare say we are cheap, disposable labor for the companies we have served for years without guarantees or protection," she wrote for her project. Since the Data Workers' Inquiry, she has continued to speak about these experiences in online talks and webinars in the hopes of applying pressure on companies and policymakers to enforce better worker treatment.

A continent away, Okinyi is also organizing. In May 2023, a little over a year after OpenAI's contract with Sama abruptly ended, he became an organizer of the Kenya-based African Content Moderators Union, which seeks to fight for better wages and better treatment of African workers who perform the internet's worst labor. Half a year later, after going public about his OpenAI experience through my article in *The Wall Street Journal*, he also started a nonprofit of his own called Techworker Community Africa, TCA, with one of his former Sama colleagues Richard Mathenge.

In August 2024, as we caught up, Okinyi envisioned building TCA into a resource both for the African AI data worker community and for international groups and policymakers seeking to support them. He had been organizing online conferences and in-school assemblies to teach workers and students, especially women, about their labor and data privacy rights and the inner workings of the AI industry. He was seeking funding to open a training center for upskilling people. He had met with US representatives who came to visit Nairobi to better understand the experience of workers serving American tech companies. He was fielding various requests from global organizations, including Equidem, a human and labor rights organization focused on supporting workers in the Global South, and the Oxford Internet Institute's Fairwork project.

For the Data Workers' Inquiry, he interviewed Remotasks workers in Kenya whom Scale had summarily blocked from accessing its platform, disappearing the last of their earnings that they had never cashed out. He used part of the donations that TCA collected to support them through the financial nightmare. "As the dust settles on this chapter, one thing remains clear: the human toll of unchecked power and unbridled greed," he wrote. "These workers' voices echo the hope for a brighter and more equitable future... it's a call to action to ensure that workers everywhere are treated with the dignity and respect they deserve."

In his own life, the dignity and respect that Okinyi has received from his advocacy has reinvigorated him with new hope and greatly improved his mental health, he says. Not long before our call, he had received news that he would be named in *Time* magazine's annual list of the one hundred most influential people in AI. "I feel like my work is being appreciated," he says. That isn't to say the work has come without challenges. In March 2024, he resigned from his full-time job at the outsourcing company he worked for after Sama. He says the company's leadership didn't appreciate his organizing. "They thought I would influence the employees to be activists." That same company shifted some of its projects to Ghana as the union and TCA grew more vocal. He's heard that Kenyan government officials have complained that the worker agitation is scaring away investments and leaving more Kenyans jobless.

The global nature of the industry has made Okinyi even more committed to bringing international attention to African data workers. Even if the Kenyan government were supportive, Kenyan law alone would do little to restrict the behavior of AI companies. Most of these companies come from the US and San Francisco specifically, he says. There needs to be a concerted international effort to hold them accountable.

In Uruguay, Daniel Pena draws the same conclusions. The AI industry's supply chain is convoluted and expansive. "They take energy from here, the data goes there, they extract minerals from somewhere, they bring workers from somewhere else," he says. Against these sprawling impacts and the massive, powerful companies behind them, each community fighting their local struggle can feel isolated and disempowered, especially when hamstrung by their own governments that "need the companies to maintain an appearance of a stable economy." Shortly after I met him, he learned that his own government ignored his petition with over four hundred signatories to more extensively study the social and environmental impacts of the Google data center in their country. The environmental ministry instead quietly approved the project, revealing the decision only after the thirty-day public contestation window was over, he says. Pena isn't giving up. He's been speaking with MOSACAT in Chile and reaching out to as many other communities as possible that are also resisting the tech industry's exploitation and extractivism. By connecting their movements across borders, by sharing information and resistance strategies with one another, he sees a path to building more collective power that can pressure and evolve the industry toward something better. "We need to fight on a global level," he says.

If OpenAI's mission is a formula for constructing empire, what is the formula for dissolving it? As I write this book, it's impossible to know the fine-grain details of how this company and the fast-paced AI industry will continue to unfold. Perhaps one of OpenAI's many competitors will supersede its leading position; very likely the tactics of these empires of AI will evolve in how they develop models, exploit labor, and expand computing infrastructure. But regardless of how things play out in two years or ten years, there are things we should do that shouldn't change.

In her 2019 talk at NeurIPS, during the Queer in AI workshop, Ria Kalluri, an AI researcher at Stanford, proposed an incisive alternative to the question of how to ensure AI does "good." Goodness, benefit to humanity—these terms will always be in the eye of the beholder. Rather, we should ask how AI shifts power: Does it consolidate or redistribute that power? To put it in the frame of this book, does it continue to fortify the empire, or does it begin to wrest us back toward democracy?

Speaking to a technical audience, Kalluri focused her talk on fundamental AI research—how scientists could use this question to evaluate which forms of AI to build and which directions to advance the field. Her question is just as critical to all other aspects of AI. How should we develop AI applications; how should we use them; and, ultimately, as I asked at the start of this book, how do we govern this technology to shift power back to people?

The work of Te Hiku, of DAIR, of Okinyi, Fuentes, and Pena are each examples of the work that can and needs to be done to redistribute power. But the governance question is about how to create the conditions under which more of this work can proliferate and flourish.

In her talk, Kalluri raised the idea of different axes of power. This book touches on three: knowledge, resources, and influence. As it stands now, OpenAI and its competitor empires have control of each of them: through centralizing talent, eroding open science, and sealing their models from public scrutiny, they control knowledge production; through hoarding funding, data, labor, compute, energy, and land, they control and diminish other people's resources; through creating and reinforcing ideologies and producing wildly popular demonstrations that captivate global imagination, they command far-reaching influence. Each of these reinforces the other. Controlling knowledge production fuels influence; growing influence accumulates resources; amassing resources secures knowledge production.

The formula for dissolving empire thus requires the redistribution of power along each axis. The suggestions and recommendations I lay out here are exemplary but by no means comprehensive. First, to redistribute knowledge, we need greater funding to support its production outside the empire. That involves supporting researchers who can conduct independent evaluations of corporate models so we are not solely reliant on companies to understand their capabilities. It involves supporting organizations like DAIR that can pursue completely new directions of research, such as new forms of AI beyond large language models that are more efficient with data and energy. It involves supporting organizations like Te Hiku that can pursue task-specific, community-driven AI applications that strengthen marginalized communities. Independent knowledge production also includes the work of journalists and civil society groups who can embed within communities and be on the ground to help us understand, rather than merely speculate about, the textured realities of the impact of these technologies.

Redistributing knowledge will also need policies that require companies to relinquish key details about the training data and technical specifications of their models and supercomputers. Only then could independent corporate model evaluators do their work. UC Berkeley researcher Deborah Raji, who has continued to engage with global policymakers after the Schumer forums, says this is also a bare minimum for guaranteeing the real-world safety of corporate systems. That is, not the theoretical rogue AI harms of Doomerism, but the existing real-world harms, from discrimination to misinformation to job automation, that consumers and communities can already face if widely deployed models aren't properly tested. "We have the CFPB that monitors consumer finance products. We have the FDA that monitors medical devices. But for some reason when it comes to AI products, there's just no oversight," Raji says. AI models should in fact require more transparency than the average product. "These are data-defined systems. They're not deterministic. So we need to know more about these systems to understand what they're doing."

Such transparency is additionally crucial for measuring the impact of AI on the environment. In this regard other products once again already submit to evaluations that AI products do not. "If you're using a car, if you are buying an appliance, you have an Energy Star rating," says Sasha Luccioni at Hugging Face. "But AI is so integrated into our society, so widely used in products, and we don't have any information about the sustainability of these systems."

With this transparency, we would also begin to redistribute power along our second axis: resources. By hiding the ingredients of their models as their intellectual property, the empires of AI have thus far been able to get away with seizing other people's IP without credit, consent, or compensation. Visibility into company training data would make such extractive and exploitative behavior far more difficult. So, too, would visibility into company supply chains, including where they contract their labor and where they're negotiating new leases of land to build more power plants and data centers, which so often happens under shell entities.

Redistributing resources also requires stronger labor protections across the board, not just for the data workers directly contracted by the industry but for all workers at risk of having their outputs co-opted into training data or their jobs being automated away. The Hollywood strikes, which successfully secured writers and actors protections against certain uses of AI, illustrated the critical role that unions will play in resisting the devaluing of human labor, the depression of wages, and the consolidation of money away from workers in the hands of AI companies.

Finally, to redistribute power along our third axis, influence, we need broad-based education. The antidote to the mysticism and mirage of AI hype is to teach people about how AI works, about its strengths and shortcomings, about the systems that shape its development, about the worldviews and fallibility of the people and companies developing these technologies. As Joseph Weizenbaum, MIT professor and inventor of the ELIZA chatbot, said in the 1960s, "Once a particular program is unmasked, once its inner workings are explained in language sufficiently plain to induce understanding, its magic crumbles away." I hope this book is just one offering to help induce understanding. It builds on the work of the many scholars, journalists, activists, and educators before me who have dedicated themselves to public education. May it be a new ground upon which many more after will rise up and build.

ACKNOWLEDGMENTS

A core theme of this book is belief. Belief in deep learning. Belief in AGI. Self-belief. How belief mobilizes and incites. Who is and isn't to be believed. Belief is a powerful and intoxicating thing. And in my own career, it has been the belief of so many people in me and my work that has been my greatest enabler.

Thank you first and foremost to the people who believed in this book project. I especially owe so much to all my sources. Many spoke to me despite legal or other risks because they believed in truth, transparency, and accountability. Many were also extremely generous with their time—inviting me to their homes, showing me around their communities, or sitting for upward of ten hours of interviews across multiple sessions. I will never take for granted the leap of faith someone makes to open up their heart and mind to a journalist. It is an absolute honor to tell your stories. Thank you. Without you, this book simply wouldn't exist.

My sincere gratitude to David Doerrer, my wonderful agent, who was first to commit his time and support to me to develop my ideas for this book before I had anything worthy to show him. It was through his patience, his probing questions, and, most importantly, his ability to kindly and firmly tell me when something just wasn't working that I began to see how my scattered collection of thoughts could form the basis of a book.

To Scott Moyers at Penguin Press, any writer's dream of an editor, who immediately understood my vision and whose unfailing support ever since allowed me to pursue it in its most ambitious form. Not only was he a moral compass and cheerleader throughout my reporting and writing process, providing incisive and illuminating feedback, he and Ann Godoff also committed Penguin Press's financial and legal resources to support the book's creation. Reporting and writing a book like this is expensive, sensitive, and time-consuming. Among many other things, it requires hiring researchers and fact-checkers; paying for flights, accommodations, local collaborators, translators, and drivers to spend time on the ground embedded within communities. Scott and Ann made it possible for me to do all of that and to work on this book full time.

To Mia Council at Penguin Press, whose sharp and compassionate edits gave me the prompting and security I needed to take risks in and push my writing to the next level. Her masterful coordination behind the scenes made the entire editing and production process feel so seamless. I am sure I didn't even see half of the logistical chaos that Mia so expertly contained. Thank you also to the rest of the stellar Penguin Press and Penguin Random House US and UK teams, including Yuki Hirose, Gail Brussel, Juli Kiyan, Danielle Plafsky, Laura Stickney, Kim Walker, Rosie Brown, Lotte Hall, Karen Dziekonski, and the many others with whom I didn't interface directly but were critical to the process.

To my incredible fact-checking team: Lindsay Muscato, Matt Mahoney, Rima Parikh, and Muriel Alarcón. All four of them fastidiously combed through the draft, cross-checking the labyrinth of details against documents and sources, and stress-testing my word choices. Matt also supported early research in my book, and Lindsay fielded many calls from me to serve as the most patient sounding board, while Rima somehow turned her fact-checking notes into standup comedy. They are all lifesavers.

Muriel was also my reporting partner extraordinaire in Chile and Uruguay. She is a one-woman wonder: She conducted research, coordinated interviews, chased down sources, and played both translator and driver across two weeks of nonstop reporting, all with the most beautiful, joyous energy. We had so many great laughs and adventures.

Thank you to everyone else who supported me in my reporting trips, especially Stephen Thuo Kiguru, my intrepid guide through Nairobi who was there for anything I needed and whose humor and relentless optimism remained unflappable even when someone called the cops on him after one

ACKNOWLEDGMENTS

of our excursions due to a gross miscommunication. That is a story for another time.

To my dear friends and mentors: Angela Chen, Gideon Lichfield, Roger McNamee, Brenda Guadalupe López Alatorre, Jose Manuel Rodriguez Moreno, Bina Venkataraman, and Tate Ryan-Mosley, who generously read early drafts of the book or various excerpts and gave me wise and invaluable feedback. To Oren Etzioni, who graciously reviewed my recounting of AI history and all of my technical explanations of AI research to ensure they were correct and appropriately nuanced. To Ria Kalluri, whose friendship was a source of strength and joy long before we both began investigating the colonial nature of AI development, and whose intellectual and moral clarity on the subject has been a guiding light.

This book also draws upon reporting and work I did throughout my journalism career. I would be remiss not to thank all of the people who supported me along the way. All my gratitude to Janet Guyon, my editor at *Quartz*, who was the first person who knew what she was talking about to tell me she believed I could make a great journalist. To Gideon, then the editor in chief of *MIT Technology Review*, who made a crazy bet to give me my first full-time job in journalism, to cover artificial intelligence, no less, putting me on a yearslong journey I could have never imagined. To Niall Firth, my editor at *Tech Review*, who said to me one day, *Why don't you profile OpenAI?* when I had never profiled a company before. For whatever reason, Niall believed I could do it. And I worked harder and pushed myself further to prove him right.

When I started realizing and turning my attention to the vast global inequality that AI was perpetuating, Niall was also instrumental in supporting my new line of inquiry, as was Angela, the colleague I referenced in chapter 4 who identified the phrase "data colonialism" in existing scholarship and helped point me in the right direction. Mat Honan, who took over from Gideon as chief editor, quickly understood the importance of my investigations and wholeheartedly supported me in pursuing them further.

In late 2021, I went on leave from *Tech Review* to pursue a six-month reporting project about "AI colonialism" with the wonderful support of MIT's Knight Science Journalism fellowship and a Pulitzer Center AI Accountability grant. I am indebted to Deborah Blum and Ashley Smart at

ACKNOWLEDGMENTS

MIT KSJ and Marina Walker Guevara and Boyoung Lim at the Pulitzer Center for giving me the funding to pursue such an expansive project in the middle of the pandemic. The result, a four-part series with stories from South Africa, Venezuela, Indonesia, and New Zealand, laid the groundwork for the thesis and, ultimately, the title of this book. Thank you to my incredible collaborators on those stories: Heidi Swart, Andrea Paola Hernández, and Nadine Freischlad, whose reporting expertise, language skills, and deep local and cultural context made those stories richer than I ever could have alone. It was under Marina's visionary leadership in global journalistic collaborations that I connected with Heidi, Andrea, and Nadine in the first place, and learned a new approach for tackling globe-spanning reporting projects. The Pulitzer Center's AI Accountability Network, which brings together journalists from around the world to advance AI accountability reporting, has since become one of my most important professional communities.

At *The Wall Street Journal*, it was my editor Josh Chin who first encouraged me to pitch what ultimately became a front-page story about Mophat Okinyi and the Kenyan workers who contracted for OpenAI; Drew Dowell and Jason Dean helped secure my reporting trip to make it happen. At *The Atlantic*, my editor Damon Beres needed no convincing when I pitched him a wonky story about the environmental impacts of the computing infrastructure behind AI, nor did Paul Bisceglio or Adrienne LaFrance, who green-lit another reporting trip to Arizona. Thank you also to Bradley Olson, Deepa Seetharaman, Daniel Engber, and Matteo Wong for helping me bring those stories to fruition. Working alongside my inimitable colleagues at both publications gave me a whole new understanding of what it means to report and write stories at the highest levels of mastery.

Finally, my deepest love and gratitude to my family. To my mom, who saw my love of writing as a little girl and poured everything she ever had into helping me achieve my dreams. To my dad, who never once questioned doing whatever he could to support me. To my 奶奶, my unending source of inspiration. To my in-laws, who wisely remind me to savor the process and celebrate the wins. To my husband: best friend, life partner, moral compass, cheerleader, number one fan, early reader, sounding board, advice giver, endless romantic. Loving you and being loved by you is my foundation for everything.

NOTES

Epigraphs

vii **"It is said"**: Joseph Weizenbaum, "ELIZA—a Computer Program for the Study of Natural Language Communication Between Man and Machine," *Communications of the ACM* 9, no. 1 (January 1966): 36–45, doi.org/10.1145/365153.365168.

vii **"Successful people create companies"**: Sam Altman, "Successful People," *Sam Altman* (blog), March 7, 2013, blog.samaltman.com/successful-people.

Prologue: A Run for the Throne

1 **"How can I help"**: Tripp Mickle, Cade Metz, Mike Isaac, and Karen Weise, "Inside OpenAI's Crisis over the Future of Artificial Intelligence," *New York Times*, December 9, 2023, nytimes.com/2023/12/09/technology/openai-altman-inside-crisis.html.

1 **Altman, still confused:** Trevor Noah, host, *What Now? with Trevor Noah*, season 1, episode 5, "Sam Altman Speaks Out about What Happened at OpenAI," Spotify Podcasts, December 7, 2023, open.spotify.com/show/122imavATqSE7eCyXIcqZL.

2 **The public announcement went up:** OpenAI, "OpenAI Announces Leadership Transition," *OpenAI* (blog), November 17, 2023, openai.com/index/openai-announces-leadership-transition.

3 **Shocked employees learned:** Unless otherwise noted, the insider accounts of the employees', board directors', and leadership's experiences throughout the board crisis are based on eleven people who were present across each of the scenes recounted.

4 **"Was there a specific incident"**: All dialogue from the all-hands meeting is from an audio recording of the meeting, November 17, 2023.

6 **Right before the event:** A screenshot of the alert, November 17, 2023.

6 **Microsoft's Nadella, who:** Hannah Miller, Brad Stone, Shirin Ghaffary, and Ashlee Vance, "Silicon Valley Boardroom Coup Leads to Ouster of an AI Champion," *Bloomberg*, November 17, 2023, bloomberg.com/news/articles/2023-11-18/openai-altman-ouster-followed-debates-between-altman-board.

7 **Riled up by Sutskever's:** Keach Hagey, Deepa Seetharaman, and Berber Jin, "Behind the Scenes of Sam Altman's Showdown at OpenAI," *Wall Street Journal*, November 22, 2023, wsj.com/tech/ai/altman-firing-openai-520a3a8c.

8 **The next day, Saturday:** Kate Clark, Natasha Mascarenhas, and Anissa Gardizy, "If

Sam Altman Returns to OpenAI, Board Will Go," *The Information*, November 18, 2023, theinformation.com/articles/altman-decision-looms-as-sequoia-tiger-negotiate-behind-scenes.

8 **"We are still working":** Erin Woo, Anissa Gardizy, and Amir Efrati, "OpenAI 'Optimistic' It Can Bring Back Sam Altman, Greg Brockman," *The Information*, November 18, 2023, theinformation.com/articles/openai-optimistic-it-can-bring-back-sam-altman-greg-brockman?rc=ot38so.

9 **A source relayed the playbook:** Alex Konrad and David Jeans, "OpenAI Investors Plot Last-Minute Push with Microsoft to Reinstate Sam Altman as CEO," *Forbes*, November 18, 2023, forbes.com/sites/alexkonrad/2023/11/18/openai-investors-scramble-to-reinstate-sam-altman-as-ceo.

9 **"The board firmly stands":** All quotes from OpenAI's Slack are pulled from screenshots.

10 **Anna Brockman, Greg's wife:** Deepa Seetharaman, Berber Jin, and Keach Hagey, "OpenAI Investors Keep Pushing for Sam Altman's Return," *Wall Street Journal*, November 21, 2023, wsj.com/tech/openai-employees-threaten-to-quit-unless-board-resigns-bbd5cc86.

11 **In the office, the company's:** Photos of the setup in the office.

12 **At some point, someone:** Amir Efrati, Anissa Gardizy, and Erin Woo, "Altman Agrees to Internal Investigation upon Return to OpenAI," *The Information*, November 21, 2023, theinformation.com/articles/breaking-sam-altman-to-return-as-openai-ceo.

12 **He tweeted it with:** Greg Brockman (@gdb), "we are so back," Twitter (now X), November 21, 2024, x.com/gdb/status/1727230819226583113.

17 **As Baidu raced to develop:** Raffaele Huang and Karen Hao, "Baidu Hurries to Ready China's First ChatGPT Equivalent Ahead of Launch," *Wall Street Journal*, March 9, 2023, wsj.com/articles/baidu-scrambles-to-ready-chinas-first-chatgpt-equivalent-ahead-of-launch-bf359ca4.

18 **Since ChatGPT, the six:** Parmy Olson and Carolyn Silverman, "ChatGPT's $8 Trillion Birthday Gift to Big Tech," *Bloomberg*, November 29, 2024, bloomberg.com/opinion/articles/2024-11-29/chatgpt-turns-2-and-gives-8-trillion-birthday-gift-to-big-tech.

18 **In June 2024, a Goldman:** *Gen AI: Too Much Spend, Too Little Benefit?*, Goldman Sachs, June 27, 2024, goldmansachs.com/insights/top-of-mind/gen-ai-too-much-spend-too-little-benefit.

18 **The following month, a survey:** "Upwork Study Finds Employee Workloads Rising Despite Increased C-Suite Investment in Artificial Intelligence," Upwork, July 23, 2024, investors.upwork.com/news-releases/news-release-details/upwork-study-finds-employee-workloads-rising-despite-increased-c.

18 **the data "raises an uncomfortable":** Olson and Silverman, "ChatGPT's $8 Trillion Birthday Gift."

19 **In a September 2024 blog post:** Sam Altman, "The Intelligence Age," *Sam Altman* (blog), September 23, 2024, ia.samaltman.com.

Chapter 1: Divine Right

23 **Everyone else had arrived:** Cade Metz, *Genius Makers: The Mavericks Who Brought AI to Google, Facebook, and the World* (Dutton, 2021), 161.

23 **It was the summer:** Various accounts of this meeting have been reported over the years, including in Cade Metz's *Genius Makers*, *Wired*, and *The Atlantic*. Greg Brockman also wrote his account in two blog posts: Greg Brockman, "My Path to OpenAI," *Greg Brockman* (blog), May 3, 2016, blog.gregbrockman.com/my-path-to-openai; and Greg Brockman, "#define CTO OpenAI," *Greg Brockman* (blog), January 9, 2017, blog.gregbrockman.com/define-cto-openai.

23 **It was as if, Musk:** Musk's views on Altman, Musk's experience cofounding OpenAI, and the evolution of Musk's views on AI are largely based on a lawsuit Musk filed against Altman, Brockman, and OpenAI on February 29, 2024, and refiled on August 5, 2024: Musk v. Altman, No. 4:24-cv-04722, CourtListener (N.D. Cal. August 5, 2024). Additional color comes primarily from Maureen Dowd, "Elon Musk's Future Shock," *Vanity Fair*, April 2017, archive.vanityfair.com/article/2017/4/elon-musks-future-shock; and Walter Isaacson, *Elon Musk* (Simon & Schuster, 2023), 239–44, Kindle.

NOTES

23 **For Altman's part:** Lex Fridman, host, *Lex Fridman Podcast*, podcast, episode 367, "Sam Altman: OpenAI CEO on GPT-4, ChatGPT, and the Future of AI," March 25, 2023, lexfridman.com/podcast.
23 **"The thing that sticks":** Sam Altman, "How to Be Successful," *Sam Altman* (blog), January 24, 2019, blog.samaltman.com/how-to-be-successful.
24 **Later, at a recurring AI:** Author interview with Timnit Gebru, August 2023.
25 **"Murdering all competing":** Tad Friend, "Sam Altman's Manifest Destiny," *New Yorker*, October 3, 2016, newyorker.com/magazine/2016/10/10/sam-altmans-manifest-destiny.
25 **"The future of AI":** Isaacson, *Elon Musk*, 241.
25 **As part of the evaluation:** "Decoding Google Gemini with Jeff Dean," posted September 11, 2024, by Google DeepMind, YouTube, 55 min., 55 sec., youtu.be/lH74gNeryhQ; author correspondence with Google spokesperson, November 2024.
25 **The meeting convinced Musk:** Based on the recollections and characterizations of four people who spoke with Musk or were present when he expressed his views, as well as *Musk*, CourtListener, ECF No. 32, Exhibit 13.
26 **"It seemed a little":** A Google DeepMind spokesperson also rejected Musk's characterization of Hassabis. Author correspondence with Google DeepMind spokesperson, November 2024.
26 **Given a simple objective:** Nick Bostrom, *Superintelligence: Paths, Dangers, Strategies* (Oxford University Press, 2014), 149–52, Kindle.
26 **To his far-reaching Twitter:** Elon Musk (@elonmusk), "Worth reading Superintelligence by Bostrom. We need to be super careful with AI. Potentially more dangerous than nukes," Twitter (now X), August 3, 2014, x.com/elonmusk/status/495759307346952192.
26 **Bostrom would apologize:** Nick Bostrom, "Apology for an Old Email," Nick Bostrom's Home Page, January 9, 2023, nickbostrom.com/oldemail.pdf.
27 **Two years later Altman would:** Olivia Carville, "The Super Rich of Silicon Valley Have a Doomsday Escape Plan in New Zealand," *Bloomberg*, September 5, 2018, bloomberg.com/features/2018-rich-new-zealand-doomsday-preppers.
27 **"probably the greatest threat":** Sam Altman, "Machine Intelligence, Part 1," *Sam Altman* (blog), February 25, 2015, blog.samaltman.com/machine-intelligence-part-1.
27 **"Been thinking a lot":** All email correspondence between Musk and Altman in this chapter are from Musk's lawsuit as exhibits attached to document number 32: *Musk*, CourtListener, ECF No. 32.
28 **"I am now very much":** Melia Russell and Julia Black, "He's Played Chess with Peter Thiel, Sparred with Elon Musk and Once, Supposedly, Stopped a Plane Crash: Inside Sam Altman's World, Where Truth Is Stranger Than Fiction," *Business Insider*, April 27, 2023, businessinsider.com/sam-altman-openai-chatgpt-worldcoin-helion-future-tech-2023-4.
28 **"You could parachute him":** Paul Graham, "A Fundraising Survival Guide," *Paul Graham* (blog), accessed November 21, 2024, paulgraham.com/fundraising.html.
28 **"Sam is extremely good":** Friend, "Sam Altman's Manifest Destiny."
29 **Jerry, the son of a:** "Megan O'Neill Is Wed to Jerold D. Altman," *New York Times*, July 24, 1977, nytimes.com/1977/07/24/archives/megan-oneill-is-wed-to-jerold-d-altman.html.
29 **"You always help people":** Berber Jin and Keach Hagey, "The Contradictions of Sam Altman, AI Crusader," *Wall Street Journal*, March 31, 2023, wsj.com/tech/ai/chatgpt-sam-altman-artificial-intelligence-openai-b0e1c8c9.
30 **From a young age:** The account of Altman's early childhood is based largely on three main profiles of him: Friend, "Sam Altman's Manifest Destiny"; Elizabeth Weil, "Sam Altman Is the Oppenheimer of Our Age," *New York*, September 25, 2023, nymag.com/intelligencer/article/sam-altman-artificial-intelligence-openai-profile.html; and Ellen Huet, host, *Foundering: The OpenAI Story*, podcast, season 5, episode 1, "The Most Silicon Valley Man Alive," Bloomberg Podcasts, June 5, 2024, bloomberg.com/news/articles/2024-06-05/foundering-sam-altman-s-rise-to-openai.
30 **When his grandmother:** "Sam Altman: How to Build the Future," posted September 27, 2016, by Y Combinator, YouTube, 20 min., 9 sec., youtu.be/sYMqVwsewSg.
30 **"I remember thinking":** Huet, "The Most Silicon Valley Man Alive."
30 **He loved to push:** Parmy Olson, *Supremacy: AI, ChatGPT, and the Race that Will Change the World* (St. Martin's Press, 2024), 5.
31 **"Either you have tolerance":** Weil, "Sam Altman Is the Oppenheimer of Our Age."

NOTES

31 **As his star rose:** Friend, "Sam Altman's Manifest Destiny."
31 **He would grow so panicked:** Joe Hudson and Brett Kistler, hosts, *The Art of Accomplishment Podcast*, podcast, episode 39, "Sam Altman—Leading with Crippling Anxiety, Discovering Meditation, and Building Intelligence with Self-Awareness," January 14, 2022, artofaccomplishment.com/podcast.
31 **After spending many hours:** Friend, "Sam Altman's Manifest Destiny."
31 **"I realized that the world":** "Office Hours with Sam Altman," posted January 11, 2017, by Y Combinator, YouTube, 24 min., 34 sec., youtu.be/45BvnJgwYjk.
32 **He dug deep into assignments:** Russell and Black, "He's Played Chess with Peter Thiel."
32 **After learning that phones:** Deepa Seetharaman, Keach Hagey, Berber Jin, and Kate Linebaugh, "Sam Altman's Knack for Dodging Bullets—with a Little Help from Bigshot Friends," *Wall Street Journal*, December 24, 2023, https://www.wsj.com/tech/ai/sam-altman-openai-protected-by-silicon-valley-friends-f3efcf68.
32 **"Work really hard":** "Sam Altman Startup School Video," posted July 26, 2017, by Waterloo Engineering, YouTube, 1 hr., 18 min., 19 sec., youtu.be/4SlNgM4PjvQ.
32 **By late 2005, he:** "Paper Chase," *Venture Capital Journal*, December 1, 2006, venturecapitaljournal.com/paper-chase.
32 **After a seven-year run:** Annie Massa and Vernal Galpotthawela, "Sam Altman Is Worth $2 Billion—That Doesn't Include OpenAI," *Bloomberg*, March 1, 2024, bloomberg.com/news/articles/2024-03-01/sam-altman-is-a-billionaire-thanks-to-vc-funds-startups.
33 **"The response has been tremendous":** "First Look: Loopt Provides More Incentives to Try Location-Based Services with Loopt Star," posted May 31, 2010, by Robert Scoble, YouTube, 15 min., 24 sec., youtu.be/P5izvkusAMM.
33 **"It's a ridiculous distinction":** "Why Loopt Partnered with Facebook," posted November 3, 2010, by CNN Business, YouTube, 2 min., 41 sec., youtu.be/tMO0Gm6yxWc.
33 **"He didn't just want":** Jessica E. Lessin, "This Is How Sam Altman Works the Press and Congress. I Know from Experience," *The Information*, June 7, 2023, theinformation.com/articles/this-is-how-sam-altman-works-the-press-and-congress-i-know-from-experience.
34 **Right as Loopt was getting:** Russell and Black, "He's Played Chess with Peter Thiel."
34 **Altman would become a Reddit:** Christine Lagorio-Chafkin, "Inside Reddit's Long, Complicated Relationship with OpenAI's Sam Altman," *Inc.*, March 8, 2024, inc.com/christine-lagorio/inside-reddits-long-complicated-relationship-with-openais-sam-altman.html.
34 **"Usain Bolt of fundraising":** Author interview with Geoff Ralston, March 2024.
35 **Twice during his time running Loopt:** Seetharaman et al., "Sam Altman's Knack for Dodging Bullets."
35 **Jobs had been worth:** Walter Isaacson, *Steve Jobs* (Simon & Schuster, 2011), 104.
35 **He'd collect luxury:** Katie Notopoulos, "Sam Altman Is Seen Driving a Car That Can Cost $5 Million. Everyone Is Thanking Him for Helping Them Pass Their Tests," *Business Insider*, July 12, 2024, businessinsider.com/sam-altman-koenigsegg-regera-expensive-sports-car-video-openai-musk-2024-7.
36 **Of particular importance:** Elizabeth Dwoskin, Marc Fisher, and Nitasha Tiku, "'King of the Cannibals': How Sam Altman Took Over Silicon Valley," *Washington Post*, December 23, 2023, washingtonpost.com/technology/2023/12/23/sam-altman-openai-peter-thiel-silicon-valley.
37 **This was not a flaw:** Eric Newcomer, "YC's Paul Graham: The Complete Interview," December 26, 2013, *The Information*, theinformation.com/articles/yc-s-paul-graham-the-complete-interview.
37 **"Loopt is probably the most":** Paul Graham, "A Student's Guide to Startups," *Paul Graham* (blog), October 2006, paulgraham.com/mit.html.
37 **Altman quickly inspired Graham:** Paul Graham, "What We Look for in Founders," *Paul Graham* (blog), October 2010, paulgraham.com/founders.html.
37 **"Sam is, along with Steve":** Paul Graham, "Five Founders," *Paul Graham* (blog), April 2009, paulgraham.com/5founders.html.
37 **When Graham asked:** Friend, "Sam Altman's Manifest Destiny."
38 **Their bond was once described:** Dwoskin et al., "'King of the Cannibals.'"
38 **"The first piece of startup":** Sam Altman, "Growth and Government," *Sam Altman* (blog), March 4, 2013, blog.samaltman.com/growth-and-government.

NOTES

- 39 **"The thing that people":** "Sam Altman Startup School Video," Waterloo Engineering.
- 39 **"Sustainable economic growth is":** Tyler Cowen, host, *Conversations with Tyler*, podcast, episode 61, "Sam Altman on Loving Community, Hating Coworking, and the Hunt for Talent," Mercatus Center Podcasts, February 27, 2019.
- 39 **Monopolies are good:** "Competition Is for Losers with Peter Thiel (How to Start a Startup 2014: 5)," posted March 22, 2017, by Y Combinator, YouTube, 50 min., 27 sec., youtu.be/3Fx5Q8xGU8k.
- 40 **"I've heard a lot":** Sam Altman, "How Things Get Done," *Sam Altman* (blog), July 17, 2013, blog.samaltman.com/how-things-get-done.
- 40 **"For startups I think":** "Sam Altman Startup School Video," Waterloo Engineering.
- 41 **Over time he accumulated:** Berber Jin, Tom Dotan, and Keach Hagey, "The Opaque Investment Empire Making OpenAI's Sam Altman Rich," *Wall Street Journal*, June 3, 2024, wsj.com/tech/ai/openai-sam-altman-investments-004fc785.
- 41 **During the 2023 Silicon Valley Bank:** Author interview with Matt Krisiloff, April 2024.
- 41 **"It's an extremely rare trait":** Author interview with Lachy Groom, February 2024.
- 41 **For a time, the political:** Sam Altman, "The 2016 Election," *Sam Altman* (blog), October 17, 2016, blog.samaltman.com/the-2016-election.
- 42 **He published a manifesto:** Sam Altman, "The United Slate," *Sam Altman* (blog), July 12, 2017, blog.samaltman.com/the-united-slate.
- 42 **He'd built eighteen pounds:** Dwoskin et al., "'King of the Cannibals.'"
- 43 **In 2016, it was Ashton:** Friend, "Sam Altman's Manifest Destiny."
- 43 **Three years later:** A photo of Schumer's visit to the Pioneer Building, March 8, 2019.
- 43 **Altman's climb would also:** Author interviews with Annie Altman, March–November 2024.
- 44 **In a public statement:** Sam Altman (@sama), "My sister has filed a lawsuit against me. Here is a statement from my mom, brothers, and me:," Twitter (now X), January 7, 2025, x.com/sama/status/1876780763653263770.
- 44 **In response to my requests:** Author correspondence with Connie Gibstine, October 2024.
- 45 **She would subsequently file:** Altman v. Altman, No. 4:25-cv-00017, CourtListener (E.D. Mo. Jan 06, 2025) ECF No. 1.

Chapter 2: A Civilizing Mission

- 46 **He had grown up:** Author interviews with Greg Brockman, August 2019.
- 47 **Lean and wiry, he:** Sutskever's education and early background is based partly on his various media interviews, including: "Interview with Dr. Ilya Sutskever, Co-founder of OPEN AI—at the Open University Studios—English," posted September 13, 2023, by The Open University of Israel, YouTube, 50 min., 28 sec., youtu.be/H1YoNlz2LxA; Nina Haikara, "This U of T Alum Is Leading AI research at $1 Billion Non-profit Backed by Elon Musk," U of T News, March 28, 2017, utoronto.ca/news/u-t-alum-leading-ai-research-1-billion-non-profit-backed-elon-musk; and Varsity Contributor, "Neural Networking," *The Varsity*, October 25, 2010, thevarsity.ca/2010/10/25/neural-networking.
- 47 **Where every other team struggled:** The breakthrough results, which happened in 2012, were published in a journal five years later: Alex Krizhevsky, Ilya Sutskever, and Geoffrey E. Hinton, "ImageNet Classification with Deep Convolutional Neural Networks," *Communications of the ACM* 60, no. 6 (May 2017): 84–90, doi.org/10.1145/3065386.
- 47 **"We thought we were":** Author interview with Geoff Hinton, August 2023.
- 48 **Even to Sutskever, who secretly:** Cade Metz, *Genius Makers: The Mavericks Who Brought AI to Google, Facebook, and the World* (Dutton, 2021), 289.
- 48 **"I knew it was going to work":** Greg Brockman, "#define CTO OpenAI," *Greg Brockman* (blog), January 9, 2017, blog.gregbrockman.com/define-cto-openai.
- 49 **Altman would later extol:** Sam Altman, "Greg," *Sam Altman* (blog), March 7, 2017, blog.samaltman.com/greg.
- 49 **"AGI might be far away":** Author interview with Pieter Abbeel, August 2019.
- 49 **Undeterred, he invited his ten:** Metz, *Genius Makers*, 163.
- 49 **"I hope for us to":** All correspondence among OpenAI and Tesla leadership in this chapter are from Musk's lawsuit as exhibits attached to document number 32: Musk v. Altman, No.

4:24-cv-04722, CourtListener (N.D. Cal. November 14, 2024) ECF No. 32; and OpenAI's responses on the company's blog: OpenAI, "OpenAI and Elon Musk," *OpenAI* (blog), March 5, 2024, openai.com/index/openai-elon-musk; OpenAI, "Elon Musk Wanted an OpenAI For-Profit," *OpenAI* (blog), December 13, 2024, openai.com/index/elon-musk-wanted-an-openai-for-profit/#summer-2017-we-and-elon-agreed-that-a-for-profit-was-the-next-step-for-openai-to-advance-the-mission.

50 **To all of the other founding:** *Musk, CourtListener*, ECF No. 32, Exhibit 7.
50 **To Sutskever, the lab had instead:** "Openai Inc," ProPublica Nonprofit Explorer, accessed August 25, 2024, projects.propublica.org/nonprofits/organizations/810861541/201703459349300445/full.
50 **Even then, Google had offered:** Metz, *Genius Makers*, 164.
50 **Musk and Altman delayed:** Metz, *Genius Makers*, 164.
51 **To preempt any other counteroffers:** *Musk, CourtListener*, ECF No. 32, Exhibit 7.
51 **Musk would later recount facing:** Walter Isaacson, *Elon Musk* (Simon & Schuster, 2023), 243, Kindle.
51 **Automated software being sold:** An early, seminal contribution to the understanding of how AI leads to discrimination comes from Solon Barocas and Andrew D. Selbst, "Big Data's Disparate Impact," *California Law Review* 104, no. 3 (2016): 671–732, ssrn.com/abstract=2477899. Here's also a story that dives more into how this discrimination plays out in practice: Karen Hao, "The Coming War on the Hidden Algorithms that Trap People in Poverty," *MIT Technology Review*, December 4, 2020, technologyreview.com/2020/12/04/1013068/algorithms-create-a-poverty-trap-lawyers-fight-back.
52 **precipitated ethnic cleansing:** Alexandra Stevenson, "Facebook Admits It Was Used to Incite Violence in Myanmar," *New York Times*, November 6, 2018, nytimes.com/2018/11/06/technology/myanmar-facebook.html.
52 **The capabilities, employees said:** Eric Lipton, "As A.I.-Controlled Killer Drones Become Reality, Nations Debate Limits," *New York Times*, November 21, 2023, nytimes.com/2023/11/21/us/politics/ai-drones-war-law.html.
52 **"It was a beacon":** Author interview with Chip Huyen, August 2019.
52 **All week the Stanford University:** Author interview with Timnit Gebru, March 2021.
53 **"Hello from Timnit":** Copy of the email, provided by Gebru.
54 **Some years later, Brockman would:** Author interview with Brockman, August 2019.
55 **"How could that *be*?":** Interview with Brockman.
55 **"a failure of imagination":** Arthur C. Clarke, *Profiles of the Future: An Inquiry into the Limits of the Possible* (Bantam Books, 1962), 30–39.
55 **"the problem of accidents":** Dario Amodei, Chris Olah, Jacob Steinhardt, Paul Christiano, John Schulman, and Dan Mané, "Concrete Problems in AI Safety," preprint, arXiv, July 25, 2016, 1–29, doi.org/10.48550/arXiv.1606.06565.
56 **By November 2024, it had:** Author correspondence with Open Philanthropy spokesperson, November 2024.
56 **Around the same time Amodei:** Julia Angwin, Jeff Larson, Surya Mattu, and Lauren Kirchner, "Machine Bias," *ProPublica*, May 23, 2016, propublica.org/article/machine-bias-risk-assessments-in-criminal-sentencing.
56 **Deborah Raji, an AI accountability:** The paper was presented at the International Conference of Learning Representations in a workshop called "Machine Learning in Real Life" on April 26, 2020, sites.google.com/nyu.edu/ml-irl-2020/home and posted on arXiv a few years later: Inioluwa Deborah Raji and Roel Dobbe, "Concrete Problems in AI Safety, Revisited," arXiv, December 18, 2023: 1–6, doi.org/10.48550/arXiv.2401.10899.
57 **"There are twenty to thirty":** Tad Friend, "Sam Altman's Manifest Destiny," *New Yorker*, October 3, 2016, newyorker.com/magazine/2016/10/10/sam-altmans-manifest-destiny.
57 **Thereafter, Open Phil would:** "OpenAI—General Support," Open Philanthropy, accessed November 27, 2024, openphilanthropy.org/grants/openai-general-support.
58 **"We have a long":** Author interview with Greg Brockman and Daniela Amodei, August 2019.
59 **In 2016, OpenAI spent:** ProPublica Nonprofit Explorer, "Openai Inc."
59 **So in March 2017:** Author interview with Brockman and Ilya Sutskever, August 2019.
60 **Around the same time, Amodei:** Interview with Brockman and Sutskever.

NOTES

433

60 **In the last six years:** OpenAI, "AI and Compute," *Open AI* (blog), May 16, 2018, openai.com/index/ai-and-compute.
61 **They briefly considered merging:** *Musk*, CourtListener, ECF No. 32; Id., ECF No. 32, Exhibit 11.
62 **So did Musk:** Id., ECF No. 32, Exhibit 13; OpenAI, "OpenAI and Elon Musk"; OpenAI, "Elon Musk Wanted an OpenAI For-Profit."
63 **Brockman and Sutskever continued:** Interviews with Brockman, August 2019.
63 **He called Reid Hoffman:** OpenAI, "OpenAI and Elon Musk."
63 **He considered launching:** *Musk*, CourtListener, ECF No. 32, Exhibit 15.
63 **Previously, with Musk's firm backing:** *Id.*, ECF No. 32, Exhibit 7.
64 **Altman became president:** ProPublica Nonprofit Explorer, "Openai Inc."
65 **Of the $1 billion commitment:** *Musk*, CourtListener, ECF No. 1, at *46–48.
65 **The intern was later commemorated:** Berber Jin and Keach Hagey, "The Contradictions of Sam Altman, AI Crusader Behind ChatGPT," *Wall Street Journal*, March 31, 2023, wsj.com/tech/ai/chatgpt-sam-altman-artificial-intelligence-openai-b0e1c8c9.
66 **Professionals were hired, and Brockman:** Correspondence with Jennifer 8. Lee, a coproducer on the documentary: *Artificial Gamer*, directed by Chad Herschberger, featuring Pieter Abbeel, Greg Brockman, and Noam Brown, released on September 24, 2021, artificialgamerfilm.com. The documentary team retained editorial independence of the film.
67 **In April 2018, OpenAI:** OpenAI, "OpenAI Charter," *Open AI* (blog), accessed August 25, 2024, openai.com/charter.
67 **That summer, as the *Dota*:** Jin and Hagey, "The Contradictions of Sam Altman."
68 **"Microsoft Research and OpenAI are":** Author interview with Xuedong Huang, July 2023.
68 **To keep the deal secret:** *Musk*, CourtListener, ECF No. 1, at *5.
68 **Around the same time, Altman:** Elizabeth Dwoskin and Nitasha Tiku, "Altman's Polarizing Past Hints at OpenAI Board's Reason for Firing Him," *Washington Post*, November 22, 2023, washingtonpost.com/technology/2023/11/22/sam-altman-fired-y-combinator-paul-graham.
69 **He had proposed the idea:** Deepa Seetharaman, Keach Hagey, and Berber Jin, "Sam Altman's Knack for Dodging Bullets—with a Little Help from Bigshot Friends," *Wall Street Journal*, December 24, 2023, wsj.com/tech/ai/sam-altman-openai-protected-by-silicon-valley-friends-f3efcf68.
70 **A payband structure:** Copy of the payband document.
70 **Executives also wrote up:** Karen Hao, "The Messy, Secretive Reality Behind OpenAI's Bid to Save the World," *MIT Technology Review*, February 17, 2020, technologyreview.com/2020/02/17/844721/ai-openai-moonshot-elon-musk-sam-altman-greg-brockman-messy-secretive-reality.
70 **"So someone who invests":** @windowshopping, "I was buying it until he said that profit is 'capped' at 100x of initial investment. So someone who invests $10 million has their investment 'capped' at $1 billion. Lol. Basically unlimited unless the company grew to a FAANG-scale market value," Hacker News, March 11, 2019, news.ycombinator.com/item?id=19360709.
70 **Initial investments poured in:** All numerical values for investments into the LP and their profit cap throughout the book are from an OpenAI internal financial document.
70 **Hoffman was initially reluctant:** Chamath Palihapitiya, Jason Calacanis, David Sacks, and David Friedberg, hosts, *All-In*, podcast, episode 194, "In Conversation with Reid Hoffman & Robert F. Kennedy Jr.," August 30, 2024, https://allin.com/episodes.
72 **"The thing that's interesting":** Kevin Scott's and Satya Nadella's emails were released in 2024 as part of the US Department of Justice's antitrust case against Google. Jyoti Mann and Beatrice Nolan, "Read the Email to Satya Nadella and Bill Gates That Shows Microsoft's CTO Was 'Very Worried' about Google's AI Progress in 2019," *Business Insider*, May 1, 2024, businessinsider.com/satya-nadella-bill-gates-microsoft-concern-google-rivals-ai-emails-2024-5.

Chapter 3: Nerve Center

74 **In 2021, OpenAI would:** Eddie Sun, "ChatGPT's San Francisco Offices Getting Nap Rooms, a Museum for Staffers," *San Francisco Standard*, July 11, 2023, sfstandard.com/2023/07/11/chatgpt-secretive-san-francisco-offices-nap-rooms-museum-open-ai.

NOTES

74 **Altman would oversee Mayo's:** I estimated the price of the furniture by reverse image searching photos of the office with Google Images. When I tried to get the exact price from the architecture firm that worked on OpenAI's Mayo office as well as confirmation of my descriptions ("Is it accurate to say the spiral staircase in the office is made of wood and stone?"), a person replied that the firm is under an NDA and cannot speak about the project.

74 **He would add a library:** Berber Jin and Keach Hagey, "The Contradictions of Sam Altman, AI Crusader," *Wall Street Journal*, March 31, 2023, wsj.com/tech/ai/chatgpt-sam-altman-artificial-intelligence-openai-b0e1c8c9.

74 **He wanted "a water feature":** Author interview with Ben Barry, former design director at OpenAI, October 2023.

77 **In December, Climate Change AI:** Information for each event can be found at: "NeurIPS 2019 Workshop: Tackling Climate Change with Machine Learning," Workshop at NeurIPS, Vancouver Convention Center, December 14, 2019, climatechange.ai/events/neurips2019; and "ML4H: Machine Learning for Health," Workshop at NeurIPS, Vancouver Convention Center, December 13, 2019, ml4h.cc/2019/index.html.

78 **"Technologies that would address":** The original white paper was written in 2019; it was published in a peer-reviewed journal in 2022. The quoted passage was edited in the 2022 version to: "Many technological tools useful in addressing climate change have been available for years but have yet to be adopted at scale by society. While we hope that ML will be useful in accelerating effective strategies for climate action, humanity also must decide to act." David Rolnick, Priya L. Donti, Lynn H. Kaack, Kelly Kochanski, Alexandre Lacoste, Kris Sankaran et al., "Tackling Climate Change with Machine Learning," *ACM Computing Surveys (CSUR)* 55, no. 2 (February 2022): 1–96, doi.org/10.1145/3485128.

79 **A recent study from:** Emma Strubell, Ananya Ganesh, and Andrew McCallum, "Energy and Policy Considerations for Deep Learning in NLP," *Proceedings of the 57th Annual Meeting of the Association for Computational Linguistics* (July 2019): 3645–50, doi.org/10.18653/v1/P19-1355.

80 **"I think that it's fairly":** Cade Metz, *Genius Makers: The Mavericks Who Brought AI to Google, Facebook, and the World* (Dutton, 2021), 299.

81 **He was a teen:** Author interviews with Brockman, August 2019.

86 **In February 2020:** Karen Hao, "The Messy, Secretive Reality Behind OpenAI's Bid to Save the World," *MIT Technology Review*, February 17, 2020, technologyreview.com/2020/02/17/844721/ai-openai-moonshot-elon-musk-sam-altman-greg-brockman-messy-secretive-reality.

86 **Hours later, Musk replied:** Elon Musk (@elonmusk), "OpenAI should be more open imo," Twitter (now X), x.com/elonmusk/status/1229544673590599681.

86 **Afterward, Altman sent OpenAI:** Copy of the email.

Chapter 4: Dreams of Modernity

88 **The authors point to:** Daron Acemoglu and Simon Johnson, *Power and Progress: Our Thousand-Year Struggle over Technology and Prosperity* (PublicAffairs, 2023), 129–33.

89 **In 1956, six years after:** The brief of what they planned to do at the summer workshop: John McCarthy, Marvin L. Minsky, Nathaniel Rochester, and Claude E. Shannon, "A Proposal for the Dartmouth Summer Research Project on Artificial Intelligence," Stanford University, August 31, 1955, jmc.stanford.edu/articles/dartmouth/dartmouth.pdf.

89 **John McCarthy, the Dartmouth professor:** At first, John McCarthy, Claude Shannon, and others collected research papers into a compendium on the same set of ideas that would be called AI and titled it "Automata Studies," which was published in 1956. McCarthy was disappointed by the papers that people submitted and their lack of ambition. He said it was that disappointment that led him to begin using the term *artificial intelligence*.

91 **In the early 1800s:** Kate Crawford, *Atlas of AI: Power, Politics, and the Planetary Costs of Artificial Intelligence* (Yale University Press, 2021), 123.

92 **A 2007 revision:** John McCarthy, "What Is Artificial Intelligence?," John McCarthy's Home Page, *Formal Reasoning Group*, November 12, 2007, www-formal.stanford.edu/jmc/whatisai.pdf.

93 **The goalposts for AI development:** Jenna Burrell, "Artificial Intelligence and the Ever-

NOTES

Receding Horizon of the Future," *Tech Policy Press*, June 6, 2023, techpolicy.press/artificial-intelligence-and-the-ever-receding-horizon-of-the-future.

95 **In 1969, he coauthored a book:** Marvin Minsky and Seymour A. Papert, *Perceptrons: An Introduction to Computational Geometry* (MIT Press, 1969).

95 **Under the hood, though:** Joseph Weizenbaum, "ELIZA—a Computer Program for the Study of Natural Language Communication Between Man and Machine," *Communications of the ACM* 9, no. 1 (January 1966): 36–45, doi.org/10.1145/365153.365168.

95 **In a paper Weizenbaum:** Weizenbaum, "ELIZA—a Computer Program."

96 **ELIZA's subsequent success:** Ben Tarnoff, "Weizenbaum's Nightmares: How the Inventor of the First Chatbot Turned Against AI," *The Guardian*, July 25, 2023, theguardian.com/technology/2023/jul/25/joseph-weizenbaum-inventor-eliza-chatbot-turned-against-artificial-intelligence-ai.

96 **He later published a tome:** Joseph Weizenbaum, *Computer Power and Human Reason: From Judgment to Calculation* (W. H. Freeman & Co, 1976).

97 **Each time the roadblocks mounted:** There isn't one unified history of when each AI winter was. Generally speaking, the first one is considered to have been during the '70s, triggered by a 1973 British Science Research Council report from Professor Sir James Lighthill of the University of Cambridge, called "Artificial Intelligence: A General Survey." In it, Lighthill observed, "In no part of the AI field have discoveries made so far produced the major impact that was then promised." The second AI winter was roughly in the late '80s to early '90s. Some scholars argue that during that time, while funding dried up for research labeled as "AI," money was still going toward the development of relevant techniques under different names. Some researchers also point to further AI winters. Stanford professor and AI luminary Fei-Fei Li describes the late '90s as a third one in her book: Fei-Fei Li, *The Worlds I See: Curiosity, Exploration, and Discovery at the Dawn of AI* (Flatiron Books, 2023), 89–90.

100 **Tech giants were already seeing:** Author interview with Geoffrey Hinton, August 2023.

101 **But alongside these impressive advances:** Shoshana Zuboff, *The Age of Surveillance Capitalism: The Fight for a Human Future at the New Frontier of Power* (PublicAffairs, 2019), 1–704.

102 **In 2023, a group:** Pratyusha Ria Kalluri, William Agnew, Myra Cheng, Kentrell Owens, Luca Soldaini, and Abeba Birhane, "The Surveillance AI Pipeline," preprint, arXiv, October 17, 2023, 10–11, doi.org/10.48550/arXiv.2309.15084.

103 **In 2019, an NBC investigation:** Olivia Solon, "Facial Recognition's 'Dirty Little Secret': Millions of Online Photos Scraped Without Consent," NBC News, March 12, 2019, nbcnews.com/tech/internet/facial-recognition-s-dirty-little-secret-millions-online-photos-scraped-n981921.

103 **I noticed, too, how:** For example, it wouldn't be long before reports would surface about how the misguided deployment of faulty facial recognition was leading to the misidentification of suspects and wrongful arrests. As of November 2024, six of the seven known individuals who were wrongfully accused in the US, leading some to jail time, job loss, separation from their children, and disrupted relationships, have been Black. Kashmir Hill, "Wrongfully Accused by an Algorithm," *New York Times*, June 24, 2020, nytimes.com/2020/06/24/technology/facial-recognition-arrest.html; Khari Johnson, "How Wrongful Arrests Based on AI Derailed 3 Men's Lives," *Wired*, March 7, 2022, wired.com/story/wrongful-arrests-ai-derailed-3-mens-lives.

103 **"We have the first mover's":** "ISTE 2017—Most Innovative Winning Pitch," posted July 13, 2018, by Max Newlon, YouTube, 7 min., 42 sec., youtu.be/oJt6cjdMGb4.

103 **A few months after:** Jane Li, "A 'Brain-Reading' Headband Is Facing a Backlash in China," *Quartz*, November 5, 2019, qz.com/1742279/a-mind-reading-headband-is-facing-backlash-in-china.

104 **I discovered the work of:** Nick Couldry and Ulises A. Mejias, *The Costs of Connection: How Data Is Colonizing Human Life and Appropriating It for Capitalism* (Stanford University Press, 2019), 1–352. For more reading on the concept of extractivism, refer to Rosemary Collard and Jessica Dempsey, "'Extractivism' Is Destroying Nature: To Tackle It Cop15 Must Go Beyond Simple Targets," *The Guardian*, December 8, 2022, theguardian.com/environment/2022/dec/08/extractivism-is-destroying-nature-to-tackle-it-cop15-must-go-beyond-simple-targets; and one of the foundational texts that defined the concept: Eduardo Gudynas, "Diez tesis urgentes sobre el nuevo extractivismo: Contextos y demandas bajo el

progresismo sudamericano actual," in *Extractivismo, Política y Sociedad*, eds. CAAP and CLAES (2009), 187, rosalux.org.ec/pdfs/extractivismo.pdf.

104 **The following year, a paper:** Shakir Mohamed, Marie-Therese Png, and William Isaac, "Decolonial AI: Decolonial Theory as Sociotechnical Foresight in Artificial Intelligence," *Philosophy and Technology* 33 (July 12, 2020): 659–84, doi.org/10.1007/s13347-020-00405-8.

104 **Not long after, in 2021:** Karen Hao and Heidi Swart, "South Africa's Private Surveillance Machine Is Fueling a Digital Apartheid," *MIT Technology Review*, April 19, 2022, technology review.com/2022/04/19/1049996/south-africa-ai-surveillance-digital-apartheid.

105 **From 2013 to 2022, corporate:** Nestor Maslej, Loredana Fattorini, Raymond Perrault, Vanessa Parli, Anka Reuel, Erik Brynjolfsson et al., *AI Index Report 2024*, Institute for Human-Centered AI, Stanford University, April 2024, 242, aiindex.stanford.edu/report.

105 **In 2021, Alphabet and Meta:** Steven Rosenbush, "Big Tech Is Spending Billions on AI Research. Investors Should Keep an Eye Out," *Wall Street Journal*, March 8, 2022, wsj.com/articles/big-tech-is-spending-billions-on-ai-research-investors-should-keep-an-eye-out-11646740800.

105 **By contrast, the US government:** Nur Ahmed, Muntasir Wahed, and Neil C. Thompson, "The Growing Influence of Industry in AI Research," *Science* 379, no. 6635 (March 2, 2023): 884–86, doi.org/10.1126/science.ade2420.

106 **From 2006 to 2020:** Ahmed et al., "The Growing Influence of Industry."

106 **Many were initially whisked away:** In 2017, Tom Eck, the CTO of industry platforms at IBM, famously said, "The top-tier A.I. researchers are getting paid the salaries of NFL quarterbacks, which tells you the demand and the perceived value." Dan Butcher, "If You really Know About Artificial Intelligence, You Could Earn As Much As an NFL Quarterback," eFinancialCareers, July 13, 2017, efinancialcareers.com/news/2017/07/top-talent-earns-high-ai-salaries-nfl-quarterbacks.

106 **In 2015, Uber infamously:** Mike Ramsey and Douglas MacMillan, "Carnegie Mellon Reels After Uber Lures Away Researchers," *Wall Street Journal*, May 31, 2015, wsj.com/articles/is-uber-a-friend-or-foe-of-carnegie-mellon-in-robotics-1433084582.

106 **In another study from Kalluri:** Abeba Birhane, Pratyusha Kalluri, Dallas Card, William Agnew, Ravit Dotan, and Michelle Bao, "The Values Encoded in Machine Learning Research," in *FAccT '22: Proceedings of the 2022 ACM Conference on Fairness, Accountability, and Transparency* (Association for Computing Machinery, 2022): 173–84, doi.org/10.1145/3531146.3533083.

107 **It might learn to associate:** You can watch a video of this happening: "Tesla FSD Beta—What-the-Hell Moments," posted January 20, 2022, by The Outspoken Nomad, YouTube, 15 min., 55 sec., youtu.be/RVkLI9pPd24?t=166.

107 **Experts concluded that:** *Collision Between Vehicle Controlled by Developmental Automated Driving System and Pedestrian, Tempe, Arizona, March 18, 2018*, Highway Accident Report, NTSB/HAR-19/03, PB2019-101402, National Transportation Safety Board, November 19, 2019, ntsb.gov/investigations/AccidentReports/Reports/HAR1903.pdf.

107 **Six years later, in April:** A crash analysis of accidents involving Tesla Autopilot: "Additional Information Regarding EA22002," National Highway Traffic Safety Administration, April 25, 2024, 1–6, static.nhtsa.gov/odi/inv/2022/INCR-EA22002-14496.pdf.

107 **In 2019, white hat hackers:** Karen Hao, "Hackers Trick a Tesla into Veering into the Wrong Lane," *MIT Technology Review*, April 1, 2019, technologyreview.com/2019/04/01/65915/hackers-trick-teslas-autopilot-into-veering-towards-oncoming-traffic.

108 **Dawn Song, a professor:** Will Knight, "How Malevolent Machine Learning Could Derail AI," *MIT Technology Review*, March 25, 2019, technologyreview.com/2019/03/25/1216/emtech-digital-dawn-song-adversarial-machine-learning.

108 **In 2019, researchers:** Benjamin Wilson, Judy Hoffman, and Jamie Morgenstern, "Predictive Inequity in Object Detection," preprint, arXiv, February 21, 2019, 1–13, doi.org/10.48550/arXiv.1902.11097.

108 **In 2024, researchers at Peking:** Xinyue Li, Zhenpeng Cheng, Jie M. Zhang, Federica Sarro, Ying Zhang, and Xuanzhe Liu, "Bias Behind the Wheel: Fairness Analysis of Autonomous Driving Systems," *ACM Transactions on Software Engineering and Methodology* (November 2024), doi.org/10.1145/3702989.

108 **Early in her career:** Author interview with Deborah Raji, April 2020.

NOTES

109 **"The human brain has"**: Author interview with Hinton at *MIT Technology Review*'s annual event, EmTech MIT, October 20, 2020. A write-up of the conversation is in: Karen Hao, "AI Pioneer Geoff Hinton: 'Deep Learning Is Going to Be Able to Do Everything,'" *MIT Technology Review*, November 3, 2020, technologyreview.com/2020/11/03/1011616/ai-godfather-geoffrey-hinton-deep-learning-will-do-everything.

110 **"We actually need both approaches"**: Author interview with Gary Marcus, September 2019. A write-up of that interview is in: Karen Hao, "We Can't Trust AI Systems Built on Deep Learning Alone," *MIT Technology Review*, September 27, 2019, technologyreview.com/2019/09/27/65250/we-cant-trust-ai-systems-built-on-deep-learning-alone.

111 **In February 2023, at the height**: OpenAI, "Planning for AGI and Beyond," *OpenAI* (blog), February 24, 2023, openai.com/index/planning-for-agi-and-beyond.

112 **When Microsoft unveiled**: Kevin Roose, "Bing's A.I. Chat: 'I Want to Be Alive.'," *New York Times*, February 16, 2023, nytimes.com/2023/02/16/technology/bing-chatbot-transcript.html.

112 **Roose's experience may have**: Pierre-François Lovens, "Sans ces conversations avec le chatbot Eliza, mon mari serait toujours là," *La Libre*, March 28, 2023, lalibre.be/belgique/societe/2023/03/28/sans-ces-conversations-avec-le-chatbot-eliza-mon-mari-serait-toujours-la-LVSLWPC5WRDX7J2RCHNWPDST24.

113 **The problem only gets harder**: There are several papers that have found this, including one cowritten by Jacob Hilton, who was an OpenAI researcher at the time. Hilton and his coauthors found that "the largest models were generally the least truthful." Stephanie Lin, Jacob Hilton, and Owain Evans, "TruthfulQA: Measuring How Models Mimic Human Falsehoods," in *Proceedings of the 60th Annual Meeting of the Association for Computational Linguistics* 1 (2021): 3214–52, doi.org/10.18653/v1/2022.acl-long.229. Additionally, for an excellent explanation of why developers have become less and less aware of the composition of their training data, read: Christo Buschek and Jer Thorp, "Models All the Way Down," Knowing Machines, March 26, 2024, knowingmachines.org/models-all-the-way.

113 **"bogus judicial decisions"**: Benjamin Weiser, "Here's What Happens When Your Lawyer Uses ChatGPT," *New York Times*, May 27, 2023, nytimes.com/2023/05/27/nyregion/avianca-airline-lawsuit-chatgpt.html.

114 **One 2023 study found that**: Katharina Jeblick, Balthasar Schachtner, Jakob Dexl, Andreas Mittermeier, Anna Theresa Stüber, Johanna Topalis et al., "ChatGPT Makes Medicine Easy to Swallow: An Exploratory Case Study on Simplified Radiology Reports," *European Radiology* 34 (October 2024): 2817–25, doi.org/10.1007/s00330-023-10213-1.

114 **They found that prompting**: Lily Hay Newman and Andy Greenberg, "Security News This Week: ChatGPT Spit Out Sensitive Data When Told to Repeat 'Poem' Forever," *Wired*, December 2, 2023, wired.com/story/chatgpt-poem-forever-security-roundup.

114 **And generative AI models amplify**: Leonardo Nicoletti and Dina Bass, "Humans Are Biased. Generative AI Is Even Worse," *Bloomberg*, June 9, 2023, bloomberg.com/graphics/2023-generative-ai-bias; Victoria Turk, "How AI Reduces the World to Stereotypes," *Rest of World*, October 10, 2023, restofworld.org/2023/ai-image-stereotypes; and Nitasha Tiku, Kevin Schaul, and Szu Yu Chen, "This Is How AI Image Generators See the World," *Washington Post*, November 1, 2023, washingtonpost.com/technology/interactive/2023/ai-generated-images-bias-racism-sexism-stereotypes.

114 **"Doctors in Africa"**: Carmen Drahl, "AI Was Asked to Create Images of Black African Docs Treating White Kids. How'd It Go?," *Goats and Soda*, NPR, October 6, 2023, npr.org/sections/goatsandsoda/2023/10/06/1201840678/ai-was-asked-to-create-images-of-black-african-docs-treating-white-kids-howd-it-.

115 **In April 2024, Dario Amodei**: Ezra Klein, host, *The Ezra Klein Show*, podcast, "What if Dario Amodei Is Right About A.I.?," April 12, 2024, *New York Times* Opinion, nytimes.com/column/ezra-klein-podcast.

Chapter 5: Scale of Ambition

117 **"How about now?"**: Cade Metz, *Genius Makers: The Mavericks Who Brought AI to Google, Facebook, and the World* (Dutton, 2021), 93; "Geoffrey Hinton | On Working with Ilya,

Choosing Problems, and the Power of Intuition," posted May 20, 2024, by Sana, YouTube, 45 min., 45 sec., youtu.be/n4IQOBka8bc.
117 **He stunned Hinton:** Author interview with Geoffrey Hinton, November 2023.
117 **At times he grew:** Metz, *Genius Makers*, 94.
117 **"One doesn't bet":** Will Douglas Heaven, "Rogue Superintelligence and Merging with Machines: Inside the Mind of OpenAI's Chief Scientist," *MIT Technology Review*, October 26, 2023, technologyreview.com/2023/10/26/1082398/exclusive-ilya-sutskever-openais-chief-scientist-on-his-hopes-and-fears-for-the-future-of-ai.
117 **"Success is guaranteed":** "NIPS: Oral Session 4—Ilya Sutskever," posted August 19, 2016, by Microsoft Research, YouTube, 23 min., 14 sec., youtu.be/-uyXE7dY5H0.
118 **Sutskever brought his die-hard belief:** Sutskever often speaks about how his belief in deep learning is really a *belief*. In September 2023, he said, "The creation of OpenAI was already an expression of this bet, of the idea that deep learning can do it. You just need to believe. And in fact, I would argue that a lot of, you know, deep learning research, at least in the past decade, maybe a bit less now, has been about faith." "Interview with Dr. Ilya Sutskever, Cofounder of OPEN AI—at the Open University Studios—English," posted September 13, 2023, by The Open University of Israel, YouTube, 50 min., 28 sec., youtu.be/H1YoNlz2LxA.
118 **His faith rested:** "What AI Is Making Possible | Ilya Sutskever and Sven Strohband," posted July 18, 2023, by Khosla Ventures, YouTube, 25 min., 26 sec., youtu.be/xym5f0XYlSc; "Ilya Sutskever: 'Sequence to Sequence Learning with Neural Networks: What a Decade,'" posted December 14, 2024, by seremot, YouTube, 24 min., 36 sec., youtu.be/1yvBqasHLZs.
118 **"Anything non–deep learning":** Author interview with Pieter Abbeel, October 2023.
118 **The intelligence of different species:** "Ilya Sutskever: 'Sequence to Sequence Learning with Neural Networks.'"
119 **"Flat out, we were wrong":** James Vincent, "OpenAI Co-founder on Company's Past Approach to Openly Sharing Research: 'We Were Wrong,'" *The Verge*, March 15, 2023, theverge.com/2023/3/15/23640180/openai-gpt-4-launch-closed-research-ilya-sutskever-interview.
119 **"it may be that today's":** Ilya Sutskever (@ilyasut), "it may be that today's large neural networks are slightly conscious," Twitter (now X), February 9, 2022, x.com/ilyasut/status/1491554478243258368.
119 **One DeepMind scientist specialized:** Murray Shanahan (@mpshanahan), "...in the same sense that it may be that a large field of wheat is slightly pasta," Twitter (now X), February 10, 2022, x.com/mpshanahan/status/1491715721289678848.
120 **The following year, Sutskever would:** Nirit Weiss-Blatt, "What Ilya Sutskever Really Wants," *AI Panic*, September 16, 2023, aipanic.news/p/what-ilya-sutskever-really-wants.
120 **That fall, he would declare:** Ilya Sutskever (@ilyasut), "In the future, once the robustness of our models will exceed some threshold, we will have *wildly effective* and dirt cheap AI therapy. Will lead to a radical improvement in people's experience of life. One of the applications I'm most eagerly awaiting.," Twitter (now X), September 27, 2023, x.com/ilyasut/status/1707027536150929689.
120 **Sutskever would get up:** A photo of Sutskever at the event.
120 **In August 2017, that changed:** Ashish Vaswani, Noam Shazeer, Niki Parmar, Jakob Uszkoreit, Llion Jones, Aidan N. Gomez et al., "Attention Is All You Need," in *NIPS '17: Proceedings of the 31st International Conference on Neural Information Processing Systems* (December 2017): 6000–10, dl.acm.org/doi/10.5555/3295222.3295349.
121 **But Sutskever, who had focused:** Sutskever's PhD thesis work focused on recurrent neural networks. RNNs, like Transformers, are designed to process sequential data, which can be widely applicable. For example: An English sentence is a sequence of words, an image is a sequence of pixels, a video is a sequence of images. His PhD thesis can be found at: Ilya Sutskever, "Training Recurrent Neural Networks" (PhD diss., University of Toronto, 2013), 1–101, cs.utoronto.ca/~ilya/pubs/ilya_sutskever_phd_thesis.pdf.
121 **Radford trained Google's neural network:** Alec Radford, Karthik Narasimhan, Tim Salimans, and Ilya Sutskever, "Improving Language Understanding by Generative Pre-Training," preprint, OpenAI, June 11, 2018, 1–12, cdn.openai.com/research-covers/language-unsupervised/language_understanding_paper.pdf. The original dataset from which Radford pulled the over seven thousand unpublished books comes from: Yukun Zhu, Ryan

NOTES

Kiros, Rich Zemel, Ruslan Salakhutdinov, Raquel Urtasun, Antonio Torralba et al., "Aligning Books and Movies: Towards Story-Like Visual Explanations by Watching Movies and Reading Books," in *Proceedings: 2015 IEEE International Conference on Computer Vision* (Institute of Electrical and Electronics Engineers, 2015): 19–27, doi.org/10.1109/ICCV.2015.11. It's not uncommon in AI research for one group to scrape together a dataset and post it and for other groups to reuse it for their own separate purposes.

121 **The company explained:** OpenAI, "Generative Models," *Open AI* (blog), June 16, 2016, openai.com/index/generative-models.

122 **In 2017, one of Amodei's:** Paul Christiano, Jan Leike, Tom B. Brown, Miljan Martic, Shane Legg, and Dario Amodei, "Deep Reinforcement Learning from Human Preferences," in *NIPS '17: Proceedings of the 31st International Conference on Neural Information Processing Systems* (December 2017): 4302–10, dl.acm.org/doi/10.5555/3294996.3295184.

123 **OpenAI touted the technique:** OpenAI, "Learning from Human Preferences," *Open AI* (blog), June 13, 2017, openai.com/index/learning-from-human-preferences.

123 **Amodei wanted to move:** Author interview with Dario Amodei, August 2019.

123 **They set their sights:** Alec Radford, Jeffrey Wu, Rewon Child, David Luan, Dario Amodei, and Ilya Sutskever, "Language Models Are Unsupervised Multitask Learners," preprint, OpenAI, February 14, 2019, 1–24, cdn.openai.com/better-language-models/language_models_are_unsupervised_multitask_learners.pdf.

123 **His team called them collectively:** Jared Kaplan, Sam McCandlish, Tom Henighan, Tom B. Brown, Benjamin Chess, Rewon Child et al., "Scaling Laws for Neural Language Models," preprint, arXiv, January 23, 2020, 1–30, doi.org/10.48550/arXiv.2001.08361.

124 **Fed a few words:** Interview with Amodei, August 2019.

124 **After GPT-2 generated a tirade:** The full tirade is in OpenAI, "Better Language Models and Their Implications," *Open AI* (blog), February 14, 2019, openai.com/index/better-language-models.

125 **Amodei, who had by then:** Author interview with Jack Clark, August 2019.

125 **"I'm like AI Wikipedia":** Interview with Clark.

126 **He, Amodei, and several others:** OpenAI, "Better Language Models."

126 **"It's very clear that if":** Will Knight, "An AI That Writes Convincing Prose Risks Mass-Producing Fake News," *MIT Technology Review*, February 14, 2019, technologyreview.com/2019/02/14/137426/an-ai-tool-auto-generates-fake-news-bogus-tweets-and-plenty-of-gibberish.

127 **"If we're right, and it":** Interview with Clark, August 2019.

127 **This was frequently discussed:** Karen Hao, "The Messy, Secretive Reality Behind OpenAI's Bid to Save the World," *MIT Technology Review*, February 17, 2020, technologyreview.com/2020/02/17/844721/ai-openai-moonshot-elon-musk-sam-altman-greg-brockman-messy-secretive-reality.

128 **"the strongest endorsement":** I attended this policy team meeting when I was embedded in the office in August 2019.

128 **Before long, it had:** Helen Toner, "GPT-2 Kickstarted the Conversation About Publication Norms in the AI Research Community," CSET, May 1, 2020, cset.georgetown.edu/article/gpt-2-kickstarted-the-conversation-about-publication-norms-in-the-ai-research-community/; PAI Staff, "Managing the Risks of AI Research: Six Recommendations for Responsible Publication," Partnership on AI, May 6, 2021, partnershiponai.org/paper/responsible-publication-recommendations.

129 **"a portfolio of bets":** Interview with Amodei, August 2019.

129 **Where Amodei did see continued:** Interview with Amodei.

129 **In company documents:** Copies of two of those documents.

129 **"Language of some form":** The quoted discussion is from one of the aforementioned documents.

130 **GPT-2 had demonstrated how easy:** Tom Simonite, "OpenAI Said Its Code Was Risky. Two Grads Re-Created It Anyway," *Wired*, August 26, 2019, wired.com/story/dangerous-ai-open-source.

133 **Amodei wanted to use all:** OpenAI didn't release the number of chips it used in its original paper on GPT-3, but after a Google controversy recounted in chapter 7, it gave the number ten thousand to Google researchers, who published it in the following paper: David

440 NOTES

Patterson, Joseph Gonzalez, Quoc Le, Chen Liang, Lluis-Miquel Munguia, Daniel Rothchild et al., "Carbon Emissions and Large Neural Network Training," preprint, arXiv, April 23, 2021, 6, doi.org/10.48550/arXiv.2104.10350.

135 **This had produced:** Details of the training data used for GPT-2 can be found in OpenAI's paper about the model: Radford et al., "Language Models Are Unsupervised Multitask Learners."

135 **So Nest expanded the data:** OpenAI is not alone in this regard. In 2023, Alex Reisner, a writer and programmer, would confirm that companies including Meta and Bloomberg had trained their models on yet another books dataset called Books3, which his analysis found contains upward of 170,000 published books. In 2024, Reisner also confirmed that the same companies, along with Anthropic, Nvidia, Apple, and others, were similarly training their models on a dataset called OpenSubtitles of the dialogue in more than 53,000 movies and 85,000 TV episodes. Alex Reisner, "Revealed: The Authors Whose Pirated Books Are Powering Generative AI," *The Atlantic*, August 19, 2023, theatlantic.com/technology/archive/2023/08/books3-ai-meta-llama-pirated-books/675063/; Alex Reisner, "There's No Longer Any Doubt That Hollywood Writing Is Powering AI," *The Atlantic*, November 18, 2024, theatlantic.com/technology/archive/2024/11/opensubtitles-ai-data-set/680650.

135 **OpenAI would respond:** Authors Guild v. OpenAI Inc., No. 1:23-cv-08292, CourtListener (S.D.N.Y. May 6, 2024) ECF No. 143, Exhibit D, at *2.

135 **So Nest turned finally:** Details of the training data used for GPT-3 can be found in OpenAI's paper about the model: Tom B. Brown, Benjamin Mann, Nick Ryder et al., "Language Models Are Few-Shot Learners," in *NIPS '20: Proceedings of the 34th International Conference on Neural Information Processing Systems* (December 2020): 1877–901, dl.acm.org/doi/abs/10.5555/3495724.3495883.

137 **"There's a big paradigm shift":** Author interview with Ryan Kolln, October 2023.

137 **In a 2023 paper, Abeba:** Abeba Birhane, Vinay Prabhu, Sang Han, and Vishnu Naresh Boddeti, "On Hate Scaling Laws for Data-Swamps," preprint, arXiv, June 28, 2023, 1–27, doi.org/10.48550/arXiv.2306.13141.

137 **Later that year, a Stanford:** David Thiel, *Identifying and Eliminating CSAM in Generative ML Training Data and Models* (Stanford Internet Observatory, 2023), 1–19, purl.stanford.edu/kh752sm9123.

137 **Among its tactics:** Billy Perrigo, "Exclusive: OpenAI Used Kenyan Workers on Less Than $2 Per Hour to Make ChatGPT Less Toxic," *Time*, January 18, 2023, time.com/6247678/openai-chatgpt-kenya-workers.

137 **It would also employ:** Karen Hao and Deepa Seetharaman, "Cleaning Up ChatGPT Takes Heavy Toll on Human Workers," *Wall Street Journal*, July 24, 2023, wsj.com/articles/chatgpt-openai-content-abusive-sexually-explicit-harassment-kenya-workers-on-human-workers-cf191483; copy of OpenAI's RLHF instructions.

138 **Psychologically harmful material:** Author interview with Hito Steyerl, September 2023.

Chapter 6: Ascension

141 **Early in his career, Altman:** Tad Friend, "Sam Altman's Manifest Destiny," *New Yorker*, October 3, 2016, newyorker.com/magazine/2016/10/10/sam-altmans-manifest-destiny.

141 **"The thing that I'm most":** "Advice to Entrepreneurs | Sam Altman & Jack Altman," posted August 1, 2019, by Khosla Ventures, YouTube, 30 min., 10 sec., youtu.be/NAaRhXQCt9o.

142 **"My sort of crazy":** "Competition Is for Losers with Peter Thiel (How to Start a Startup 2014: 5)," posted March 22, 2017, by Y Combinator, YouTube, 50 min., 27 sec., youtu.be/3Fx5Q8xGU8k.

142 **"If your iteration cycle":** "Sam Altman Startup School Video," posted July 26, 2017, by Waterloo Engineering, YouTube, 1 hr., 18 min., 19 sec., youtu.be/4SlNgM4PjvQ.

142 **"And we will, over time":** Tyler Cowen, host, *Conversations with Tyler*, podcast, episode 61, "Sam Altman on Loving Community, Hating Coworking, and the Hunt for Talent," Mercatus Center Podcasts, February 27, 2019.

142 **"Sam was the first person":** Author interview with Geoff Ralston, March 2024.

142 **In a memo he sent:** Copy of the memo.

NOTES

144 **The Amodei siblings, meanwhile:** Stephanie Palazzolo, Erin Woo, and Amir Efrati, "How Anthropic Got Inside OpenAI's Head," *The Information*, December 12, 2024, theinformation.com/articles/how-anthropic-got-inside-openais-head.

147 **Altman himself was paranoid:** Details of Altman's and Sutskever's paranoias and the way the company ramped up digital and physical security come from the recollections and contemporaneous notes of people who spoke with Altman or had knowledge of the measures and recordings of those measures being either tested or discussed. Altman's emphasis on security is also referenced in the aforementioned memo. Every detail (e.g., the focus on insider threat, the palm scanner, the distress passwords) is corroborated by at least two people, contemporaneous notes, a recording, or the memo.

150 **As they had done:** Tom B. Brown, Benjamin Mann, Nick Ryder et al., "Language Models Are Few-Shot Learners," in *NIPS '20: Proceedings of the 34th International Conference on Neural Information Processing Systems* (2020): 1877–901, dl.acm.org/doi/abs/10.5555/3495724.3495883.

151 **impressive technical milestone:** In addition to interviews with sources, the idea of using code-generation models to accelerate OpenAI's research comes up in two of the internal company memos for which I have copies.

152 **a faster rise in unemployment:** Rakesh Kochhar, "Unemployment Rose Higher in Three Months of COVID-19 Than It Did in Two Years of the Great Recession," Pew Research Center, June 11, 2020, pewresearch.org/short-reads/2020/06/11/unemployment-rose-higher-in-three-months-of-covid-19-than-it-did-in-two-years-of-the-great-recession.

153 **Google had published:** Daniel Adiwardana, Minh-Thang Luong, David R. So, Jamie Hall, Noah Fiedel, Romal Thoppilan et al., "Towards a Human-Like Open-Domain Chatbot," preprint, arXiv, February 27, 2020, 1–38, doi.org/10.48550/arXiv.2001.09977.

153 **Google's executives determined:** Miles Kruppa and Sam Schechner, "How Google Became Cautious of AI and Gave Microsoft an Opening," *Wall Street Journal*, March 7, 2023, wsj.com/articles/google-ai-chatbot-bard-chatgpt-rival-bing-a4c2d2ad.

154 **At NeurIPS that year:** The paper won one of the Best Paper Awards at NeurIPS in 2020. Hsuan-Tien Lin, Maria Florina Balcan, Raia Hadsell, and Marc'Aurelio Ranzato, "Announcing the NeurIPS 2020 Award Recipients," Neural Information Processing Systems Conference, December 8, 2020, neuripsconf.medium.com/announcing-the-neurips-2020-award-recipients-73e4d3101537.

155 **At one point, Welinder:** *Simple Sabotage Field Manual* (Office of Strategic Services: 1944), cia.gov/static/5c875f3ec660e092cf893f60b4a288df/SimpleSabotage.pdf.

Chapter 7: Science in Captivity

158 **Shortly after joining DeepMind:** Copy of that memo.

159 **But executives weren't interested:** Karen Hao, Salvador Rodriguez, and Deepa Seetharaman, "Mark Zuckerberg Was Early in AI. Now Meta Is Trying to Catch Up," *Wall Street Journal*, June 17, 2023, wsj.com/articles/mark-zuckerberg-was-early-in-ai-now-meta-is-trying-to-catch-up-94a86284.

159 **In China, GPT-3 similarly:** Jeffrey Ding and Jenny W. Xiao, *Recent Trends in China's Large Language Model Landscape*, Centre for the Governance of AI, April 28, 2023, 1–14, cdn.governance.ai/Trends_in_Chinas_LLMs.pdf.

159 **By providing evidence:** Raffaele Huang and Karen Hao, "Baidu Hurries to Ready China's First ChatGPT Equivalent Ahead of Launch," *Wall Street Journal*, March 9, 2023, wsj.com/articles/baidu-scrambles-to-ready-chinas-first-chatgpt-equivalent-ahead-of-launch-bf359ca4.

159 **In June 2019, Emma:** Emma Strubell, Ananya Ganesh, and Andrew McCallum, "Energy and Policy Considerations for Deep Learning in NLP," *Proceedings of the 57th Annual Meeting of the Association for Computational Linguistics* (July 2019): 3645–50, doi.org/10.18653/v1/P19-1355.

160 **consuming 1,287 megawatt-hours:** David Patterson, Joseph Gonzalez, Quoc Le, Chen Liang, Lluis-Miquel Munguia, Daniel Rothchild et al., "Carbon Emissions and Large Neural Network Training," preprint, arXiv, April 23, 2021, doi.org/10.48550/arXiv.2104.10350.

NOTES

161 **This included a groundbreaking:** Joy Buolamwini and Timnit Gebru, "Gender Shades: Intersectional Accuracy Disparities in Commercial Gender Classification," in *Proceedings of the 1st Conference on Fairness, Accountability and Transparency* (2018): 77–91, proceedings.mlr.press/v81/buolamwini18a.html.

161 **Buolamwini would subsequently:** The follow-on paper: Inioluwa Deborah Raji and Joy Buolamwini, "Actionable Auditing: Investigating the Impact of Publicly Naming Biased Performance Results of Commercial AI Products," in *AIES '19: Proceedings of the 2019 AAAI/ACM Conference on AI, Ethics, and Society* (January 2019): 429–35, doi.org/10.1145/3306618.3314244; the US government audit: Patrick Grother, Mei Ngan, and Kayee Hanaoka, *Face Recognition Vendor Test (FRVT) Part 3: Demographic Effects*, NISTIR 8280, National Institute of Standards and Technology, December 2019, doi.org/10.6028/NIST.IR.8280.

161 **Two years later, widespread:** The full story of Buolamwini's research and advocacy is recounted in her bestselling memoir: Joy Buolamwini, *Unmasking AI: My Mission to Protect What Is Human in a World of Machines* (Random House Trade Paperbacks, 2024); and the Netflix documentary: *Coded Bias*, directed by Shalini Kantayya (2020; Brooklyn, NY: 7th Empire Media), Netflix. For more on the wide-reaching impacts of "Gender Shades" and "Actionable Auditing," see: "Celebrating 5 Years of Gender Shades," Algorithmic Justice League, accessed on January 15, 2025, gs.ajl.org/.

161 **Black in AI sparked:** Karen Hao, "Inside the Fight to Reclaim AI from Big Tech's Control," *MIT Technology Review*, June 14, 2021, technologyreview.com/2021/06/14/1026148/ai-big-tech-timnit-gebru-paper-ethics.

162 **had approached Gebru:** Author interview with Timnit Gebru, August 2023.

162 **In 2017, a Facebook:** Alex Hern, "Facebook Translates 'Good Morning' into 'Attack Them,' Leading to Arrest," *The Guardian*, October 24, 2017, theguardian.com/technology/2017/oct/24/facebook-palestine-israel-translates-good-morning-attack-them-arrest.

162 ***Algorithms of Oppression* by Safiya:** Safiya Umoja Noble, *Algorithms of Oppression: How Search Engines Reinforce Racism* (NYU Press, 2018), 1–248.

163 **OpenAI had simply admitted:** In the GPT-3 paper, under Section 6.2 Fairness, Bias, and Representation, it discusses several different types of bias found in the model, and then reads, "We have presented this preliminary analysis to share some of the biases we found in order to motivate further research." Tom B. Brown, Benjamin Mann, Nick Ryder, Melanie Subbiah, Jared Kaplan, Prafulla Dhariwal et al., "Language Models Are Few-Shot Learners," in *NIPS '20: Proceedings of the 34th International Conference on Neural Information Processing Systems*, no. 159 (2020): 1877–901, dl.acm.org/doi/abs/10.5555/3495724.3495883.

163 **Gebru chimed in:** The account of Gebru's experiences around the "Stochastic Parrots" paper comes primarily from author interviews with Gebru, 2020–24, including one day after her ouster, as well as a detailed account in Tom Simonite, "What Really Happened When Google Ousted Timnit Gebru," *Wired*, June 8, 2021, wired.com/story/google-timnit-gebru-ai-what-really-happened.

164 **If not, she would be:** Dialogue between Gebru and Emily M. Bender pulled from screenshots of exchanges, provided by Bender.

164 **"Our goal with these initial":** Copy of email, provided by Bender.

165 **"Definitely not my area":** Simonite, "What Really Happened."

165 **In total, it presented four:** Emily M. Bender, Timnit Gebru, Angelina McMillan-Major, and Shmargaret Shmitchell [Meg Mitchell], "On the Dangers of Stochastic Parrots: Can Language Models Be Too Big? 🦜" in *FAccT '21: Proceedings of the 2021 ACM Conference on Fairness, Accountability, and Transparency* (March 2021): 610–23, doi.org/10.1145/3442188.3445922. Because Google would not let Meg Mitchell publish the paper for the reasons detailed in this chapter, she listed her name on the paper as Shmargaret Shmitchell and created a corresponding email address. As her affiliation, she hailed from "the Aether."

167 **On another internal LISTSERV:** Casey Newton, "The Withering Email That Got an Ethical AI Researcher Fired at Google," Platformer, December 3, 2020, platformer.news/the-withering-email-that-got-an-ethical.

168 **"We, the undersigned":** Google Walkout for Real Change, "Standing with Dr. Timnit Gebru—#ISupportTimnit #BelieveBlackWomen," Medium, December 3, 2020, https://googlewalkout.medium.com/standing-with-dr-timnit-gebru-isupporttimnit-believeblackwomen-6dadc300d382.

NOTES

169 **A few hours later, I:** Karen Hao, "We Read the Paper That Forced Timnit Gebru out of Google. Here's What It Says," *MIT Technology Review*, December 4, 2020, technologyreview.com/2020/12/04/1013294/google-ai-ethics-research-paper-forced-out-timnit-gebru.

169 **On December 9, as protests:** Ina Fried, "Scoop: Google CEO Pledges to Investigate Exit of Top AI Ethicist," Axios, December 9, 2020, axios.com/2020/12/09/sundar-pichai-memo-timnit-gebru-exit.

169 **On December 16, representatives:** Karen Hao, "Congress Wants Answers from Google About Timnit Gebru's Firing," *MIT Technology Review*, December 17, 2020, technologyreview.com/2020/12/17/1014994/congress-wants-answers-from-google-about-timnit-gebrus-firing.

169 **For more than a year, the protests:** Ina Fried, "Google Fires Another AI Ethics Leader," Axios, February 19, 2021, axios.com/2021/02/19/google-fires-another-ai-ethics-leader.

169 **Google said she had violated:** Sam Shead, "New Google Union 'Concerned' After a Senior A.I. Ethics Researcher Is Reportedly Locked Out of Her Account," CNBC, January 21, 2021, cnbc.com/2021/01/21/margaret-mitchell-google-investigating-ai-researcher-awu-concerned.html.

169 **The company sought to:** Sepi Hejazi Moghadam, "Marian Croak's Vision for Responsible AI at Google," *The Keyword*, February 18, 2021, blog.google/technology/ai/marian-croak-responsible-ai.

169 **"This was a painful":** Author correspondence with Google spokesperson, November 2024.

170 **It was a warning:** Mohamed Abdalla and Moustafa Abdalla, "The Grey Hoodie Project: Big Tobacco, Big Tech, and the Threat on Academic Integrity," in *AIES '21: Proceedings of the 2021 AAAI/ACM Conference on AI, Ethics, and Society* (July 2021): 287–97, doi.org/10.1145/3461702.3462563.

170 **As one of Google's earliest:** James Somers, "The Friendship That Made Google Huge," *New Yorker*, December 3, 2018, newyorker.com/magazine/2018/12/10/the-friendship-that-made-google-huge.

170 **saying his objections:** Simonite, "What Really Happened."

171 **Strubell felt it was more:** Author interview with Emma Strubell, November 2023.

172 **A Google spokesperson said Strubell:** Correspondence with Google spokesperson, November 2024.

172 **The blog post Patterson:** David Patterson, "Good News About the Carbon Footprint of Machine Learning Training," *Google Research* (blog), February 15, 2022, research.google/blog/good-news-about-the-carbon-footprint-of-machine-learning-training.

173 **It was then that OpenAI:** Correspondence with Google spokesperson, November 2024.

173 **Nearly all of the companies:** Nitasha Tiku and Gerrit De Vynck, "Google Shared AI Knowledge with the World—Until ChatGPT Caught Up," *Washington Post*, May 4, 2023, washingtonpost.com/technology/2023/05/04/google-ai-stop-sharing-research.

173 **All ten of the companies:** Rishi Bommasani, Kevin Klyman, Shayne Longpre, Sayash Kapoor, Nestor Maslej, Betty Xiong et al., *The Foundation Model Transparency Index* (Stanford Center for Research on Foundation Models, October 2023), crfm.stanford.edu/fmti/October-2023/index.html.

Chapter 8: Dawn of Commerce

175 **With new consensus:** Copy of the road map.

178 **A year later, Google:** This is colloquially called the "Chinchilla paper": Jordan Hoffmann, Sebastian Borgeaud, Arthur Mensch, Elena Buchatskaya, Trevor Cai, Eliza Rutherford et al., "Training Compute-Optimal Large Language Models," preprint, arXiv, March 29, 2022, 1–36, arxiv.org/abs/2203.15556.

179 **OpenAI called this process:** The first use of this term in the AI context comes from the paper Miles Brundage, Shahar Avin, Jasmine Wang, Haydn Belfield, Gretchen Krueger, Gillian Hadfield et al., "Toward Trustworthy AI Development: Mechanisms for Supporting Verifiable Claims," preprint, arXiv, April 20, 2020, 2, doi.org/10.48550/arXiv.2004.07213.

180 **Khlaaf, who worked with OpenAI:** Khlaaf has written a paper that analyzes the differences between red teaming in AI and security. Heidy Khlaaf, "Toward Comprehensive

NOTES

Risk Assessments and Assurance of AI-Based Systems," Trail of Bits, March 7, 2023, 1–30, trailofbits.com/documents/Toward_comprehensive_risk_assessments.pdf.

180 **The company had partnered:** Lex Fridman, host, *Lex Fridman Podcast*, podcast, episode 121, "Eugenia Kuyda: Friendship with an AI Companion," September 5, 2020, lexfridman.com/podcast.

180 **Latitude had already been using:** Tom Simonite, "It Began as an AI-Fueled Dungeon Game. It Got Much Darker," *Wired*, May 5, 2021, wired.com/story/ai-fueled-dungeon-game-got-much-darker.

182 **Microsoft executives directed:** Charles Duhigg, "The Inside Story of Microsoft's Partnership with OpenAI," *New Yorker*, December 1, 2023, newyorker.com/magazine/2023/12/11/the-inside-story-of-microsofts-partnership-with-openai.

184 **Microsoft would get its moment:** Nat Friedman, "Introducing GitHub Copilot: Your AI Pair Programmer," GitHub, June 29, 2021, github.blog/news-insights/product-news/introducing-github-copilot-ai-pair-programmer.

184 **OpenAI would then release:** OpenAI, "OpenAI Codex," *Open AI* (blog), August 10, 2021, openai.com/index/openai-codex.

184 **The arrangement would:** Tiernan Ray, "Microsoft Has Over a Million Paying Github Copilot Users: CEO Nadella," ZDNet, October 25, 2023, zdnet.com/article/microsoft-has-over-a-million-paying-github-copilot-users-ceo-nadella.

185 **"If you could wave":** "Advice to Entrepreneurs | Sam Altman & Jack Altman," posted August 1, 2019, by Khosla Ventures, YouTube, 30 min., 10 sec., youtu.be/NAaRhXQCt9o.

185 **The venture was a dedicated:** Ellen Huet and Gillian Tan, "Sam Altman Wants to Scan Your Eyeball in Exchange for Cryptocurrency," *Bloomberg*, June 29, 2021, bloomberg.com/news/articles/2021-06-29/sam-altman-s-worldcoin-will-give-free-crypto-for-eyeball-scans.

185 **At YC he had started:** Sarah Holder and Shirin Ghaffary, "Sam Altman–Backed Group Completes Largest US Study on Basic Income," *Bloomberg*, July 22, 2024, bloomberg.com/news/articles/2024-07-22/ubi-study-backed-by-openai-s-sam-altman-bolsters-support-for-basic-income.

185 **In July 2024, OpenResearch:** OpenResearch, "Key Findings: Spending," *OpenResearch* (blog), July 21, 2024, openresearchlab.org/findings/key-findings-spending.

185 **Tools for Humanity's main product:** Huet and Tan, "Sam Altman Wants to Scan Your Eyeball."

186 **An extensive investigation:** Eileen Guo and Adi Renaldi, "Deception, Exploited Workers, and Cash Handouts: How Worldcoin Recruited Its First Half a Million Test Users," *MIT Technology Review*, April 6, 2022, technologyreview.com/2022/04/06/1048981/worldcoin-cryptocurrency-biometrics-web3.

186 **In July 2023, Worldcoin:** Anita Nkonge, "Worldcoin Suspended in Kenya as Thousands Queue for Free Money," BBC, August 3, 2023, bbc.com/news/world-africa-66383325.

186 **"I basically just took":** Antonio Regalado, "Sam Altman Invested $180 Million into a Company Trying to Delay Death," *MIT Technology Review*, March 8, 2023, technologyreview.com/2023/03/08/1069523/sam-altman-investment-180-million-retro-biosciences-longevity-death.

187 **To Antonio Regalado, cofounder:** Antonio Regalado, "A Startup Is Pitching a Mind-Uploading Service That Is '100 percent Fatal,'" *MIT Technology Review*, March 13, 2018, technologyreview.com/2018/03/13/144721/a-startup-is-pitching-a-mind-uploading-service-that-is-100-percent-fatal.

187 **"destroy the planet":** "Office Hours with Sam Altman," posted January 11, 2017, by Y Combinator, YouTube, 24 min., 34 sec., youtu.be/45BvnJgwYjk.

187 **"more than an investment":** "StrictlyVC in Conversation with Sam Altman, Part One," posted on January 16, 2023, by Connie Loizos, YouTube, 20 min., 32 sec., youtu.be/57OU18cogJI.

187 **To the astonishment:** Justine Calma, "Microsoft Just Made a Huge, Far-from-Certain Bet on Nuclear Fusion," *The Verge*, May 10, 2023, theverge.com/2023/5/10/23717332/microsoft-nuclear-fusion-power-plant-helion-purchase-agreement.

187 **That May, he launched:** Information can be found at its own website, openai.fund.

188 **Altman's net worth:** Berber Jin, Tom Dotan, and Keach Hagey, "The Opaque Investment Empire Making OpenAI's Sam Altman Rich," *Wall Street Journal*, June 3, 2024, wsj.com/tech/ai/openai-sam-altman-investments-004fc785.

NOTES

Chapter 9: Disaster Capitalism

189 **In 2021, in parallel:** Karen Hao and Deepa Seetharaman, "Cleaning Up ChatGPT Takes Heavy Toll on Human Workers," *Wall Street Journal*, July 24, 2023, wsj.com/articles/chatgpt-openai-content-abusive-sexually-explicit-harassment-kenya-workers-on-human-workers-cf191483.

190 **To build the automated filter:** Copies of OpenAI's Statements of Work for the project.

190 **After six months of searching:** Author interview with OpenAI spokesperson, June 2023.

190 **OpenAI sent Sama:** Review of the email.

190 **Sama provided thorough answers:** Review of the answers.

190 **OpenAI signed four contracts:** Copy of two contracts and review of the two others.

190 **You can see the markers:** Based on the author's reporting trip to Nairobi, May 2023.

191 **Under these conditions:** Author interviews with Mercy Mutemi, the lawyer who represented the four Kenyan workers to fight for digital labor reforms in Kenya, May 2023; and Jonathan Beardsley, an executive at the time at data-annotation firm CloudFactory, May 2023.

192 **It wasn't until early 2022:** Billy Perrigo, "Inside Facebook's African Sweatshop," *Time*, February 14, 2022, time.com/6147458/facebook-africa-content-moderation-employee-treatment.

192 **Sama would defend itself:** Author correspondence with Sama spokesperson, November 2024.

192 **Nearly two hundred workers would:** Caroline Kimeu, "'A Watershed': Meta Ordered to Offer Mental Health Care to Moderators in Kenya," *The Guardian*, June 7, 2023, theguardian.com/global-development/2023/jun/07/a-watershed-meta-ordered-to-offer-mental-health-care-to-moderators-in-kenya.

192 **Under the code names PBJ1:** Based on the contracts and project documents as well as Sama's response to the author's comment request for her story in *The Wall Street Journal*: Hao and Seetharaman, "Cleaning Up ChatGPT Takes Heavy Toll."

192 **Workers had no idea:** Author interviews with four of those workers, Mophat Okinyi, Richard Mathenge, Alex Kairu, and Bill Mulinya, 2023.

192 **What they did know:** Copy of the instructions that the workers received. These categories correspond to those available in OpenAI's content moderation API, which can be viewed here: "Moderation," OpenAI Platform, OpenAI, accessed October 17, 2024, platform.openai.com/docs/guides/moderation.

193 **For one of them:** Author interviews with Mophat, his brother Albert, one of his friends, and Mutemi, 2023.

193 **In 2019, they published:** Mary L. Gray and Siddharth Suri, *Ghost Work: How to Stop Silicon Valley from Building a New Global Underclass* (Harper Business, 2019), 1–288; and author interview with Mary L. Gray, May 2019.

194 **Before generative AI:** Florian Alexander Schmidt, "Crowdsourced Production of AI Training Data—How Human Workers Teach Self-Driving Cars How to See," *Working Paper Forschungsförderung* 155 (2019), hdl.handle.net/10419/216075.

195 **But right as this new:** Author interviews with Florian Alexander Schmidt, 2022; and Julian Posada, 2021.

196 **10 million percent:** According to the International Monetary Fund.

196 **By mid-2018, hundreds:** Schmidt, "Crowdsourced Production of AI."

197 **Looking back several years later:** Julian Posada, "The Coloniality of Data Work: Power and Inequality in Outsourced Data Production for Machine Learning" (PhD diss., University of Toronto, 2022), 1–229, hdl.handle.net/1807/126388.

197 **In December 2021, I journeyed:** Karen Hao and Andrea Paola Hernández, "How the AI Industry Profits from Catastrophe," *MIT Technology Review*, April 20, 2022, technologyreview.com/2022/04/20/1050392/ai-industry-appen-scale-data-labels.

197 **Fuentes was the first:** Author interviews with Oskarina Veronica Fuentes Anaya, including at her home, 2021.

199 **Wilson Pang, Appen's CTO:** Author interview with Wilson Pang, December 2021.

201 **Fuentes taught me:** Author interviews with data-annotation workers 2021–24 in Kenya, the Philippines, Colombia, Venezuela (in partnership with Andrea Paola Hernández), North Africa, and elsewhere.

NOTES

202 **Among the crop:** The account of Scale's business practices is based on author interviews with five current and former Scale employees, screenshots of company documents, reviews of instructions provided to workers, embedding in their Discord, as well as author interviews with nearly two dozen workers globally who have worked on the platform.
203 **"If you could be pulling":** Ashlee Vance, "Silicon Valley's Latest Unicorn Is Run by a 22-Year-Old," *Bloomberg*, August 5, 2019, bloomberg.com/news/articles/2019-08-05/scale-ai-is-silicon-valley-s-latest-unicorn.
204 **We found through a spreadsheet:** Copy of spreadsheet of worker pay.
204 **Inside Scale, Remotasks Plus:** Author correspondence with Scale spokesperson, November 2024.
204 **With nowhere to go:** Correspondence with Scale spokesperson.
204 **"Remotasks is committed":** Hao and Hernández, "How the AI Industry Profits from Catastrophe."
205 **"We care deeply":** Correspondence with Scale spokesperson, November 2024.
205 **At least one worker:** Screenshot of the worker's payments.
205 **"revolutions and protests":** Screenshot of the message in the workers' Discord channel.
206 **One such firm, CloudFactory:** Author interviews with founder Mark Sears, May 2023; and executive Jonathan Beardsley, and around a dozen CloudFactory workers; as well as a visit to the CloudFactory Nairobi headquarters, May 2023.
207 **Mophat Okinyi grew up:** Author interviews with Mophat Okinyi, May 2023; and Albert Okinyi, May and June 2023.
207 **The country's youth unemployment:** According to the Federation of Kenya Employers, which defines youth as fifteen to thirty-four years old.
207 **In 2021, the World Bank:** "Continued Rebound, but Storms Cloud the Horizon: Policies to Accelerate the Productive Economy for Inclusive Growth," *Kenya Economic Update*, no. 26 (World Bank, 2022), 1–54, hdl.handle.net/10986/38386.
207 **It felt like a miracle:** Author correspondence with a Sama spokesperson, June 2023.
208 **He had just met:** Author interviews with Mophat, May 2023; Albert, May and June 2023; and a friend of Mophat's, May 2023.
208 **Okinyi was placed:** Copy of OpenAI's Statement of Work with Sama.
208 **OpenAI's instructions split:** Copy of instructions.
209 **Others were generated:** OpenAI researchers later wrote a paper explaining some of their practices for building the content moderation filter. Section 3.3 goes into how they generated synthetic data for training. The paper further explains the categories of severity. Todor Markov, Chong Zhang, Sandhini Agarwal, Tyna Eloundou, Teddy Lee, Steven Adler et al., "A Holistic Approach to Undesired Content Detection in the Real World," in *AAAI'23/IAAI'23/EAAI'23: Proceedings of the Thirty-Seventh AAAI Conference on Artificial Intelligence and Thirty-Fifth Conference on Innovative Applications of Artificial Intelligence and Thirteenth Symposium on Educational Advances in Artificial Intelligence*, no. 1683 (2022): 15009–18, dl.acm.org/doi/10.1609/aaai.v37i12.26752.
210 **In March 2022, Sama:** Correspondence with Sama spokesperson, June 2023.
210 **The company never received:** Correspondence with Sama spokesperson.
212 **As the product went viral:** Interview with Albert Okinyi, May and June 2023.
212 **But the consistency of workers' experiences:** Milagros Miceli and Julian Posada, "The Data-Production Dispositif," in *Proceedings of the ACM on Human-Computer Interaction* 6, no. 460 (November 2022): 1–37, dl.acm.org/doi/10.1145/3555561; James Muldoon and Boxi A. Wu, "Artificial Intelligence in the Colonial Matrix of Power," *Philosophy and Technology* 36, no. 80 (December 2023), doi.org/10.1007/s13347-023-00687-8.
213 **"It's just so unbelievably ugly":** Interview with Sears, May 2023.
213 **Between the spring of 2022:** OpenAI deals based on screenshot of closed contracts between OpenAI and Scale; estimated revenue in 2023 from Cory Weinberg, "Fame, Feud and Fortune: Inside Billionaire Alexandr Wang's Relentless Rise in Silicon Valley," *The Information*, June 28, 2024, theinformation.com/articles/fame-feud-and-fortune-inside-billionaire-alexandr-wangs-relentless-rise-in-silicon-valley.
214 **Where self-driving cars:** Long Ouyang, Jeff Wu, Xu Jiang, Diogo Almeida, Carroll L. Wainwright, Pamela Mishkin et al., "Training Language Models to Follow Instructions with Human Feedback," arXiv, March 4, 2022, 1–68, doi.org/10.48550/arXiv.2203.02155.

NOTES

214 **"follow user instructions"**: OpenAI, "Aligning Language Models to Follow Instructions," *Open AI* (blog), January 27, 2022, openai.com/index/instruction-following.
215 **The company began using**: Based on copies of over a hundred pages of OpenAI's RLHF documents.
216 **"You will play the role"**: RLHF documents.
216 **To properly rank outputs**: RLHF documents.
216 **"Your goal is to provide"**: RLHF documents.
217 **during a talk at UC Berkeley**: "John Schulman—Reinforcement Learning from Human Feedback: Progress and Challenges," posted April 19, 2023, by UC Berkeley EECS, YouTube, 1 hr., 3 min., 31 sec., youtu.be/hhiLw5Q_UFg.
218 **Scale AI, whose business**: Berber Jin, "The 27-Year-Old Billionaire Whose Army Does AI's Dirty Work," *Wall Street Journal*, September 20, 2024, wsj.com/tech/ai/alexandr-wang-scale-ai-d7c6efd7.
218 **"soon companies will"**: Alexandr Wang (@alexandr_wang), "we're starting to see top companies spend the same amount on RLHF and compute in training ChatGPT-like LLMs . . . for example, OpenAI hired >1000 devs to RLHF their code models . . . crazy—but soon companies will start spending $ hundreds of Ms or $ billions on RLHF, just as w/compute," Twitter (now X), February 1, 2023, x.com/alexandr_wang/status/1620934510820093952.
218 **Scale would soon ban**: Author correspondence with Scale spokesperson, November 2024.
219 **Among the workers**: Based on visits to the homes of three Remotasks workers and four Sama workers in Nairobi, May 2023, as well as the addresses of two other Remotasks workers.
219 **the only girl**: Author interviews with Winnie and her partner, Millicent, May 2023.
220 **There was a project called**: Review of Flamingo Generation instructions.
220 **There was another project**: Review of Crab Generation instructions.
221 **Crab Paraphrase was similar**: Copy of Crab Paraphrase instructions.
222 **Kenya, they decided**: Russell Brandom, "Scale AI's Remotasks Platform Is Dropping Whole Countries Without Explanation," *Rest of World*, March 28, 2024, restofworld.org/2024/scale-ai-remotasks-banned-workers.
222 **In a great irony**: Jin, "The 27-Year-Old Billionaire."
222 **Scale downgraded Kenya**: Screenshots of group designations and an announcement of a change in groups.
222 **Scale was now recruiting**: Cory Weinberg, "Why a $14 Billion Startup Is Now Hiring PhD's to Train AI from Their Living Rooms," *The Information*, June 25, 2024, theinformation.com/articles/why-a-14-billion-startup-is-now-hiring-phds-to-train-ai-from-their-living-rooms.
223 **In her inbox**: Hilary Kimuyu, "Online Gig Site Remotasks Exits Kenya," *Business Daily*, March 13, 2024, businessdailyafrica.com/bd/corporate/technology/online-gig-site-remotasks-exits-kenya-4555340.

Chapter 10: Gods and Demons

227 **We were young**: Andrew Van Dam, "What Percent Are You?," *Economics Blog, Wall Street Journal*, March 2, 2016, wsj.com/articles/what-percent-are-you-1456922287.
228 **"Where I grew up"**: Tyler Cowen, host, *Conversations with Tyler*, podcast, episode 61, "Sam Altman on Loving Community, Hating Coworking, and the Hunt for Talent," Mercatus Center Podcasts, February 27, 2019.
229 **Core to the EA philosophy**: Émile P. Torres, "The Acronym Behind Our Wildest AI Dreams and Nightmares," *Truthdig*, June 15, 2023, truthdig.com/articles/the-acronym-behind-our-wildest-ai-dreams-and-nightmares.
229 **In a 2013 paper**: William MacAskill, "Replaceability, Career Choice, and Making a Difference," *Ethical Theory and Moral Practice* 17 (2013): 269–83, doi.org/10.1007/S10677-013-9433-4.
229 **Under the logic**: "What Is Effective Altruism?," Effective Altruism Forum, accessed October 8, 2024, effectivealtruism.org/articles/introduction-to-effective-altruism.
229 **"I and others"**: Will MacAskill, "What Are the Most Important Moral Problems of Our Time?," TED Talk, April 2018, 11 min., 45 sec., ted.com/talks/will_macaskill_what_are_the_most_important_moral_problems_of_our_time.

230 **A decade earlier, Facebook:** "About Us," Open Philanthropy, accessed October 17, 2024, openphilanthropy.org/about-us.

231 **Open Philanthropy became:** Holden Karnofsky, "The Open Philanthropy Project Is Now an Independent Organization," Open Philanthropy, June 12, 2017, openphilanthropy.org/research/the-open-philanthropy-project-is-now-an-independent-organization.

231 **Bankman-Fried, or SBF:** David Yaffe-Bellany, "A Crypto Emperor's Vision: No Pants, His Rules," *New York Times*, May 14, 2022, nytimes.com/2022/05/14/business/sam-bankman-fried-ftx-crypto.html.

231 **As he amassed his wealth:** Rebecca Ackermann, "Inside Effective Altruism, Where the Far Future Counts a Lot More Than the Present," *MIT Technology Review*, October 17, 2022, technologyreview.com/2022/10/17/1060967/effective-altruism-growth.

231 **At the start of 2022:** "Announcing the Future Fund," FTX Future Fund, archived on November 27, 2022, at web.archive.org/web/20221127183608/https://ftxfuturefund.org/announcing-the-future-fund.

231 **According to estimates compiled:** "An Overview of the AI Safety Funding Situation," Effective Altruism Forum, accessed October 8, 2024, forum.effectivealtruism.org/posts/XdhwXppfqrpPL2YDX/an-overview-of-the-ai-safety-funding-situation; author correspondence with Open Philanthropy spokesperson, November 2024.

232 **Online EA and AI safety forums:** Shazeda Ahmed, Klaudia Jaźwińska, Archana Ahlawat, Amy Winecoff, and Mona Wang, "Building the Epistemic Community of AI Safety," preprint, SSRN, December 1, 2023, 1–14, ssrn.com/abstract=4641526; "What Is Effective Altruism?," Effective Altruism Forum.

232 **The influx of members:** Most of these definitions are pulled from *LessWrong* and Effective Altruism Forum; for example: "AI Timelines," *LessWrong*, accessed on October 17, 2024, lesswrong.com/tag/ai-timelines; "Global Catastrophic Risk," Effective Altruism Forum, accessed on November 27, 2024, forum.effectivealtruism.org/topics/global-catastrophic-risk.

232 **Mixed with the tech:** Charlotte Alter, "Effective Altruism Promises to Do Good Better. These Women Say It Has a Toxic Culture of Sexual Harassment and Abuse," *Time*, February 3, 2023, time.com/6252617/effective-altruism-sexual-harassment; and Kelsey Piper, "Why Effective Altruism Struggles on Sexual Misconduct," *Vox*, February 16, 2023, vox.com/future-perfect/2023/2/15/23601143/effective-altruism-sexual-harassment-misconduct.

235 **The first, called CLIP:** Alec Radford, Jong Wook Kim, Chris Hallacy, Aditya Ramesh, Gabriel Goh, Sandhini Agarwal et al., "Learning Transferable Visual Models from Natural Language Supervision," preprint, arXiv, February 26, 2021, 1–48, doi.org/10.48550/arXiv.2103.00020.

235 **The second, DALL-E 1:** OpenAI, "DALL·E: Creating Images from Text," *Open AI* (blog), January 5, 2021, openai.com/index/dall-e.

235 **The original idea:** Jascha Sohl-Dickstein, Eric A. Weiss, Niru Maheswaranathan, and Surya Ganguli, "Deep Unsupervised Learning Using Nonequilibrium Thermodynamics," in *ICML '15: Proceedings of the 32nd International Conference on Machine Learning* 37 (July 2015): 2256–65, dl.acm.org/doi/10.5555/3045118.3045358.

235 **Five years later, Jonathan:** Jonathan Ho, Ajay Jain, and Pieter Abbeel, "Denoising Diffusion Probabilistic Models," in *NIPS '20: Proceedings of the 34th International Conference on Neural Information Processing Systems*, no. 574 (December 2020): 6840–51, dl.acm.org/doi/abs/10.5555/3495724.3496298; Anil Ananthaswamy, "The Physics Principle That Inspired Modern AI Art," *Quanta Magazine*, January 5, 2023, quantamagazine.org/the-physics-principle-that-inspired-modern-ai-art-20230105.

236 **OpenAI changed tack:** "DALL·E 2," OpenAI, accessed September 17, 2024, openai.com/index/dall-e-2.

236 **Ramesh and other researchers:** Alex Nichol, Prafulla Dhariwal, Aditya Ramesh, Pranav Shyam, Pamela Mishkin, Bob McGrew et al., "GLIDE: Towards Photorealistic Image Generation and Editing with Text-Guided Diffusion Models," in *Proceedings of the 39th International Conference on Machine Learning* (2022): 16784–804, proceedings.mlr.press/v162/nichol22a.html.

236 **Researchers outside of OpenAI:** Robin Rombach, Andreas Blattmann, Dominik Lorenz, Patrick Esser, and Björn Ommer, "High-Resolution Image Synthesis with Latent Diffusion Models," in *2022 IEEE/CVF Conference on Computer Vision and Pattern Recognition* (2022): 10674–85, doi.ieeecomputersociety.org/10.1109/CVPR52688.2022.01042.

NOTES

236 **256 Nvidia A100s:** Author interview with Björn Ommer, March 2024.
236 **With DALL-E 2's remarkable:** Fraser Kelton and Nabeel Hyatt, hosts, *Hallway Chat*, podcast, "Launch Stories of ChatGPT," December 2, 2023, hallwaychat.co/launch-stories-of-chatgpt.
238 **In December 2023:** Hayden Field, "Microsoft Engineer Warns Company's AI Tool Creates Violent, Sexual Images, Ignores Copyrights," CNBC, March 6, 2024, cnbc.com/2024/03/06/microsoft-ai-engineer-says-copilot-designer-creates-disturbing-images.html.
241 **"This is intoxicating":** Kelton and Hyatt, *Hallway Chat*.
244 **To solve OpenAI's data:** Cade Metz, Cecilia Kang, Sheera Frenkel, Stuart A. Thompson, and Nico Grant, "How Tech Giants Cut Corners to Harvest Data for A.I.," *New York Times*, April 6, 2024, nytimes.com/2024/04/06/technology/tech-giants-harvest-data-artificial-intelligence.html.
244 **OpenAI had previously:** Davey Alba and Emily Chang, "YouTube Says OpenAI Training Sora with Its Videos Would Break Rules," *Bloomberg*, April 4, 2024, bloomberg.com/news/articles/2024-04-04/youtube-says-openai-training-sora-with-its-videos-would-break-the-rules.
244 **He then used a speech-recognition tool:** OpenAI, "Introducing Whisper," *OpenAI* (blog), September 21, 2022, openai.com/index/whisper.
244 **Then, with several others:** "GPT-4 Contributions," OpenAI, accessed October 13, 2024, openai.com/contributions/gpt-4.
245 **"an idiot savant":** Bill Gates, host, *Unconfuse Me with Bill Gates*, podcast, episode 2, "Sal Khan," Gates Notes, August 10, 2023, gatesnotes.com/podcast.
245 **AP Bio because:** Bill Gates, "The Age of AI Has Begun," GatesNotes, March 21, 2023, gatesnotes.com/The-Age-of-AI-Has-Begun.
246 **This showcase, Gates said:** Bill Gates has since said this many times publicly, including in Gates, "The Age of AI Has Begun."
247 **Brockman and Fraser Kelton:** Kelton and Hyatt, *Hallway Chat*.
252 **The jokes delighted:** Will Hurd, "Should 4 People Be Able to Control the Equivalent of a Nuke?," *Politico*, January 30, 2024, politico.com/news/magazine/2024/01/30/will-hurd-ai-regulation-00136941.
252 **"The CEO is supposed":** "Sam Altman Startup School Video," posted July 26, 2017, by Waterloo Engineering, YouTube, 1 hr., 18 min., 19 sec., youtu.be/4SlNgM4PjvQ.
253 **"The board is a nonprofit":** Bilawal Sidhu, host, *The TED AI Show*, podcast, "What Really Went Down at OpenAI and the Future of Regulation w/ Helen Toner," May 28, 2024, ted.com/talks/the_ted_ai_show_what_really_went_down_at_openai_and_the_future_of_regulation_w_helen_toner.
253 **"Who am I":** Rebecca Heilweil, "Why Silicon Valley Is Fertile Ground for Obscure Religious Beliefs," *Vox*, June 30, 2022, vox.com/recode/2022/6/30/23188222/silicon-valley-blake-lemoine-chatbot-eliza-religion-robot.
253 **When company executives:** Nitasha Tiku, "The Google Engineer Who Thinks the Company's AI Has Come to Life," *Washington Post*, June 11, 2022, washingtonpost.com/technology/2022/06/11/google-ai-lamda-blake-lemoine.
254 **But despite enormous:** Tom Hartsfield, "Koko the Impostor: Ape Sign Language Was a Bunch of Babbling Nonsense," *Big Think*, May 11, 2022, bigthink.com/life/ape-sign-language.
254 **In conversations with Hinton:** Author interview with Geoff Hinton.

Chapter 11: Apex

256 **For a photo:** The photo in question, October 2022.
257 **Financial documents released:** United States v. Samuel Bankman-Fried, No. 1:22-cr-00673, CourtListener (S.D.N.Y. March 15, 2024) ECF No. 410, at *12–13. The pertinent section reads: "From late 2021 through the first quarter of 2022, Bankman-Fried directed billions of dollars in spending, which used FTX customers' money. Those expenditures included . . . Anthropic PBC (an artificial intelligence company)."
258 **A judge would rule:** Zack Abrams, "FTX Offloads Remaining Anthropic Shares as Bankruptcy Cost Surpasses $500 Million," The Block, June 1, 2024, theblock.co/post/298010/ftx-offloads-remaining-anthropic-shares-as-bankruptcy-cost-surpasses-700-million.

NOTES

259 **The instant runaway:** Will Douglas Heaven, "The Inside Story of How ChatGPT Was Built from the People Who Made It," *MIT Technology Review*, March 3, 2023, technologyreview.com/2023/03/03/1069311/inside-story-oral-history-how-chatgpt-built-openai.
260 **"one order of magnitude less":** "StrictlyVC in Conversation with Sam Altman, Part Two," posted on January 17, 2023, by Connie Loizos, YouTube, 38 min., 58 sec., youtu.be/bjkD1Om4uw.
260 **numbering just over:** Erin Woo and Stephanie Palazzolo, "OpenAI Overhauls Content Moderation Efforts as Elections Loom," *The Information*, December 18, 2023, theinformation.com/articles/openai-overhauls-content-moderation-efforts-as-elections-loom.
261 **The severe shortage:** "Behind the Scenes Scaling ChatGPT—Evan Morikawa at LeadDev West Coast 2023," posted October 26, 2023, by LeadDev, YouTube, 27 min., 12 sec., youtu.be/PeKMEXUrlq4.
261 **In an attempt to leverage:** OpenAI, "Using GPT-4 for Content Moderation," *Open AI* (blog), August 15, 2023, openai.com/index/using-gpt-4-for-content-moderation.
261 **As he'd expected:** Nico Grant and Cade Metz, "A New Chat Bot Is a 'Code Red' for Google's Search Business," *New York Times*, December 21, 2022, nytimes.com/2022/12/21/technology/ai-chatgpt-google-search.html.
262 **"We are now":** Copy of the memo.
264 **The way in which Microsoft:** The account of the climate at Microsoft is based on author interviews with ten current and former Microsoft employees and executives as well as copies of several emails that executives sent to employees.
265 **Nadella implemented a new strategy:** Copy of email referencing the new strategy.
266 **"Azure OpenAI Service":** Each of the Microsoft emails cited are based on copies of those emails.
266 **In January 2023, it had:** Growth in inferencing requests is based on copies of the above emails as well as screenshots of an internal dashboard.
266 **"We have stopped":** Author correspondence with Microsoft spokesperson, November 2024, who provided this quote from a transcript of the meeting.
267 **Still numbering fewer:** Woo and Palazzolo, "OpenAI Overhauls Content Moderation."
267 **By the end of that:** Woo and Palazzolo, "OpenAI Overhauls Content Moderation."
268 **latched on to ChatGPT:** Copy of the document.
268 **After ChatGPT went viral:** Dylan Patel and Afzal Ahmad, "The Inference Cost of Search Disruption—Large Language Model Cost Analysis," *SemiAnalysis*, February 9, 2023, semianalysis.com/p/the-inference-cost-of-search-disruption.
269 **Arrakis felt like:** Jon Victor and Aaron Holmes, "OpenAI Dropped Work on New 'Arrakis' AI Model in Rare Setback," *The Information*, October 17, 2023, theinformation.com/articles/openai-dropped-work-on-new-arrakis-ai-model-in-rare-setback.
269 **There was also a new:** Tom Dotan and Deepa Seetharaman, "The Awkward Partnership Leading the AI Boom," *Wall Street Journal*, June 13, 2023, wsj.com/articles/microsoft-and-openai-forge-awkward-partnership-as-techs-new-power-couple-3092de51.
270 **Nadella would tell:** Karen Weise and Cade Metz, "How Microsoft's Satya Nadella Became Tech's Steely Eyed A.I. Gambler," *New York Times*, July 14, 2026, nytimes.com/2024/07/14/technology/microsoft-ai-satya-nadella.html.
270 **To fulfill that aggressive:** Anissa Gardizy and Amir Efrati, "Microsoft and OpenAI Plot $100 Billion Stargate AI Supercomputer," *The Information*, March 29, 2024, theinformation.com/articles/microsoft-and-openai-plot-100-billion-stargate-ai-supercomputer; Anissa Gardizy, Aaron Holmes, and Amir Efrati, "OpenAI Leaders Say Microsoft Isn't Moving Fast Enough to Supply Servers," *The Information*, October 8, 2024, theinformation.com/articles/openai-eases-away-from-microsoft-data-centers.

Chapter 12: Plundered Earth

271 **The mountains come:** Based on author's reporting trip in Santiago and the Atacama Desert, 2024.
271 **Indigenous elders still warn:** Author interview with Sonia Ramos, an Atacameño activist, June 2024; the cutting off of tongues is also referenced in the introduction of a dictionary

NOTES

for Kunza, an Atacameño language that has largely gone extinct: Julio Vilte Vilte, *Kunza: Lengua del Pueblo Lickan Antai o Atacameño* (Codelco Chile, 2004), 11.

272 **Today nearly 60 percent:** "Chile—Country Commercial Guide: Mining," International Trade Administration, December 7, 2023, trade.gov/country-commercial-guides/chile-mining.

272 **The country has struggled:** Samo Burja, "Chile Is a Politically Disunited Resource Exporter," *Bismarck Brief*, June 19, 2024, brief.bismarckanalysis.com/p/chile-is-a-politically-disunited.

272 **Long after the Spanish:** Naomi Klein, *The Shock Doctrine: The Rise of Disaster Capitalism* (Picador, 2008), 55.

272 **In the 1950s and '60s:** Klein, *The Shock Doctrine*, 64.

272 **Friedman was a towering:** Milton Friedman, "A Friedman Doctrine—the Social Responsibility of Business Is to Increase Its Profits," *New York Times*, September 13, 1970, timesmachine.nytimes.com/timesmachine/1970/09/13/223535702.html?pageNumber=379.

272 **As Naomi Klein details:** Klein, *The Shock Doctrine*, 61.

273 **under conditions fomented:** James Doubek, "The U.S. Set the Stage for a Coup in Chile. It Had Unintended Consequences at Home," NPR, September 10, 2023, npr.org/2023/09/10/1193755188/chile-coup-50-years-pinochet-kissinger-human-rights-allende; the original Senate report detailing the CIA's heavy spending and influence campaign in Chile leading up to the coup: *Covert Action in Chile 1963–1973, Staff Report of the Select Committee to Study Governmental Operations with Respect to Intelligence Activities* (US Senate: 1975), intelligence.senate.gov/sites/default/files/94chile.pdf.

273 **Under Pinochet's rule:** Daniel Matamala, "The Complicated Legacy of the 'Chicago Boys' in Chile," *Promarket*, September 12, 2021, promarket.org/2021/09/12/chicago-boys-chile-friedman-neoliberalism.

273 **Chile is among:** "Income Share of the Richest 1%," Our World in Data, accessed October 14, 2024, ourworldindata.org/grapher/income-share-top-1-before-tax-wid?tab=chart&country=CHL.

273 **the government proudly:** Gobierno de Chile, "International InvestChile Forum: 100 Companies from 28 Countries Will Meet in the Country," Gobierno de Chile, May 16, 2024, gob.cl/en/news/international-investchile-forum-100-companies-from-28-countries-will-meet-in-the-country.

274 **"If we are going to develop":** Author interview with Martín Tironi Rodó, June 2024.

274 **The four largest hyperscalers:** Author interviews with Alan Howard, a cloud and data center analyst at the technology consultancy firm Omdia, August and September 2023.

274 **It's difficult to imagine:** Indeed it was, until the author visited the one training OpenAI's models in Arizona, September 2023.

274 **"Now football fields":** Author interview with Mél Hogan, August 2023.

275 **The equipment all together:** Bianca Bosker, "Why Everything Is Getting Louder," *The Atlantic*, November 15, 2019, theatlantic.com/magazine/archive/2019/11/the-end-of-silence/598366.

275 **Now developers use:** Rich Miller, "The Gigawatt Data Center Campus Is Coming," Data Center Frontier, April 29, 2024, datacenterfrontier.com/hyperscale/article/55021675/the-gigawatt-data-center-campus-is-coming.

275 **A rack of GPUs:** Author interviews with Hogan, August 2023; and a data center investor, March 2024.

275 **According to the International Energy:** Goldman Sachs, "AI Is Poised to Drive 160% Increase in Data Center Power Demand," Goldman Sachs, May 14, 2024, goldmansachs.com/insights/articles/AI-poised-to-drive-160-increase-in-power-demand.

275 **close to 122,000 American households:** A 150-megawatt facility can consume up to 150 megawatt-hours of energy in an hour, or 1,314,000 megawatt-hours of energy in a year. According to the US Energy Information Administration, an average American household consumed 10,791 kilowatt-hours in a year in 2022; 1,314,000 megawatt-hours divided by 10,791 kilowatt-hours equals 121,768.

275 **A single one could:** A 1,000-megawatt facility can consume up to 8,760,000 megawatt-hours of energy in a year, and a 2,000-megawatt facility, twice that. According to the California Energy Commission, San Francisco County consumed 5,120,586 megawatt-hours

NOTES

in 2022; 8,760,000 megawatt-hours divided by 5,120,586 megawatt-hours equals 1.7. Twice that is 3.4. "Electricity Consumption by County," California Energy Commission, accessed October 17, 2024, ecdms.energy.ca.gov/elecbycounty.aspx.

275 **After the last decade of flatlined:** Goldman Sachs, "AI Is Poised to Drive 160% Increase."

275 **Utility companies are now delaying:** Evan Halper, "A Utility Promised to Stop Burning Coal. Then Google and Meta Came to Town," *Washington Post*, October 12, 2024, washingtonpost.com/business/2024/10/08/google-meta-omaha-data-centers/; C Mandler, "Three Mile Island Nuclear Plant Will Reopen to Power Microsoft Data Centers," NPR, September 20, 2024, npr.org/2024/09/20/nx-s1-5120581/three-mile-island-nuclear-power-plant-microsoft-ai.

275 **By 2030, at the current:** Goldman Sachs, "AI Is Poised to Drive 160% Increase"; Ian King, "AI Computing on Pace to Consume More Energy Than India, Arm Says," *Bloomberg*, April 17, 2024, news.bloomberglaw.com/artificial-intelligence/ai-computing-on-pace-to-consume-more-energy-than-india-arm-says.

276 **AGI will solve climate change:** This claim is one that Altman has used many times, including in Sam Altman, "The Intelligence Age," *Sam Altman* (blog), September 23, 2024, ia.samaltman.com.

276 **While the last claim:** Author interview with Sasha Luccioni, August 2023.

276 **There are indeed many:** Climate Change AI details these technologies in several reports on its website, climatechange.ai, including David Rolnick, Priya L. Donti, Lynn H. Kaack, Kelly Kochanski, Alexandre Lacoste, Kris Sankaran et al., "Tackling Climate Change with Machine Learning," *ACM Computing Surveys (CSUR)* 55, no. 2 (February 2022): 1–96, doi.org/10.1145/3485128.

276 **In one paper, together:** Alexandra Sasha Luccioni, Yacine Jernite, and Emma Strubell, "Power Hungry Processing: Watts Driving the Cost of AI Deployment?," in *FAccT '24: Proceedings of the 2024 ACM Conference on Fairness, Accountability, and Transparency* (June 2024): 85–99, doi.org/10.1145/3630106.3658542.

276 **They found that producing:** These numbers are based on Table 2 in the aforementioned paper, and the EPA's estimate before January 2024 that a smartphone charge consumed 0.012 kWh of energy.

277 **Even as hyperscalers:** Transcript of meeting.

277 **build their campuses in threes:** Interview with Alan Howard, August 2023.

277 **During Hurricane Irma:** James Glanz, "How the Internet Kept Humming During 2 Hurricanes," *New York Times*, September 18, 2017, nytimes.com/2017/09/18/us/harvey-irma-internet.html.

277 **According to an estimate:** Pengfei Li, Jianyi Yang, Mohammad A. Islam, and Shaolei Ren, "Making AI Less 'Thirsty': Uncovering and Addressing the Secret Water Footprint of AI Models," preprint, arXiv, October 29, 2023, 1, doi.org/10.48550/arXiv.2304.03271.

278 **Another study found:** Md Abu Bakar Siddik, Arman Shehabi, and Landon Marston, "The Environmental Footprint of Data Centers in the United States," *Environmental Research Letters* 16, no. 6 (June 2021): 064017, doi.org/10.1088/1748-9326/abfba1.

278 **In response, data center developers:** Author interviews with six different communities facing data center expansions in Arizona, New Mexico, Virginia, two in Chile, and Uruguay, 2023–24, as well as interviews with three Microsoft employees and executives, including Noelle Walsh, corporate vice president of cloud operations and innovation, who oversees all of the company's data center expansions, about the company's practices from their perspective, 2023–24.

278 **In one case in Virginia:** Author interview with Roger Yackel, a Virginia resident leading protests against the data center expansion, March 2024.

278 **"We need a mole":** Copy of the email.

278 **"In AI, whoever has":** Author interview with Greg Brockman, August 2019.

278 **Altman began referring:** The code names, numbers, and location of Phases 1, 2, and 3 are pulled from an internal OpenAI document. The locations and cost of Phases 4 and 5 come from Anissa Gardizy and Amir Efrati, "Microsoft and OpenAI Plot $100 Billion Stargate AI Supercomputer," *The Information*, March 29, 2024, theinformation.com/articles/microsoft-and-openai-plot-100-billion-stargate-ai-supercomputer.

279 **Equipped with ten thousand:** Matt O'Brien and Hannah Fingerhut, "Artificial Intelligence

NOTES

Technology Behind ChatGPT Was Built in Iowa—with a Lot of Water," AP, September 9, 2023, apnews.com/article/chatgpt-gpt4-iowa-ai-water-consumption-microsoft-f551fde980 83d17a7e8d904f8be822c4.

279 **the company also invested:** Author correspondence with Microsoft spokesperson, November 2024.

279 **After carefully cultivating:** Author interviews with Barbara Chappell, the city of Goodyear's water services director, October 2023; two community members, September 2023; the three aforementioned Microsoft sources, 2023–24; and copies of the Goodyear city council's meeting minutes and other government documents and correspondence, obtained through public records requests, 2023–24. Those interviews and additional reporting produced the following story: Karen Hao, "AI Is Taking Water from the Desert," *The Atlantic*, March 1, 2024, theatlantic.com/technology/archive/2024/03/ai-water-climate-microsoft /677602.

280 **In Microsoft and OpenAI's design:** The 5,000 megawatt estimate for Stargate comes from Gardizy and Efrati, "Microsoft and OpenAI Plot $100 Billion Stargate"; and according to the NYC Mayor's Office of Climate and Environmental Justice, the city used on average of about 5,500 megawatts of power in 2022: "Systems," NYC Mayor's Office of Climate and Environmental Justice, accessed October 17, 2024, climate.cityofnewyork.us/subtopics /systems.

280 **Altman had recused himself:** Berber Jin, Tom Dotan, and Keach Hagey, "The Opaque Investment Empire Making OpenAI's Sam Altman Rich," *Wall Street Journal*, June 3, 2024, wsj.com/tech/ai/openai-sam-altman-investments-004fc785.

280 **Microsoft's data centers had consumed:** According to the West Des Moines Water Works, as cited by: O'Brien and Fingerhut, "Artificial Intelligence Technology Behind ChatGPT."

281 **the company is working to increase:** Correspondence with Microsoft spokesperson, November 2024.

281 **In 2022, as Microsoft:** A. Park Williams, Benjamin I. Cook, and Jason E. Smerdon, "Rapid Intensification of the Emerging Southwestern North American Megadrought in 2020–2021," *Nature Climate Change* 12, no. 3 (March 2022): 232–34, doi.org/10.1038/s41558-022-01290-z.

281 **Without drastic action:** This refers to a condition called "deadpooling," as explained in Christopher Flavelle and Mira Rojanasakul, "As the Colorado River Shrinks, Washington Prepares to Spread the Pain," *New York Times*, January 27, 2023, nytimes.com/2023/01 /27/climate/colorado-river-biden-cuts.html.

281 **over six hundred dead:** Kira Caspers, "645 People Died Due to Heat in Metro Phoenix in 2023. Here's What Is Changing This Year," AZ Central, March 15, 2024, azcentral.com /story/news/local/phoenix/2024/03/15/heat-deaths-maricopa-county/72980594007.

281 **"All things," says Tom Buschatzke":** Author interview with Tom Buschatzke, October 2023.

281 **Meta would come out:** Kevin Lee, Adi Gangidi, Mathew Oldham, "Building Meta's GenAI Infrastructure," Engineering at Meta, March 12, 2024, engineering.fb.com/2024/03/12 /data-center-engineering/building-metas-genai-infrastructure.

281 **She was born into:** Interview with Sonia Ramos, June 2024.

281 **In 1957, a part:** "Tres muertos y treinta heridos en explosión de una mina en Chuquicamata," *El Mercurio*, September 6, 1957.

282 **That displaced rock:** "The Battle for Chile's Critical Minerals," posted July 22, 2022, by Sky News, YouTube, 13 min., 54 sec., youtu.be/oywE0mQnWI0

282 **The mining has also:** Author interview with Cristina Dorador, a Chilean scientist who studies the Atacama Desert's ecosystems, June 2024.

282 **Less visible are the trails:** "The Battle for Chile's Critical Minerals," Sky News; interview with Dorador.

282 **The shift has plunged:** Author visits and interviews with three Atacameños leaders, including Sonia Ramos and Sergio Cubillos, June 2024.

283 **Instead, many are forced:** Visits and interviews with the three Atacameños leaders; and visit to an industry-sponsored health clinic, June 2024.

283 **Lithium is a more recent:** Author interviews with Dorador, June 2024; and SQM, a Chilean mining company and the world's largest lithium producer; as well as an on-site tour of SQM's lithium mines in Atacama, June 2024.

283 **Chile produces roughly a third:** Govind Bhutada, "This Chart Shows Which Countries Pro-

duce the Most Lithium," *World Economic Forum* (blog), January 5, 2023, weforum.org/stories/2023/01/chart-countries-produce-lithium-world.

283 **The material is primarily:** Author interviews with Dorador, June 2024; SQM, June 2024; and architect and researcher Marina Otero Verzier, May 2024, who has a talk about the connection between lithium extraction, data center development, Chile's colonial history, and global technology futures here: "Marina Otero Verzier-Data Mourning," posted March 1, 2023, by Columbia GSAPP, YouTube, 1 hr., 30 min., youtu.be/vbFPaNBNB-M.

283 **Now the flamingos are gone:** Visit and interview with Cubillos, the Peine leader, June 2024.

283 **In 2022, as the European:** Interview with SQM, June 2024.

283 **"Local people never have":** Interview with Dorador, June 2024.

283 **The accelerated copper:** Paul R. La Monica, "Move Over, Nvidia. Copper Is Getting a Big AI Boost Too," *Barron's*, May 22, 2024, barrons.com/articles/copper-price-ai-microsoft-utilities-c99058b7.

284 **In Brazil, a 2023 art exhibition:** "Artificial Intelligence, Art and Indigeneity," accessed October 2, 2024, aei.art.br/aiai/en/the-research.

284 **central to Indigenous demands:** Visits and interviews with the three Atacameños leaders, June 2024.

285 **"the largest infrastructure buildout":** Dylan Patel and Myron Xie, "Microsoft Infrastructure—AI & CPU Custom Silicon Maia 100, Athena, Cobalt 100," *SemiAnalysis*, November 15, 2023, semianalysis.com/p/microsoft-infrastructure-ai-and-cpu.

285 **Google, meanwhile, said:** Alphabet, "2024 Q3 Earnings Call," Alphabet Investor Relations, October 29, 2024, abc.xyz/2024-q3-earnings-call.

285 **Meta said it would likely:** Meta, "Meta Reports Third Quarter 2024 Results," Meta Investor Relations, October 30, 2024, investor.fb.com/investor-news/press-release-details/2024/Meta-Reports-Third-Quarter-2024-Results/default.aspx.

285 **Less than a thirty-minute drive:** Based on author's visit to Quilicura, June 2024.

286 **When I ask the company's:** Author correspondences with Google Chile spokesperson, June and November 2024.

286 **Arancibia had just started:** Author interviews with Alexandra Arancibia, June 2024.

286 **Only two decades ago:** Author interviews with Arancibia, June 2024; Rodrigo Vallejos, June 2024; Lorena Antiman, another environmental activist in Quilicura, June 2024; and Miguel Mora, a Quilicura-based teacher who studies its wetlands, and Felipe Gonzalez, who heads the Environmental Management Unit of Quilicura, June 2024.

287 **The data center—as activists:** Interviews with Arancibia; Vallejos; and Antiman.

288 **It announced a project:** Subsecretaría de Telecomunicaciones, "Gobierno de Chile escoge ruta mediante Nueva Zelanda y hasta Australia para implementar el Cable Transoceánico," Subsecretaría de Telecomunicaciones, July 27, 2020, subtel.gob.cl/gobierno-de-chile-escoge-ruta-mediante-nueva-zelanda-y-hasta-australia-para-implementar-el-cable-transoceanico.

288 **Google backed the partnership:** Google, "Announcing Humboldt, the First Cable Route Between South America and Asia-Pacific," *Google Cloud* (blog), January 11, 2024, cloud.google.com/blog/products/infrastructure/announcing-humboldt-the-first-cable-route-between-south-america-and-asia-pacific.

288 **From the 1930s:** Josefa Silva González, "A más de 20 años de Miño: La estancada lucha contra el asbestos," *La Voz de Maipú*, February 18, 2022, lavozdemaipu.cl/la-estancada-lucha-contra-el-asbesto.

288 **That summer, as Google:** Interviews with Arancibi; Vallejos; Gonzalez; and author interview with Tania Rodriguez, June 2024.

288 **In other words, the data:** The Google environmental impact report to SEA stated that the data center could use 169 liters of potable water a second, or 5,329,584,000 liters a year. According to the water service authority in Cerillos, the municipality consumed 5,097,946 liters in all of 2019, the year Google sought to come in; 5,329,584,000 liters a year divided by 5,097,946 liters a year equals 1,045.

289 **Chile was already nine years:** "Persistent Drought Is Drying Out Chile's Drinking Water," Reuters, March 20, 2024, reuters.com/world/americas/persistent-drought-is-drying-out-chiles-drinking-water-2024-03-20.

289 **MOSACAT was founded:** The account of MOSACAT's activism against Google is based on

NOTES

author interviews with Rodriguez and eight other MOSACAT members, June 2024. Additional details are from Chilean media coverage, primarily Alberto Arellano, Lucas Cifuentes, and Cristóbal Ríos, "Las zonas oscuras de la evaluación ambiental que autorizó 'a ciegas' el megaproyecto de Google en Cerrillos," Ciper, May 25, 2020, ciperchile.cl/2020/05/25/las-zonas-oscuras-de-la-evaluacion-ambiental-que-autorizo-a-ciegas-el-megaproyecto-de-google-en-cerrillos.

291 **Cumulatively, they take:** All details about Antel's operations are based on author's visit to an Antel data center and an interview with its manager, Javier Echeverria, June 2024.

292 **Some cheekily call it:** Author interview with Marcos Umpiérrez, a professor at the University of the Republic in Uruguay, and his colleagues, June 2024.

292 **The park even looks somewhat:** Based on a visit to Parque de las Ciencias, June 2024.

292 **The water shortage was:** Author interview with Marcelo Fozati, an Uruguayan agronomist and farmer who heads an organization to protect local farmers and crops, and Daniel Pena, an Uruguayan researcher who studies the environmental extractivism of multinationals in his country, June 2024; as well as "Uruguay: Drought Losses Estimated at USD 1.200 million, Minister Says," MercoPress, February 2, 2023, en.mercopress.com/2023/02/02/uruguay-drought-losses-estimated-at-usd-1.200-million-minister-says; and Guillermo Garat, "My City Has Run Out of Fresh Water. Will Your City Be Next?," *New York Times*, July 19, 2023, nytimes.com/2023/07/19/opinion/drinking-water-montevideo.html.

293 **Those who couldn't drank:** Author interviews with three Uruguayan residents and water activists: Fabiana, June 2024; Noelia Lagos, June 2024; and Carmen Sosa, June 2024.

293 **Where Silicon Valley had ascended:** "Google's and Microsoft's Profits Soar as Pandemic Benefits Big Tech," *New York Times*, October 18, 2021, nytimes.com/live/2021/04/27/business/stock-market-today.

293 **Fabiana, the boisterous head:** Interview with Fabiana.

293 **The water crisis emerged:** Grace Livingstone, "'It's Pillage': Thirsty Uruguayans Decry Google's Plan to Exploit Water Supply," *Guardian*, July 11, 2023, theguardian.com/world/2023/jul/11/uruguay-drought-water-google-data-center.

293 **Most such farms:** Interview with the Fozati and Pena.

293 **Their activities deplete:** "Fertilizer Use Per Capita, 1961 to 2019," Our World in Data, accessed October 17, 2024, ourworldindata.org/grapher/fertilizer-per-capita?tab=table.

294 **He drives around the country:** Details about Pena's research and activism are based on author interviews with Pena, May and June 2024, including a day spent in his truck traveling through some of the poorest parts of Montevideo and its outskirts.

294 **Now, in a bitter irony:** Interviews with Pena, June 2024; and Sosa, June 2024.

294 **The environmental ministry revealed:** Livingstone, "'It's Pillage.'"

295 **"This is not drought":** Livingstone, "'It's Pillage.'"

295 **The Google Chile spokesperson said:** Author correspondence with Google Chile spokesperson, November 2024.

296 **In 2022, Microsoft finalized:** Dan Swinhoe, "Microsoft Files Plans for Chilean Data Center Region," Data Center Dynamics, January 24, 2022, datacenterdynamics.com/en/news/microsoft-files-plans-for-chilean-data-center-region.

296 **In his victory speech:** Matamala, "The Complicated Legacy of the 'Chicago Boys.'"

297 **"It is deeply striking":** Rodrigo Vallejos Calderón, "Los costos de estar conectados: Datacenters y el consume hídrico," *Bits* 23 (2022), 28–33, revistasdex.uchile.cl/index.php/bits/issue/view/1049.

297 **Vallejos caught the attention:** Author interviews with Marina Otero Verzier, May 2024; and Serena Dambrosio and Nicolás Díaz Bejarano, June 2024.

299 **The students designed:** Photos of the mock-ups.

299 **But in fairness, the coalition:** Interviews with Martín Tironi and Aisén Etcheverry, the head of the Ministry of Science.

Chapter 13: The Two Prophets

302 **"Would you be qualified":** "Watch: OpenAI CEO Sam Altman Testifies Before Senate Judiciary Committee," *PBS News*, May 16, 2023, pbs.org/newshour/politics/watch-live-openai-ceo-sam-altman-testifies-before-senate-judiciary-committee.

NOTES

302 **Marcus would later backtrack:** Gary Marcus, "OpenAI's Sam Altman Is Becoming One of the Most Powerful People on Earth. We Should Be Very Afraid," *Guardian*, August 3, 2024, theguardian.com/technology/article/2024/aug/03/open-ai-sam-altman-chatgpt-gary-marcus-taming-silicon-valley.

302 **Altman's prep team:** Hasan Chowdhury, "Insiders Say Sam Altman's AI World Tour Was a Success," *Business Insider*, June 24, 2023, businessinsider.com/sam-altman-world-tour-ai-chatgpt-openai-2023-6.

302 **For months, with or without:** Cecilia Kang, "How Sam Altman Stormed Washington to Set the A.I. Agenda," *New York Times*, June 7, 2023, nytimes.com/2023/06/07/technology/sam-altman-ai-regulations.html.

302 **By early June, Altman:** Kang, "How Sam Altman Stormed Washington."

302 **On the day of Altman's:** Author interviews with Karla Ortiz, December 2023 and April 2024; and Rachel Meinerding and Nicole Hendrix Herman, the cofounders and coleaders of the Concept Art Association, April 2024.

303 **Those jobs that were:** Interview with Meinerding and Hendrix Herman; CVL Economics, *Future Unscripted: The Impact of Generative Artificial Intelligence on Entertainment Industry Jobs* (2024), 1–58, animationguild.org/wp-content/uploads/2024/01/Future-Unscripted-The-Impact-of-Generative-Artificial-Intelligence-on-Entertainment-Industry-Jobs-pages-1.pdf.

303 **Altman was attending:** Kang, "How Sam Altman Stormed Washington."

303 **The same narrative Altman:** Karen Hao, "The New AI Panic," *The Atlantic*, October 11, 2023, theatlantic.com/technology/archive/2023/10/technology-exports-ai-programs-regulations-china/675605.

304 **"If you'd told me":** Alex W. Palmer, "'An Act of War': Inside America's Silicon Blockade Against China," *New York Times*, July 12, 2023, nytimes.com/2023/07/12/magazine/semiconductor-chips-us-china.html.

304 **Nvidia's own maneuvering:** Jane Lee, "Exclusive: Nvidia Offers New Advanced Chip for China That Meets U.S. Export Controls," Reuters, November 7, 2022, reuters.com/technology/exclusive-nvidia-offers-new-advanced-chip-china-that-meets-us-export-controls-2022-11-08.

304 **The ban was also a lift:** Fanny Potkin and Yelin Mo, "Chinese Chip Equipment Makers Grab Market Share as US Tightens Curbs," Reuters, October 18, 2023, reuters.com/technology/chinese-chip-equipment-makers-grab-market-share-us-tightens-curbs-2023-10-18.

304 **After vigorously playing catch-up:** Khari Johnson, "Meta's Open Source Llama Upsets the AI Horse Race," *Wired*, July 26, 2023, wired.com/story/metas-open-source-llama-upsets-the-ai-horse-race.

305 **a critical building block for:** Tony Peng, "What Llama 3 Means to China, ERNIE Bot Hits 200 Million Users, and China Trails US in AI Models," Recode China AI, April 22, 2024, recodechinaai.substack.com/p/what-llama-3-means-to-china-ernie.

305 **Amid the climate:** Markus Anderljung, Joslyn Barnhart, Anton Korinek, Jade Leung, Cullen O'Keefe, Jess Whittlestone et al., "Frontier AI Regulation: Managing Emerging Risks to Public Safety," preprint, arXiv, November 7, 2023, 1–51, doi.org/10.48550/arXiv.2307.03718.

306 **But Hooker and many:** Author interviews with Sara Hooker, October 2024; Deborah Raji, August 2024; Sarah Myers West, codirector of AI Now, October 2024; and other AI policy experts, 2023–24.

306 **While scale *can* lead:** Sara Hooker, "On the Limitations of Compute Thresholds as a Governance Strategy," preprint, arXiv, July 30, 2024, 1–54, doi.org/10.48550/arXiv.2407.05694.

307 **It captured significant:** Author interviews with Myers West, September 2023; Amba Kak, the other codirector of AI Now, October 2023; Emily Weinstein, September 2023; Raji, October 2023; and two other AI policy experts, November 2023.

307 **"Parts of the administration":** Interviews with Weinstein.

307 **The white paper's ideas:** Hao, "The New AI Panic."

308 **"If we don't know":** "Watch: OpenAI CEO Sam Altman Testifies."

308 **"it's because they trained it":** Interview with Myers West, September 2023.

308 **As Commerce consulted:** US Department of Commerce, "NTIA Solicits Comments on

NOTES

Open-Weight AI Models," press release, February 21, 2024, commerce.gov/news/press-releases/2024/02/ntia-solicits-comments-open-weight-ai-models.

308 **Facing off against:** Mozilla, "Mozilla's Response to the National Telecommunications and Information Administration's Request for Comments on Dual Use Foundation Artificial Intelligence Models with Widely Available Model Weights," *Mozilla Foundation* (blog), March 2024, blog.mozilla.org/netpolicy/files/2024/03/Mozilla-RfC-Submission-Dual-Use-Foundation-Models-With-Widely-Available-Model-Weights.pdf.

309 **Such recipes already abound:** Portions of this section appeared in different form as Hao, "The New AI Panic."

309 **In critical ways:** Cameron F. Kerry, Joshua P. Meltzer, Matt Sheehan, "Can Democracies Cooperate with China on AI Research?," Brookings, January 9, 2023, brookings.edu/articles/can-democracies-cooperate-with-china-on-ai-research.

309 **One of the most famous:** Matt Sheehan, "Who Benefits from American AI Research in China?," Macro Polo, October 21, 2019, macropolo.org/china-ai-research-resnet.

310 **the ideas championed by:** The account of how the EO came together is based on author interviews with Alondra Nelson, former OSTP director, October 2023; Suresh Venkatasubramanian, former OSTP deputy director, October 2023; and two other policy folks, November 2023.

310 **The order, one:** Portions of this section appeared in different form as Karen Hao and Matteo Wong, "The White House Is Preparing for an AI-Dominated Future," *The Atlantic*, October 30, 2023, theatlantic.com/technology/archive/2023/10/biden-white-house-ai-executive-order/675837.

311 **California governor Gavin Newsom:** Khari Johnson, "Why Silicon Valley Is Trying So Hard to Kill This AI Bill in California," CalMatters, August 12, 2024, calmatters.org/economy/technology/2024/08/ai-regulation-showdown.

311 **"It was a step":** Interview with Hooker, October 2024.

311 **Raji had found herself:** Gabby Miller, "US Senate AI 'Insight Forum' Tracker," Tech Policy Press, December 9, 2023, techpolicy.press/us-senate-ai-insight-forum-tracker.

311 **As her fellow witnesses:** Author interview with Raji, October 2023; Inioluwa Deborah Raji, "AI's Present Matters More Than Its Imagined Future," *The Atlantic*, October 4, 2023, theatlantic.com/technology/archive/2023/10/ai-chuck-schumer-forum-legislation/675540.

311 **A Schumer spokesperson would later:** Cat Zakrzewski, "Meet the Woman Who Transformed Sam Altman into the Avatar of AI," *Washington Post*, January 9, 2024, washingtonpost.com/technology/2024/01/09/openai-anna-makanju-ai-regulation.

312 **In March of that year:** Sam Altman (@sama), "i'm doing a trip in may/june to talk to openai users and developers (and people interested in AI generally). please come hang out and share feature requests and other feedback! more detail here: https://openai.com/form/openai-tour-2023 or email oai23tour@openai.com," Twitter (now X), March 29, 2023, x.com/sama/status/1641181668206858240.

312 **The model had involved:** "GPT-4 contributions," OpenAI, accessed October 13, 2024, openai.com/contributions/gpt-4.

312 **The author of the company's:** "GPT-4," OpenAI, March 14, 2023, openai.com/index/gpt-4-research.

312 **Altman had then tweeted credit:** Sam Altman (@sama), "GPT-4 was truly a team effort from our entire company, but the overall leadership and technical vision of Jakub Pachocki for the pretraining effort was remarkable and we wouldn't be here without it," Twitter (now X), March 14, 2023, x.com/sama/status/1635700851619819520.

313 **OpenAI's response:** OpenAI, "OpenAI and Journalism," *OpenAI* (blog), January 8, 2024, openai.com/index/openai-and-journalism.

313 **That same week, OpenAI's policy:** Dan Milmo, "'Impossible' to Create AI Tools like ChatGPT Without Copyrighted Material, OpenAI Says," *Guardian*, January 8, 2024, theguardian.com/technology/2024/jan/08/ai-tools-chatgpt-copyrighted-material-openai.

314 **Iterative deployment, Altman:** OpenAI, "Our Approach to AI Safety," *Open AI* (blog), April 5, 2023, openai.com/index/our-approach-to-ai-safety.

316 **The post also announced:** OpenAI, "Introducing Superalignment," *OpenAI* (blog), July 5, 2023, openai.com/index/introducing-superalignment.

316 **"i was hoping that"**: Sam Altman (@sama), "i was hoping that the oppenheimer movie would inspire a generation of kids to be physicists but it really missed the mark on that. let's get that movie made! (i think the social network managed to do this for startup founders.)," Twitter (now X), July 22, 2023, x.com/sama/status/1682809958734131200.
316 **Altman was fond:** Elizabeth Weil, "Sam Altman Is the Oppenheimer of Our Age," *New York*, September 25, 2023, nymag.com/intelligencer/article/sam-altman-artificial-intelligence-openai-profile.html.
316 **He also liked to paraphrase:** Cade Metz, "The ChatGPT King Isn't Worried, but He Knows You Might Be," *New York Times*, March 31, 2023, nytimes.com/2023/03/31/technology/sam-altman-open-ai-chatgpt.html.
317 **"The way the world was"**: Tyler Cowen, host, *Conversations with Tyler*, podcast, episode 61, "Sam Altman on Loving Community, Hating Coworking, and the Hunt for Talent," Mercatus Center Podcasts, February 27, 2019.
319 **In March 2023, he'd:** Eliezer Yudkowsky, "Pausing AI Developments Isn't Enough. We Need to Shut It All Down," *Time*, March 29, 2023, time.com/6266923/ai-eliezer-yudkowsky-open-letter-not-enough.
320 **A month later, Hoffman:** Musk v. Altman, No. 4:24-cv-04722, CourtListener (N.D. Cal. November 14, 2024) ECF No. 32, Exhibit 18.
320 **That position became:** Julia Black, "Elon Musk Had Twins Last Year with One of His Top Executives," *Business Insider*, July 6, 2022, businessinsider.com/elon-musk-shivon-zilis-secret-twins-neuralink-tesla.
320 **"This is a bait":** "Elon Musk Wanted an OpenAI For-Profit," *OpenAI* (blog), December 13, 2024, openai.com/index/elon-musk-wanted-an-openai-for-profit/#summer-2017-we-and-elon-agreed-that-a-for-profit-was-the-next-step-for-openai-to-advance-the-mission.
321 **In the announcement, Altman:** Sam Altman, "Quora," *Sam Altman* (blog), April 21, 2017, blog.samaltman.com/quora.
322 **In her most popular:** Helen Toner, "Leaning into EA Disillusionment," Effective Altruism Forum, July 22, 2022, forum.effectivealtruism.org/posts/MjTB4MvtedbLjgyja/leaning-into-ea-disillusionment.
322 **By late summer of 2023:** Cade Metz, Tripp Mickle, and Mike Isaac, "Before Altman's Ouster, OpenAI's Board Was Divided and Feuding," *New York Times*, November 21, 2023, nytimes.com/2023/11/21/technology/openai-altman-board-fight.html.
323 **After the meeting, one:** Kevin Roose, "OpenAI Insiders Warn of a 'Reckless' Race for Dominance," *New York Times*, June 4, 2024, nytimes.com/2024/06/04/technology/openai-culture-whistleblowers.html.
324 **When Altman finally handed:** Dan Primack, "Sam Altman Owns OpenAI's Venture Capital Fund," *Axios*, February 15, 2024, axios.com/2024/02/15/sam-altman-openai-startup-fund.

Chapter 14: Deliverance

326 **"Annie Altman?" Weil wrote:** Elizabeth Weil, "Sam Altman Is the Oppenheimer of Our Age," *New York*, September 25, 2023, nymag.com/intelligencer/article/sam-altman-artificial-intelligence-openai-profile.html.
327 **The final day before the:** Copy of email, which Annie posted online: Annie Altman (@anniealtman108), "Less than 24 hours before the @NYMag publishing, the first 'official' public recognition of my existence and relation. x.com/bullishdumping/bullishdumping/status/1753869400719958519," Twitter (now X), February 3, 2024, x.com/anniealtman108/status/1753881201482629258.
327 **In 2024, I would reach out:** Author interviews and visit with Annie Altman, March–November 2024.
327 **Gibstine offered a brief statement:** Author correspondence with Connie Gibstine, October 2024.
328 **In January 2025, after Annie:** Altman v. Altman, No. 4:25-cv-00017, CourtListener (E.D. Mo. Jan 06, 2025) ECF No. 1; Sam Altman (@sama), "My sister has filed a lawsuit against me. Here is a statement from my mom, brothers, and me:," Twitter (now X), January 7, 2025, x.com/sama/status/1876780763653263770.

NOTES

329 **The only other Altman:** Fact-checker correspondence with Burroughs, October 2024; author interview with James Roble, July 2024.
329 **While still in college:** Copy of Annie's Tufts medical records.
329 **In a span of six years:** Each of Annie's diagnoses are corroborated by copies of one of the following: her childhood medical records; Tufts medical records; Tufts therapy notes; adulthood diagnostic imaging scans and readouts, including an ultrasound and an MRI; an obstetrics and gynecology evaluation; and physical therapy notes. Details of the impact of these diagnoses on her mobility and quality of life are also corroborated by her Tufts therapy notes; adulthood physical therapy notes; and photos, such as of her walking boot and sweat-soaked sheets.
329 **died of a sudden heart attack:** "Jerry Altman Obituary," *St. Louis Post-Dispatch*, May 27, 2018, legacy.com/us/obituaries/stltoday/name/jerry-altman-obituary?id=1683283.
329 **She was diagnosed at a young age:** Copy of Annie's childhood medical records; copy of Annie's Tufts therapy notes.
330 **Sam has also spoken publicly:** Trevor Noah, host, *What Now? with Trevor Noah*, season 1, episode 5, "Sam Altman Speaks Out About What Happened at OpenAI," Spotify Podcasts, December 7, 2023, open.spotify.com/show/122imavATqSE7eCyXIcqZL.
331 **In May 2019, as her:** Copy of 401(k) email notification and 401(k) statement with balance.
331 **The best tax strategy:** Copy of email.
331 **Her therapist's notes:** Copy of Annie's therapist notes in LA.
331 **In December 2019, her bank account:** Copy of bank notification email.
331 **Scared and alone:** Screenshot of SeekingArrangement activation email.
331 **From late 2019 to mid-2020, Annie:** Copies of various emails and texts exchanged between Annie and her family.
332 **they agreed to cover:** Copies of various emails and texts.
332 **In May 2020, as her family's:** Copy of text exchange.
332 **Eight months after:** Copy of email.
334 **She continued her podcast:** Copy of Etsy and Patreon activation emails.
334 **A strange thing was happening:** Various screenshots.
334 **Sometimes she noticed chunks:** Annie's old Instagram stories with screenshots of her Apple podcast reviews tagging Apple podcast support about her disappearing reviews.
334 **At least twice, on both:** Screenshot of an Instagram where number of Likes is greater than number of views; and two screenshots of the same YouTube video, where the screenshot with the later time stamp has fewer views.
334 **it's possible that Annie's:** Author interviews with Olivia Snow, a researcher at UCLA focused on sex work, tech, and policy, May 2024; Val Elefante, a researcher and founding team member of the feminist social media company Reliabl, September 2024; and a former Facebook data scientist, October 2024.
334 **"Sam carried AI into the world":** Author interview with Neily Messerschmidt, November 2024.
334 **In sessions from July 2021:** Copy of Annie's therapy notes in Hawai'i.
335 **From fifteen sessions:** Annie's therapy notes.
335 **In these childhood memories:** Interview with Messerschmidt, November 2024.
335 **The victim's brain:** Author interviews with a therapist, June and August 2024, who also referenced the bestselling book: Bessel van der Kolk, M.D., *The Body Keeps the Score: Brain, Mind, and Body in the Healing of Trauma* (Penguin Books, 2015), 1–464.
335 **"I experienced sexual":** Annie Altman (@anniealtman108), "I experienced sexual, physical, emotional, verbal, financial, and technological abuse from my biological siblings, mostly Sam Altman and some from Jack Altman. (2/3)," Twitter (now X), November 13, 2021, x.com/anniealtman108/status/1459696444802142213.
336 **"Sexual, physical, emotional":** Annie Altman (@anniealtman108), "Sam and Jack, I know you remember my Torah portion was about Moses forgiving his brothers. 'Forgive them father for they know not what they've done' Sexual, physical, emotional, verbal, financial, and technological abuse. Never forgotten.," Twitter (now X), September 10, 2022, x.com/anniealtman108/status/1568689744951005185.
336 **In the three months:** Screenshot of Annie's OnlyFans income history.

336 **In July 2023, Sam:** Copy of email thread.
337 **In addition to his 401(k):** Copy of Jerry Altman's will; and copy of Jerry Altman's trust.
337 **In early 2024, with her:** Copy of email correspondence between Annie's lawyer and Gibstine's lawyer.
337 **It would allege:** *Altman*, CourtListener, ECF No. 1.
337 **In October 2024, after:** Copy of Annie's diagnosis.
338 **Borderline personality disorder is marked:** Author interviews with the aforementioned therapist and Blaise Aguirre, an assistant professor of psychiatry at Harvard Medical School, October 2024, both of whom have worked with many patients with the disorder. Neither reviewed Annie's case, only commented on the condition more broadly.
338 **the disorder usually goes away:** The most comprehensive study of borderline personality disorder is an ongoing twenty-four-year longitudinal study called McLean Study of Adult Development, which is conducted by Mary C. Zanarini and has followed over 360 individuals diagnosed with the disorder. Among the study's key findings: The disorder has a good symptomatic prognosis, and psychotropic medications are not curative. The study regularly publishes new papers, including: Mary C. Zanarini, Frances R. Frankenburg, Isabel V. Glass, and Garrett M. Fitzmaurice, "The 24-Year Course of Symptomatic Disorders in Patients with Borderline Personality Disorder and Personality-Disordered Comparison Subjects: Description and Prediction of Recovery From BPD," *The Journal of Clinical Psychiatry* 85 (2024), doi.org/10.4088/JCP.24m15570.

Chapter 15: The Gambit

343 **Born in Albania:** The account of Murati's upbringing comes primarily from Charles Duhigg, "The Inside Story of Microsoft's Partnership with OpenAI," *New Yorker*, December 1, 2023, newyorker.com/magazine/2023/12/11/the-inside-story-of-microsofts-partnership-with-openai; and Murati's appearance on Kevin Scott's podcast: Kevin Scott, host, *Behind the Tech with Kevin Scott*, "Mira Murati, Chief Technology Officer, OpenAI," Microsoft, July 11, 2023, microsoft.com/en-us/behind-the-tech/mira-murati-chief-technology-officer-openai.
343 **The shift happened:** Christopher Jarvis, "The Rise and Fall of Albania's Pyramid Schemes," *Finance & Development*, International Monetary Fund, March 2000, imf.org/external/pubs/ft/fandd/2000/03/jarvis.htm.
343 **The upheaval would leave:** Duhigg, "The Inside Story of Microsoft's Partnership with OpenAI."
345 **But the more Murati worked:** Unless otherwise noted, the account of the lead-up to the board crisis and the behind the scenes of the crisis itself in this and the next chapter is based on author interviews with eight people who were directly involved in or close to the people directly involved in the described events; their contemporaneous notes; and screenshots of Slack messages, emails, and other corroborating evidence, including the audio recording of the all-hands meeting on November 17, 2023 after the board fired Altman.
347 **In the summer, Murati:** Mike Isaac, Tripp Mickle, and Cade Metz, "Key OpenAI Executive Played a Pivotal Role in Sam Altman's Ouster," *New York Times*, March 7, 2024, nytimes.com/2024/03/07/technology/openai-executives-role-in-sam-altman-ouster.html.
350 **To anyone resisting:** Isaac et al., "Key OpenAI Executive Played a Pivotal Role."
359 **"I did not feel":** Cade Metz, Tripp Mickle, and Mike Isaac, "Before Altman's Ouster, OpenAI's Board Was Divided and Feuding," *New York Times*, November 21, 2023, nytimes.com/2023/11/21/technology/openai-altman-board-fight.html.

Chapter 16: Cloak-and-Dagger

367 ***The Wall Street Journal* would later report:** Deepa Seetharaman, Keach Hagey, Berber Jin, and Kate Linebaugh, "Sam Altman's Knack for Dodging Bullets—with a Little Help from Bigshot Friends," *Wall Street Journal*, December 24, 2023, wsj.com/tech/ai/sam-altman-openai-protected-by-silicon-valley-friends-f3efcf68.
367 **along with Reid Hoffman:** Natasha Mascarenhas, "Behind OpenAI Meltdown, Valley Heavyweight Reid Hoffman Calmed Microsoft Nerves," *The Information*, January 17, 2024,

NOTES

theinformation.com/articles/behind-openai-meltdown-valley-heavyweight-reid-hoffman-calmed-microsoft-nerves.

368 **The New York Times would later:** Mike Isaac, Tripp Mickle, and Cade Metz, "Key OpenAI Executive Played a Pivotal Role in Sam Altman's Ouster," *New York Times*, March 7, 2024, nytimes.com/2024/03/07/technology/openai-executives-role-in-sam-altman-ouster.html.

369 **"Ilya has a good":** Elon Musk (@elonmusk), "I am very worried. Ilya has a good moral compass and does not seek power. He would not take such drastic action unless he felt it was absolutely necessary.," Twitter (now X), November 19, 2023, x.com/elonmusk/status/1726376406785925566.

369 **Later, at around 2:00 a.m.:** Elon Musk (@elonmusk), Twitter (now X), November 20, 2023, x.com/elonmusk/status/1726542015087927487.

369 **But on Tuesday:** Elon Musk (@elonmusk), "This letter about OpenAI was just sent to me. These seem like concerns worth investigating. https://gist.github.com/Xe/32d7bc436e401f3323ae77e7e242f858," Twitter (now X), November 21, 2023, x.com/elonmusk/status/1727096607752282485.

369 **It was a different letter:** "Xe/openai-message-to-board.md," GitHub Gist, archived November 21, 2023, at web.archive.org/web/20231121225252/https://gist.github.com/Xe/32d7bc436e401f3323ae77e7e242f858.

370 **The first was my 2020:** Karen Hao, "The Messy, Secretive Reality Behind OpenAI's Bid to Save the World," *MIT Technology Review*, February 17, 2020, technologyreview.com/2020/02/17/844721/ai-openai-moonshot-elon-musk-sam-altman-greg-brockman-messy-secretive-reality; Karen Hao and Charlie Warzel, "Inside the Chaos at OpenAI," *The Atlantic*, November 19, 2023, theatlantic.com/technology/archive/2023/11/sam-altman-open-ai-chatgpt-chaos/676050.

371 **"current employee here":** All quotes from emails are from the screenshots that the person provided.

373 **During the board crisis, one:** Anna Tong, Jeffrey Dastin and Krystal Hu, "OpenAI Researchers Warned Board of AI Breakthrough Ahead of CEO Ouster, Sources Say," Reuters, November 23, 2023, reuters.com/technology/sam-altmans-ouster-openai-was-precipitated-by-letter-board-about-ai-breakthrough-2023-11-22.

373 **The algorithm had been a brainchild of:** Jon Victor and Amir Efrati, "OpenAI Made an AI Breakthrough Before Altman Firing, Stoking Excitement and Concern," *The Information*, November 22, 2023, theinformation.com/articles/openai-made-an-ai-breakthrough-before-altman-firing-stoking-excitement-and-concern.

374 **He would later explain:** "Ilya Sutskever: 'Sequence to Sequence Learning with Neural Networks: What a Decade,'" posted December 14, 2024, by seremot, YouTube, 24 min., 36 sec., youtu.be/1yvBqasHLZs.

374 **They siloed the company:** Anna Tong and Katie Paul, "Exclusive: OpenAI Working on New Reasoning Technology Under Code Name 'Strawberry,'" Reuters, July 15, 2024, reuters.com/technology/artificial-intelligence/openai-working-new-reasoning-technology-under-code-name-strawberry-2024-07-12.

374 **The frenetic Q* discourse:** Portions of this section appeared in different form as Karen Hao, "Why Won't OpenAI Say What the Q* Algorithm Is?," *The Atlantic*, November 28, 2023, theatlantic.com/technology/archive/2023/11/openai-sam-altman-q-algorithm-breakthrough-project/676163.

375 **OpenAI teased Sora:** OpenAI, "Creating Video from Text," *Open AI* (blog), openai.com/index/sora.

375 **In 2022, Taylor had played:** Kate Conger and Lauren Hirsch, "The Board Chair Squaring Up to Elon Musk in the Feud Over Twitter," *New York Times*, October 4, 2022, nytimes.com/2022/10/04/technology/twitter-board-elon-musk.html.

375 **Soon after, Taylor cofounded:** "OpenAI Chair's AI Startup Sierra Gets $4.5 Bln Valuation in Latest Funding Round," Reuters, October 28, 2024, reuters.com/technology/artificial-intelligence/openai-chairs-ai-startup-sierra-gets-45-bln-valuation-latest-funding-round-2024-10-28.

375 **For the OpenAI investigation:** OpenAI, "Review Completed & Altman, Brockman to Continue to Lead OpenAI," *Open AI* (blog), March 8, 2024, openai.com/index/review-completed-altman-brockman-to-continue-to-lead-openai.

376 **"We have unanimously concluded":** OpenAI, "Review Completed."
376 **"Accountability is important":** Helen Toner released the statement in a screenshot on X: Helen Toner (@hlntnr), "A statement from Helen Toner and Tasha McCauley:," Twitter (now X), March 8, 2024, x.com/hlntnr/status/1766269137628590185.

Chapter 17: Reckoning

377 **Two of them were:** Erin Woo and Stephanie Palazzolo, "OpenAI Researchers, Including Ally of Sutskever, Fired for Alleged Leaking," *The Information*, April 11, 2024, theinformation.com/articles/openai-researchers-including-ally-of-sutskever-fired-for-alleged-leaking.
377 **he had been in the process:** Edward Ludlow and Ashlee Vance, "Altman Sought Billions for Chip Venture Before OpenAI Ouster," *Bloomberg*, November 19, 2023, bloomberg.com/news/articles/2023-11-19/altman-sought-billions-for-ai-chip-venture-before-openai-ouster.
377 **In February 2024, after:** Keach Hagey and Asa Fitch, "Sam Altman Seeks Trillions of Dollars to Reshape Business of Chips and AI," *Wall Street Journal*, February 8, 2024, wsj.com/tech/ai/sam-altman-seeks-trillions-of-dollars-to-reshape-business-of-chips-and-ai-89ab3db0.
377 **Altman would later say:** Lex Fridman, host, *Lex Fridman Podcast*, podcast, episode 419, "Sam Altman: OpenAI, GPT-5, Sora, Board Saga, Elon Musk, Ilya, Power & AGI," March 18, 2024, lexfridman.com/podcast.
378 **The audio work had:** Author interview with Alexis Conneau, January 2025.
379 **Altman and Brockman had set a new deadline:** Copy of internal OpenAI memo.
379 **Scallion would be the first:** OpenAI, *Preparedness Framework (Beta)* (OpenAI, December 18, 2023), 1–27, cdn.openai.com/openai-preparedness-framework-beta.pdf.
380 **Upstream processes and:** OpenAI, *Preparedness Framework (Beta)*.
380 **the launch for Scallion:** OpenAI, "Hello GPT-4o," *OpenAI* (blog), May 13, 2024, openai.com/index/hello-gpt-4o.
380 **There had also been:** Wording confirmed independently by two people.
381 **On *The Daily Show*:** "Trump's Thirsty VP Contenders Crash Trial & ChatGPT's Flirty AI Update | The Daily Show," posted on May 15, 2024, by The Daily Show, YouTube, 9 min., 57 sec., youtu.be/eFkUOi_9140.
382 **"I think it's the best":** Audio recording of the meeting, May 15, 2024.
382 **The company was going:** Audio recording of the meeting, May 15, 2024.
382 **Altman would come to:** Audio recording of Altman expressing his regret to employees.
382 **What did Altman think:** Sarah Krouse, Deepa Seetharaman, and Joe Flint, "Behind the Scenes of Scarlett Johansson's Battle with OpenAI," *Wall Street Journal*, May 23, 2024, wsj.com/tech/ai/scarlett-johansson-openai-sam-altman-voice-fight-7f81a1aa.
383 **In March, he appeared:** Fridman, "Sam Altman."
383 **"When we just do":** "Sam Altman & Brad Lightcap: Which Companies Will Be Steamrolled by OpenAI?," posted April 15, 2024, by 20VC with Harry Stebbings, YouTube, 53 min., 6 sec., youtu.be/G8T1O81W96Y.
383 **In May, he then joined:** Julia Black, "The Besties' Revenge: How the 'All-In' Podcast Captured Silicon Valley," *The Information*, December 15, 2023, theinformation.com/articles/the-besties-revenge-how-the-all-in-podcast-ate-silicon-valley.
383 **"It feels to me like":** "In Conversation with Sam Altman," posted May 10, 2024, by All-In Podcast, YouTube, 1 hr., 43 min., 2 sec., youtu.be/nSM0xd8xHUM.
383 **"i try not to think":** Sam Altman (@sama), "i try not to think about competitors too much, but i cannot stop thinking about the aesthetic difference between openai and google," Twitter (now X), May 16, 2024, x.com/sama/status/1791183356274921568.
384 **After The Blip, the board's:** Deepa Seetharaman, "SEC Investigating Whether OpenAI Investors Were Misled," *Wall Street Journal*, February 28, 2024, wsj.com/tech/sec-investigating-whether-openai-investors-were-misled-9d90b411.
384 **Most notably, in March:** Karen Weise and Cade Metz, "How Microsoft's Satya Nadella Became Tech's Steely Eyed A.I. Gambler," *New York Times*, July 14, 2026, nytimes.com/2024/07/14/technology/microsoft-ai-satya-nadella.html.
384 **He was known to those:** Based on the recollections and characterizations of three people who worked for him.

NOTES

385 **After years of HR complaints:** Rob Copeland and Parmy Olson, "Artificial Intelligence Will Define Google's Future. For Now, It's a Management Challenge," *Wall Street Journal*, January 26, 2021, wsj.com/articles/artificial-intelligence-will-define-googles-future-for-now-its-a-management-challenge-11611676945; Giles Turner and Mark Bergen, "Google DeepMind Co-Founder Placed on Leave From AI Lab," *Bloomberg*, August 21, 2019, bloomberg.com/news/articles/2019-08-21/google-deepmind-co-founder-placed-on-leave-from-ai-lab.

385 **Later in 2024, Microsoft:** Jordan Novet, "Microsoft Says OpenAI Is Now a Competitor in AI and Search," CNBC, July 31, 2024, cnbc.com/2024/07/31/microsoft-says-openai-is-now-a-competitor-in-ai-and-search.html; Alex Heath, "Microsoft Now Lists OpenAI as a Competitor," *The Verge*, August 2, 2024, theverge.com/2024/8/2/24212370/microsoft-now-lists-openai-as-a-competitor.

385 **More were reaching out:** Ellen Huet, host, *Foundering: The OpenAI Story*, podcast, season 5, episode 1, "The Most Silicon Valley Man Alive," Bloomberg Podcasts, June 6, 2024, bloomberg.com/news/articles/2024-06-05/foundering-sam-altman-s-rise-to-openai?srnd=foundering.

386 **"It's a strangely":** "Sam Altman Talks GPT-4o and Predicts the Future of AI," posted May 14, 2024, by the Logan Bartlett Show, YouTube, 46 min., 14 sec., youtu.be/fMtbrKhXMWc.

386 **"Ilya is easily":** OpenAI, "Ilya Sutskever to Leave OpenAI, Jakub Pachocki Announced as Chief Scientist," *OpenAI* (blog), May 14, 2024, openai.com/index/jakub-pachocki-announced-as-chief-scientist.

386 **Sutskever tweeted his own:** Ilya Sutskever (@ilyasut), "After almost a decade, I have made the decision to leave OpenAI. The company's trajectory has been nothing short of miraculous, and I'm confident that OpenAI will build AGI that is both safe and beneficial under the leadership of @sama, @gdb, @miramurati and now, under the excellent research leadership of @merettm. It was an honor and a privilege to have worked together, and I will miss everyone dearly. So long, and thanks for everything. I am excited for what comes next—a project that is very personally meaningful to me about which I will share details in due time," Twitter (now X), May 14, 2024, x.com/ilyasut/status/1790517455628198322.

387 **OpenAI executives internally:** Screenshot of Slack announcement, May 14, 2024.

387 **"Being AGI ready":** All quotes about AI safety and the Superalignment team are pulled are from an audio recording of the all-hands meeting, May 15, 2024.

388 **"I have been disagreeing":** Jan Leike (@janleike), "I joined because I thought OpenAI would be the best place in the world to do this research. However, I have been disagreeing with OpenAI leadership about the company's core priorities for quite some time, until we finally reached a breaking point.," Twitter (now X), May 17, 2024, x.com/janleike/status/1791498178346549382.

388 **Kelsey Piper, a senior:** Kelsey Piper, "ChatGPT Can Talk, but OpenAI Employees Sure Can't," *Vox*, May 17, 2024, vox.com/future-perfect/2024/5/17/24158478/openai-departures-sam-altman-employees-chatgpt-release.

389 **They agreed to not:** Daniel Kokotajlo has written at length about his decision-making, including in this thread: Daniel Kokotajlo (@DKokotajlo67142), "1/15: In April, I resigned from OpenAI after losing confidence that the company would behave responsibly in its attempt to build artificial general intelligence—'AI systems that are generally smarter than humans,'" Twitter (now X), June 4, 2024, x.com/DKokotajlo67142/status/1797994238468407380; and his posts and comments on the AI Safety forum LessWrong: "Daniel Kokotajlo," LessWrong, accessed November 25, 2024, lesswrong.com/users/daniel-kokotajlo. The estimated value of his equity comes from Kevin Roose, "OpenAI Insiders Warn of a 'Reckless' Race for Dominance," *New York Times*, June 4, 2023, nytimes.com/2024/06/04/technology/openai-culture-whistleblowers.html.

389 **With Piper's story out:** All quotes from Slack pulled from screenshots.

390 **"we have never clawed back":** Sam Altman (@sama), "in regards to recent stuff about how openai handles equity: we have never clawed back anyone's vested equity, nor will we do that if people do not sign a separation agreement (or don't agree to a non-disparagement agreement). vested equity is vested equity, full stop. there was a provision about potential equity cancellation in our previous exit docs; although we never clawed anything back, it should never have been something we had in any documents or communication. this is on

me and one of the few times i've been genuinely embarrassed running openai; i did not know this was happening and i should have. the team was already in the process of fixing the standard exit paperwork over the past month or so. if any former employee who signed one of those old agreements is worried about it, they can contact me and we'll fix that too. very sorry about this," Twitter (now X), May 18, 2024, x.com/sama/status/1791936857594581428.

390 **On May 20, Scarlett:** Bobby Allyn (@BobbyAllyn), "Statement from Scarlett Johansson on the OpenAI situation. Wow:," Twitter (now X), May 20, 2024, x.com/BobbyAllyn/status/1792679435701014908.

390 **All week, on top of:** Kylie Robison, "ChatGPT Will Be Able to Talk to You Like Scarlett Johansson in Her," *The Verge*, May 13, 2024, theverge.com/2024/5/13/24155652/chatgpt-voice-mode-gpt4o-upgrades.

390 **Was she mad:** Sarah Krouse et al, "Behind the Scenes of Scarlett Johansson's Battle with OpenAI."

391 **On May 19, the company had:** OpenAI, "How the Voices for ChatGPT Were Chosen," *OpenAI* (blog), May 19, 2024, openai.com/index/how-the-voices-for-chatgpt-were-chosen.

391 **before they could find:** Krouse et al., "Scarlett Johansson's Battle with OpenAI."

391 **"In a time when we":** Allyn, "Statement from Scarlett Johansson."

392 **"We are sorry":** OpenAI, "How the Voices for ChatGPT Were Chosen."

392 **"I've seen a lot of policymakers":** Derek Robertson, "Sam Altman's Scarlett Johansson Blunder Just Made AI a Harder Sell in DC," *Politico*, May 22, 2024, politico.com/news/magazine/2024/05/22/scarlett-johansson-sam-altmans-washington-00159507.

392 **The leadership team gave:** All descriptions of and quotes from the meeting are based on an audio recording of the meeting, May 22, 2024.

394 **Published just that day:** Kelsey Piper, "Leaked OpenAI Documents Reveal Aggressive Tactics Toward Former Employees," *Vox*, May 22, 2024, vox.com/future-perfect/351132/openai-vested-equity-nda-sam-altman-documents-employees.

395 **"The situation is, I think":** All quotes from an audio recording of the meeting, May 23, 2024.

396 **Murati, Brockman, and Pachocki arrived:** Deepa Seetharaman, "Turning OpenAI into a Real Business Is Tearing It Apart," *Wall Street Journal*, September 27, 2024, wsj.com/tech/ai/open-ai-division-for-profit-da26c24b.

Chapter 18: A Formula for Empire

399 **Altman once remarked onstage:** Tyler Cowen, host, *Conversations with Tyler*, podcast, episode 61, "Sam Altman on Loving Community, Hating Coworking, and the Hunt for Talent," Mercatus Center Podcasts, February 27, 2019.

400 **"The most successful founders":** Sam Altman, "Successful People," *Sam Altman* (blog), March 7, 2013, blog.samaltman.com/successful-people.

400 **"Who will control the future of AI?":** Sam Altman, "Who Will Control the Future of AI?," Opinion, *Washington Post*, July 25, 2024, washingtonpost.com/opinions/2024/07/25/sam-altman-ai-democracy-authoritarianism-future.

401 **"We're now going to assume":** All quotes from the all-hands meeting pulled from an audio recording, May 15, 2024.

402 **On May 28, less than:** Screenshot of Slack message.

403 **On Altman's list:** Amir Efrati and Wayne Ma, "OpenAI CEO Cements Control as He Secures Apple Deal," *The Information*, May 29, 2024, theinformation.com/articles/openai-ceo-cements-control-as-he-secures-apple-deal.

403 **Altman was considering:** Aaron Holmes, Natasha Mascarenhas, and Julia Hornstein, "OpenAI CEO Says Company Could Become Benefit Corporation Akin to Rivals Anthropic, xAI," *The Information*, June 14, 2024, theinformation.com/articles/openai-ceo-says-company-could-become-benefit-corporation-akin-to-rivals-anthropic-xai.

403 **On June 4, *The New York Times*:** Kevin Roose, "OpenAI Insiders Warn of a 'Reckless' Race for Dominance," *New York Times*, June 4, 2024, nytimes.com/2024/06/04/technology/openai-culture-whistleblowers.html.

403 **In an open letter:** "A Right to Warn About Advanced Artificial Intelligence," accessed November 5, 2024, righttowarn.ai.

NOTES

403 **A month later, *The Washington Post*:** Pranshu Verma, Cat Zakrzewski, and Nitasha Tiku, "OpenAI Illegally Barred Staff from Airing Safety Risks, Whistleblowers Say," *Washington Post*, July 13, 2024, washingtonpost.com/technology/2024/07/13/openai-safety-risks-whistleblower-sec.

403 **Later that month, five US senators:** Pranshu Verma, Cat Zakrzewski, and Nitasha Tiku, "Senators Demand OpenAI Detail Efforts to Make Its AI Safe," *Washington Post*, July 23, 2024, washingtonpost.com/technology/2024/07/23/openai-senate-democrats-ai-safe.

404 **OpenAI was also bringing in:** OpenAI, "OpenAI Welcomes Sarah Friar (CFO) and Kevin Weil (CPO)," *OpenAI* (blog), June 10, 2024, openai.com/index/openai-welcomes-cfo-cpo.

404 **First to go:** John Schulman (@johnschulman2), "I shared the following note with my OpenAI colleagues today: I've made the difficult decision to leave OpenAI. This choice stems from my desire to deepen my focus on AI alignment, and to start a new chapter of my career where I can return to hands-on technical work. I've decided to pursue this goal at Anthropic, where I believe I can gain new perspectives and do research alongside people deeply engaged with the topics I'm most interested in . . . ," Twitter (now X), August 5, 2024, x.com/johnschulman2/status/1820610863499509855.

404 **Brockman announced that he was:** Greg Brockman (@gdb), "I'm taking a sabbatical through end of year. First time to relax since co-founding OpenAI 9 years ago. The mission is far from complete; we still have a safe AGI to build.," Twitter (now X), August 5, 2024, x.com/gdb/status/1820644694264791459.

404 **The following month, on September 25:** Mira Murati (@miramurati), "I shared the following note with the OpenAI team today.," Twitter (now X), September 25, 2024, x.com/miramurati/status/1839025700009030027.

404 **Within hours, two more key leaders:** Bob McGrew (@bobmcgrewai), "I just shared this with OpenAI:," Twitter (now X), September 25, 2024, x.com/bobmcgrewai/status/1839099787423134051; Barret Zoph (@barret_zoph), "I posted this note to OpenAI.," September 25, 2024, x.com/barret_zoph/status/1839095143397515452.

404 **the shipping of OpenAI's latest model:** OpenAI, "Introducing OpenAI o1," *OpenAI* (blog), accessed January 6, 2025, openai.com/o1.

404 **Musk was expanding:** Dara Kerr, "How Memphis Became a Battleground over Elon Musk's xAI Supercomputer," NPR, September 11, 2024, npr.org/2024/09/11/nx-s1-5088134/elon-musk-ai-xai-supercomputer-memphis-pollution.

404 **Anthropic's latest version:** Stephanie Palazzolo, Erin Woo, and Amir Efrati, "How Anthropic Got Inside OpenAI's Head," *The Information*, December 12, 2024, theinformation.com/articles/how-anthropic-got-inside-openais-head; Kevin Roose, "How Claude Became Tech Insiders' Chatbot of Choice," *New York Times*, December 13, 2024, nytimes.com/2024/12/13/technology/claude-ai-anthropic.html.

405 **Sutskever had officially formed:** Kenrick Cai, Krystal Hu, and Anna Tong, "Exclusive: OpenAI Co-Founder Sutskever's New Safety-Focused AI Startup SSI Raises $1 Billion," Reuters, September 4, 2024, reuters.com/technology/artificial-intelligence/openai-co-founder-sutskevers-new-safety-focused-ai-startup-ssi-raises-1-billion-2024-09-04.

405 **OpenAI was still struggling:** Stephanie Palazzolo, Erin Woo, and Amir Efrati. "OpenAI Shifts Strategy as Rate of 'GPT' AI Improvements Slows," *The Information*, November 9, 2024, theinformation.com/articles/openai-shifts-strategy-as-rate-of-gpt-ai-improvements-slows; Deepa Seetharaman, "The Next Great Leap in AI Is Behind Schedule and Crazy Expensive," *Wall Street Journal*, December 20, 2024, wsj.com/tech/ai/openai-gpt5-orion-delays-639e7693.

405 **OpenAI's latest fundraise:** OpenAI, "New Funding to Scale the Benefits of AI," *OpenAI* (blog), October 2, 2024, openai.com/index/scale-the-benefits-of-ai.

405 **The post was titled:** Sam Altman, "The Intelligence Age," *Sam Altman* (blog), September 23, 2024, ia.samaltman.com.

405 **During an all-hands, Murati:** Audio recording of the meeting, September 26. 2024.

405 **"Mira, Bob, and Barret made":** Sam Altman (@sama), "i just posted this note to openai: Hi All– Mira has been instrumental to OpenAI's progress and growth the last 6.5 years; she has been a hugely significant factor in our development from an unknown research lab to an important company. When Mira informed me this morning that she was leaving, I was saddened but of course support her decision. For the past year, she has been building

out a strong bench of leaders that will continue our progress. I also want to share that Bob and Barret have decided to depart OpenAI. Mira, Bob, and Barret made these decisions independently of each other and amicably, but the timing of Mira's decision was such that it made sense to now do this all at once, so that we can work together for a smooth handover to the next generation of leadership.," Twitter (now X), September 25, 2024, x.com/sama/status/1839096160168063488.

406 **Musk, allied now with:** Musk v. Altman, No. 4:24-cv-04722, CourtListener (N.D. Cal. November 14, 2024) ECF No. 32.

407 **"OpenAI's conduct could have":** Jessica Toonkel, Keach Hagey, Meghan Bobrowsky, "Meta Urges California Attorney General to Stop OpenAI from Becoming For-Profit," *Wall Street Journal*, December 13, 2024, wsj.com/tech/ai/elon-musk-open-ai-lawsuit-response-c1f415f8.

407 **Late in the year, nestled:** OpenAI, "Why OpenAI's Structure Must Evolve to Advance Our Mission," *OpenAI* (blog) December 27, 2024, openai.com/index/why-our-structure-must-evolve-to-advance-our-mission.

407 **"We are now confident":** Sam Altman, "Reflections," *Sam Altman* (blog), January 5, 2025, blog.samaltman.com/reflections.

Epilogue: How the Empire Falls

409 **In 2021, I came across:** Karen Hao, "A New Vision of Artificial Intelligence for the People," *MIT Technology Review*, April 22, 2022, technologyreview.com/2022/04/22/1050394/artificial-intelligence-for-the-people.

410 **Large language models accelerate:** Author interviews with Kathleen Siminyu, November 2021; Michael and Caroline Running Wolf, November 2021; Kevin Scannell, December 2021; Vukosi Marivate, April 2023; and Pelonomi Moiloa and Jade Abbott, April 2023; Matteo Wong, "The AI Revolution Is Crushing Thousands of Languages," *The Atlantic*, April 12, 2024, theatlantic.com/technology/archive/2024/04/generative-ai-low-resource-languages/678042.

410 **Among the over seven thousand:** "Kevin Scannell on 'Language from Below: Grassroots Efforts to Develop Language Technology for Minoritized Languages' 24.S96 Special Seminar: Linguistics & social justice," posted on November 17, 2021, by MIT-Haiti Initiative, Facebook, 2 hr., 56 min., 46 sec., facebook.com/mithaiti/videos/1060463734714819; OpenAI, "GPT-4," OpenAI, March 14, 2023, openai.com/index/gpt-4-research.

410 **It was up against:** Author interviews with Keoni Mahelona, October, November, and December 2021; and Peter-Lucas Jones, November, December 2021, and January 2022.

411 **This is where:** Author interviews with Mahelona; Jones; Caleb Moses, a data scientist who worked on the project, November 2021; and several others engaged in Te Hiku's language preservation work, November 2021–January 2022.

412 **"Data is the last frontier":** Interview with Mahelona, October 2021.

412 **That data pool paled:** Alec Radford, Jong Wook Kim, Tao Xu, Greg Brockman, Christine McLeavey, Ilya Sutskever, "Robust Speech Recognition via Large-Scale Weak Supervision," preprint, arXiv, December 6, 2022, 1–2, arxiv.org/pdf/2212.04356.

413 **a free speech-recognition model:** Mozilla, "About DeepSpeech," Mozilla GitHub, accessed December 16, 2024, mozilla.github.io/deepspeech-playbook/DEEPSPEECH.html.

414 **After Timnit Gebru was ousted:** Author interviews with Timnit Gebru, August 2024; and Milagros Miceli, August 2024.

414 **she founded a nonprofit:** "About Us," Distributed AI Research, accessed December 16, 2024, dair-institute.org/about.

414 **"Our research is intended to":** "Research Philosophy," Distributed AI Research, accessed December 16, 2024, dair-institute.org/research-philosophy.

415 **She created the Data Workers':** "Data Workers' Inquiry," Data Workers' Inquiry, accessed December 16, 2024, data-workers.org.

415 **For her project, Fuentes:** Oskarina Veronica Fuentes Anaya, "Life of a Latin American Data Worker," Data Workers' Inquiry, accessed December 16, 2024, data-workers.org/oskarina; author correspondence with Fuentes, July 2024.

415 **A continent away:** Author interview with Mophat Okinyi, August 2024.

NOTES

416 **he also started a nonprofit:** "Our Story," Techworker Community Africa, accessed December 16, 2024, techworkercommunityafrica.org/About.html.
416 **"As the dust settles":** Mophat Okinyi, "Impact of Remotasks' Closure on Kenyan Workers," Data Workers' Inquiry, accessed December 16, 2024, data-workers.org/mophat.
416 **he would be named:** Billy Perrigo, "Mophat Okinyi," *Time*, September 5, 2024, time.com/7012787/mophat-okinyi.
417 **In Uruguay, Daniel Pena:** Author interview with Daniel Pena, May 2024.
418 **In her 2019 talk:** Ria Kalluri, "The Values of Machine Learning," conference talk, December 9, 2019, posted December 9, 2019, by NIPS 2019, SlidesLive, 28 min., 51 sec., slideslive.com/38923453/the-values-of-machine-learning.
419 **UC Berkeley researcher:** Author interview with Deborah Raji, August 2024.
420 **"If you're using a car":** Author interview with Sasha Luccioni, August 2024.
420 **As Joseph Weizenbaum, MIT professor:** Joseph Weizenbaum, "ELIZA—a Computer Program for the Study of Natural Language Communication Between Man and Machine," *Communications of the ACM* 9, no. 1, (January 1966): 36–45, doi.org/10.1145/365153.365168.

INDEX

Abbeel, Pieter, 49, 118, 235
Abbott, Andy, 30
acceleration risk, 232, 249
Acemoglu, Daron, 88–89
Achiam, Joshua, 406
African Content Moderators Union, 416
AGI (artificial general intelligence), 47–48, 76–79, 129–31, 232, 388–89
   Google and, 24–25
   OpenAI and Altman, 7–8, 12–13, 19, 31, 47–48, 49, 62, 65, 67, 75, 111, 121–22, 142–43, 183, 240, 253, 254–55, 301, 319, 357, 400–402, 405
   use of term, 76–77, 93–94
Agnew, William, 102, 106, 161
agriculture, 229, 292–93
Aguirre, Blaise, 338
AI (artificial intelligence)
   AGI compared with, 76–77
   anthropomorphizing, 90–91, 111
   author's reporting, 14–16
   benefits of, 13, 16, 19, 76, 77–78, 84–85, 88–89, 90, 333–34, 400, 418
   commercialization of, 14–15, 51, 75, 101–15, 150–52
   definition of intelligence, 90–94
   empires of, 16–20, 197, 222–23, 270, 414, 418, 420
   funding, 101–6, 110, 132
   model training, 4, 61, 98, 134–37, 163, 244–45, 278–81, 307
   regulatory policy, 25, 27, 84, 86, 134, 136, 265, 272, 301, 303–4, 306–7, 311–12, 357, 358, 384

   research and development, 13, 14, 17–18, 64, 89–90, 101–6, 110
   paper conventions and peer review, 15, 15n
   research faculty exodus, 105–6, 134
   risks and harms of, 16–19, 23–27, 55–58, 78–81, 106–10, 380
   scraping, 102–3, 114, 134–38, 151–52, 182–84, 384
   theories of, 94–101
   timeline, 93, 133, 232–33, 260, 388–89
   total corporate investments in, 105
   use of term, 90, 91, 400
AI alignment, 26, 248
   misalignment, 55, 86, 124, 145–46, 320, 347
   OpenAI, 54, 70, 86, 122–23, 164, 240, 248, 250, 262, 315–18, 347
   Superalignment, 316–17, 353, 387–88
AI Index, 105
AI Insight Forums, 311
AI Now Institute, 308
Airbnb, 36, 41, 136, 150, 202, 367
air pollution, 286
AI safety, 55–58, 122–32, 301–12, 316–24, 419. *See also* data privacy; existential risks
   alignment and, 122–23, 124, 145–46, 316–18
   effective altruism and, 55–56, 230–34, 321–22
   Frontier Model Forum, 305–6, 309
   Senate Judiciary Hearing, 301–3, 307–9, 314–15
   thresholds, 301–2, 305–8, 310–11
AI Scientist, 183, 318–19, 325, 347, 375
"AI takeoff," 232

INDEX

"AI winter," 97, 435n
Alameda Research, 231
Algorithmic Justice League, 161
algorithms, 51–52, 56, 373–74
Algorithms of Oppression (Noble), 162
Alibaba, 15, 159
Alignment Manhattan Project, 315–18
Allen & Company, 67–68
Alphabet, 105
AlphaFold, 309–10
AlphaGo, 59, 93
Altman, Annie, 43–45, 326–40, 352–55, 406, 458–59n
  appeals to family for financial help, 327, 331–32
  death of father, 329–31
  early life and education of, 29, 30, 328–29
  mental health struggles of, 44–45, 329–30, 331–32, 339–40
  New York magazine article, 326–27, 328–29, 332–33, 336–40, 343, 352
  physical health struggles of, 329, 332–33
  sexual abuse allegations, 3, 44–45, 327–28, 334–38, 352–53, 406
  sex work of, 326, 332–36
Altman, Jack, 29, 30, 35–36, 41, 69, 185, 327–28, 331, 336
Altman, Jerold "Jerry," 29–31, 44, 329–31, 332
Altman, Max, 29, 30, 36, 326, 327–28, 331
Altman, Sam
  AI chip company plan, 3, 377–78
  background of, 23, 29–30
  benefits of AGI, 19, 405
  birth and early life of, 29, 30–31
  board of directors and, 40, 252–53, 320–25, 375–76
    leadership questions, 345–65
  business structure of OpenAI, 13–14, 61–64, 66–67, 86, 402–3, 407
  ChatGPT, 260, 261, 262, 280, 346
  commercialization plan, 66–67, 150–51
  compute phases, plan, 278–81
  conflicts and rifts at OpenAI, 149, 150–51, 233–34, 313–16, 396
  congressional testimony of, 301–3, 314–15
  education of, 30–32
  effective altruism ideology and, 233–34
  equity crisis and, 388–90, 392–96
  firing and reinstatement of, 1–12, 14, 364–73
    the investigation, 369–70, 375–76, 377, 392
  founding of OpenAI, 12–13, 26–28, 46, 47–51, 53–54
  fundraising, 61–62, 65–68, 71–72, 132, 141, 156, 262, 320–21, 331, 367, 377, 405
  GPT-3, 133–34, 278–79
  GPT-4, 246, 248–52, 279, 346, 383–84, 386, 390–91
  Graham and, 28, 32, 36–39, 40, 69
  "Intelligence Age," 19, 405
  Jobs comparisons with, 2, 34, 35, 37
  Johansson crisis, 382, 390–92, 393
  leadership of, 64–65, 69–70, 75, 141–44, 243–44, 354–55, 403–4
  leadership behavior, 345–60, 361–65, 382–83, 385–86
  Loopt and, 32–37, 43, 68
  Manhattan Project, 146–47, 315–17
  Mayo's office design and, 74
  media relations of, 33, 34, 383
  mission of OpenAI, 5, 400–402
  MIT Technology Review and, 86–87
  on Napoleon, 399–400
  net worth of, 35, 44, 188, 389, 390
  other investment projects of, 3, 185–88
  paranoia of, 147–48
  personality of, 31, 34, 42–45, 333, 346
  politics of, 41–42, 43, 62
  research road map, 59, 175–78
  retreat of October 2022, 256–57
  Scallion, 379–80, 380, 382
  sexuality of, 31, 41
  sister Annie and, 43–45, 326–40, 385–86, 406
    sexual abuse allegations, 3, 44–45, 327–28, 334–38, 352–53, 356
  success formula of, 32–35, 37, 142–44
  vision for OpenAI, 9, 83, 142–43, 262
  World Tour of, 312, 313, 337
  at Y Combinator (YC), 23, 27–28, 32, 34, 36–38, 39, 43, 68–69, 75, 141, 142, 185, 186, 187–88, 321
altruism, 13, 14, 400. *See also* effective altruism
Amazon, 41, 46, 142, 161
  data centers, 274–75, 277, 287
  Mechanical Turk, 194
American Sign Language, 254
Amodei, Daniela, 55–56, 58, 144–45, 156, 157, 230
Amodei, Dario, 55–58
  AI safety and risks, 55–56, 57–58, 87, 122–27, 131, 133, 134, 145–46, 147, 149–52, 156–57, 362
  Altman's firing, 366
  at Anthropic, 58, 60, 115, 128, 157, 213–14, 230
  background of, 55
  The Divorce, 57–58, 156–57, 181, 213, 230, 233, 242, 353
  Dota 2, 129, 144–45
  founding of OpenAI, 28, 55
  GPT-2, 125, 129–32, 150
  GPT-3, 133–34, 134–35, 144–45, 156
  Nest, 134–35, 144–45, 150, 151, 156, 244
  promotion to director of research, 125, 133
  scaling, 129–33, 156–57
Android, 100, 239
"anonymous crowd work" model, 206

INDEX

Antel, 291-92
Anthropic, 6, 60, 115, 157, 233
    Claude, 261, 358, 379, 400, 404-5, 406
    founding of, and The Divorce, 58, 128, 157, 213, 230
    Frontier Model Forum, 305-6, 309
    FTX bankruptcy and, 257-58
    Leike joins, 388
    valuation, 18
AP Bio, 245-46
APEC CEO Summit, 2
APIs (application programming interfaces), 150-51. *See also specific APIs*
Apollo 11 (movie), 317
Apollo program, 317
Appen, 137, 195, 197-202
Apple, 30, 202, 334, 402
Arancibia, Alexandra, 285-87, 296-99, 300
Arizona, 15, 279, 281, 292
arms race, 16-17
Arrakis, 269, 374
arsenic, 282
artificial general intelligence. *See* AGI
artificial intelligence. *See* AI
arXiv, 15*n*
asbestos, 288
Asimov, Isaac, 83
Atacama Desert, 271-72, 284-87
atomic bomb, 316-17
authoritarianism, 71, 147, 195-96, 400
Authors Guild, 135
automata studies, 89-90, 434*n*
autonomous weapons, 52, 310, 380
Azure AI, 68, 72, 75, 156, 266, 279

babbage, 150
Babbage, Charles, 150
backpropagation, 97-98
Baidu, 15, 17, 55, 159, 413
Bankman-Fried, Samuel, 231-32, 233, 257-58, 380
Beckham, David, 1
Bell Labs, 55
Bender, Emily M., 164-69, 253-54
    "On the Dangers of Stochastic Parrots," 164-73, 254, 276, 414
Bengio, Samy, 161-62, 165, 166-67, 169
Bengio, Yoshua, 105, 162
Bezos, Jeff, 41
Biden, Joe, 115-16, 310
Bing, 112, 113, 247, 264, 355
biological viruses, 27
biological weapons, 305, 309, 310, 380
Birhane, Abeba, 102, 106, 137-38
"black box," 107
Black in AI, 52, 53, 161
blacklists, 222
Black Lives Matter, 152-53, 162-63, 167
blind spots, 88

Blip, The, 375, 377, 384, 386, 396, 397-98
board of directors, of OpenAI
    Altman's firing and reinstatement, 1-12, 14, 336, 364-73, 375-76, 384, 386, 396, 402
    author's reporting, 370-73
    the investigation, 369-70, 375-76, 377, 392
    Murati as interim CEO, 1-2, 8, 357, 364-65, 366
    open letter, 10-11, 367-68
    Altman's leadership behavior, 324-25, 345-65, 385
    members departing and joining, 11, 57-58, 58, 320-23, 375
    oversight questions, 322-25
Bolt, Usain, 34
Books2, 135
Books3, 440*n*
Boomers (Boomerism), 233-34, 250, 305-6, 314, 315, 387, 396, 402, 403-4
bootstrapping, 49
borderless science, 308-11
borderline personality disorder, 338, 460*n*
Boric Font, Gabriel, 296-97, 299-300
Bostrom, Nick, 26-27, 55-56, 57, 122-23
bot tax, 200
bottleneck, 47, 78, 244-45, 280, 309
Boyd, Eric, 266
Brady, Tom, 231
brain-scale AI, 60
Bridgewater Associates, 230
Brin, Sergey, 249
Brockman, Anna, 10, 256-57, 333, 338
Brockman, Greg
    Altman and, 243-44, 349, 355, 395-96, 406-7
    firing and reinstatement, 2, 6, 8-12, 345-46, 366
    leadership behavior, 34, 363-64
    author's 2019 interview, 74-81, 84-85, 159-60, 278
    background of, 46
    board of directors and, 240
    board of directors and oversight, 322-23
    commercialization plan, 150-51
    computing infrastructure, 278-79
    culture and mission of OpenAI, 53-54, 84-85
    departure of, 404
    Dota 2, 66, 144-45
    founding of OpenAI, 28, 46-51
    governance structure of OpenAI, 61-63
    GPT-4, 244-48, 250-51, 252, 257, 260, 346
    Latitude, 180-81
    leadership of OpenAI, 58-59, 61-62, 63-65, 69, 70, 83, 84-85, 243-44
    Omnicrisis, 396-98
    recruitment efforts of, 48-49, 53-54, 57-58

Brockman, Greg (*cont.*)
  research road map, 59–61
  retreat of October 2022, 256–57
  Scallion, 379–80
  Stripe, 41, 46, 55, 58, 73, 82
Brundage, Miles, 248, 250, 314, 388, 406
Buolamwini, Joy, 161
Burning Man, 35, 263
Burrell, Jenna, 93
Buschatzke, Tom, 281

California Senate Bill 1047, 311
cancers, 192, 282, 288, 293, 301, 378
capped-profit structure, 70, 72, 75, 322, 370–71, 401
carbon emissions, 79–80, 159–60, 171–73, 275–78, 295, 309
Carnegie Mellon University, 97, 106, 172
Carr, Andrew, 385
Carter, Ashton, 43
CBRN weapons, 301, 380
Center for AI Safety, 322
Center for Security and Emerging Technology (CSET), 7, 307, 321, 357, 358
Center on Long-Term Risk, 388
Centre for the Governance of AI, 321–22
Cerrillos, Chile, 288–91, 296, 297
CFPB (Consumer Financial Protection Bureau), 419–20
chatbots, 17, 112–14, 189–90, 217–18, 220
  ELIZA, 95–97, 111, 420–21
  GPT-3, 217–18
  GPT-4, 258–59
  LaMDA, 153, 253–54
  Meena, 153
  Tay, 153
ChatGPT, 258–62, 267, 280
  connectionist tradition of, 95
  GPT-3.5 as basis, 217–18, 258
  hallucinations problem, 113, 114, 268
  release, 2, 58, 101, 111, 120, 158, 159, 212, 220, 258–62, 264, 265–66, 268, 302
  sign-up incentive, 267
  voice mode, 378–79, 380–81, 391
Chauvin, Derek, 152–53
Chen, Mark, 381, 405–6
Chesky, Brian, 41, 367
Chicago Boys (Chicago school of economics), 272–73, 296
child sex abuse material (CSAM), 137, 180–81, 189, 192, 208, 237–39, 241, 242
Chile, 15, 271–81
  data centers, 285–91, 295–99
  extractivism, 272, 273–74, 281–85, 296–99, 417
Chilean coup d'état of 1973, 273
Chilean protests of 2019-2022, 291, 296–97
Chile Project, 272–73

China
  AI chips, 115–16, 304
  AI development, 55, 103, 132, 146, 159, 191, 301, 303–4, 305, 307, 309–10, 311
  mass surveillance, 103–4
Chuquicamata mine collapse of 1957, 281–82
CIA (Central Intelligence Agency), 155, 273, 321
Clarifai, 108, 238
Clark, Jack, 76, 81, 125–28, 154, 156–57, 311
Clarke, Arthur C., 55
Claude, 261, 358, 379, 400, 404–5, 406
clawback clause, 389, 393–96
climate change, 24, 52, 76–80, 93, 165, 196, 276, 281, 292–95, 301
Climate Change AI, 77–78, 276
CLIP, 235, 236
closed-domain questions, 268
closed systems, 308–11
CloudFactory, 206–7, 212–13
code generation, 151–53, 181–84, 318
Codex, 184, 243, 247, 269, 318
cofounders, overview of, 48
Cogito, 242
cognition, 109, 119–20
cognitive dissonance, 227–28
Cohere, 306–7
Coinbase, 136
Collard, Rosemary, 104*n*
Colombia, 15, 103
Colorado River and water usage, 281
Commerce Department, U.S., 304, 307, 308
Common Crawl, 135–36, 137, 151, 163
companion bots, 179, 180
"compositional generation," 238
compression, 122, 235
compute, 59–61, 115–16, 278–81, 387
  efficiency, 175–77, 268–69, 375, 419
  threshold, 98, 301–2, 305–8, 310–11
Conception, 41
Conneau, Alexis, 378–79
connectionism, 94–100, 105, 109–10, 117–18
content moderation, 136–37, 155, 179–81, 189–90, 238–39. *See also* data annotation
  Sama, 190–92, 206–13, 218–19
Copilot, 238–39, 247–48, 264
copper, 272, 273, 277, 281, 282–84, 291
copyright infringement, 90–91, 102, 135, 301, 308, 313, 384
"costly signals," 357–58
cotton gin, 88–89
Couldry, Nick, 104
COVID-19 pandemic, 54, 74, 149, 152, 181–82, 192, 203, 205, 206, 208, 213, 218, 293, 323
Cowen, Tyler, 399
Crab Generation, 220–21
Creative Commons, 182
cryogenics, 186–87

cryptocurrencies, 63, 80, 185–86
CSAM. *See* child sex abuse material
CUDA (Compute Unified Device Architecture), 61
curie, 150
Curie, Marie, 150
Curry, Steph, 231
cybersecurity, 114, 147, 148, 179–80, 380
Cyc, 97

DAIR (Distributed AI Research Institute), 414–15, 419
Dalí, Salvador, 234
DALL-E, 11, 114, 234–39, 241–42, 258–59, 269
  avocado armchair, 235, 237–38
Damon, Matt, 317–18
D'Angelo, Adam, 321
  Altman's firing, 7, 11, 366, 367
  Altman's leadership behavior, 324–25, 352, 357, 359–60, 361–62
Dartmouth Summer Research Project (1856), 89–90, 94
data annotation, 15, 178, 189–90, 192–223, 414–17
  Kenya workers, 15, 18, 190–92, 206–13, 415–17
  Scale AI, 202–6, 213–14
  self-driving cars, 193–95, 202–6, 214–15
  Venezuela workers, 195–96, 198–202, 203–4, 218
data centers, 15, 274–78
  Altman's compute phases, 278–81
  carbon emissions, 79–80, 159–60, 171–73
  in Chile, 285–91, 295–99
  energy usage, 77, 80, 274–78, 280–81, 288–90, 294
  Google, 274–75, 285–91, 295–96
  in Uruguay, 291–96
"data colonialism," 103–4
data filtering, 137, 155, 177–78
Dataluna, 289–90
data privacy, 19–20, 33, 56, 103, 136, 186, 301, 308, 310, 413, 416
data scraping, 102–3, 114, 134–38, 151–52, 182–84, 384
"data swamps," 137–38, 212–13
Data Workers' Inquiry, 415–17
davinci, 150
da Vinci, Leonardo, 150
Dean, Jeff, 25, 158, 161–62, 163–65, 170–72
deepfakes, 79–80, 239, 391
deep learning, 98–101
  discriminatory impacts of, 57, 108–9
  ImageNet, 47, 100–101, 117–18, 259
  limitations and risks of, 106–10
DeepMind, 6, 17, 24–26, 48, 66, 158–59, 261–62, 384–85
  AlphaFold, 309–10
  AlphaGo, 59, 93

OpenAI and ChatGPT, 114, 119–20, 132, 159, 261–62
  scaling, 132, 158–59
Democratic Party, 41, 231
Dempsey, Jessica, 104*n*
dense neural networks, 177–78
Deployment Safety Board (DSB), 248, 323–24, 346, 350, 362, 363
Desmond-Hellmann, Sue, 376
Díaz Bejarano, Nicolás, 297–99
diffusion, 235–36, 375
  Stable Diffusion, 114, 137, 236, 242, 284
Digital Realty, 274
disaster capitalism, 189–223
discriminatory impact, 51–52, 57, 108–9, 114, 137, 161–64, 179, 310, 419, 432*n*
dissolving empire, 418–19
distillation, 177, 307
distress passwords, 149
Divorce, The, 156–57, 181, 213, 230, 233, 242
DNNresearch, 47, 50, 98–99, 100
Doctor Strange (movie), 303
Doomers (Doomerism), 233–34, 250, 267–68, 305–6, 308, 310, 311, 314, 315, 317–18, 319, 377, 387, 388–90, 396, 402, 403–4, 419
doomsday scenario, 26–27
Dorador, Cristina, 283
Dota 2, 66–67, 71, 129, 144–45, 244–45
Dowling, Steve, 154, 256, 382–83
doxing, 303
drinking water. *See* water resources
DUST, 269
Du, Yilun, 121

"earn to give," 229, 231
economic growth, 38–39
edge cases, 112
Edison, Thomas, 54, 55
education, 420–21
effective accelerationism (e/acc), 233
effective altruism (EA), 55–56, 228–33, 321–22, 388–89
Effective Ventures Foundation, 321–22
Ehlers-Danlos syndrome, 257, 338
election of 2016, 38, 42, 51–52, 321
ELIZA, 95–97, 111, 420–21
empires of AI, 16–20, 197, 222–23, 270, 414, 418, 420
energy usage, 77, 80, 160, 171, 173, 186–87, 275–78, 280–81, 288–90, 294, 295, 419, 451*n*
Enigma, 91
environmental impact, 20–21, 57, 79–80, 84, 89, 134, 165, 170–71, 309, 417, 420. *See also* extractivism; water resources
  plundered Earth, 271–300
Equinix, 274
Estallido Social, 291, 296
Etcheverry, Aisén, 300

European Commission, 105
European Union (EU), 283
    AI Act, 311
Evolved Transformers, 160, 171–73
Executive Order 14110, 310
existential risks, 24–25, 26, 55–56, 97, 125, 145, 229–32, 314, 410. *See also* Doomers
    p(doom) (probability of doom), 232, 250, 317, 319–20, 377
expected values, 229–30
expert systems, 94–95
Exploratory Research, 149, 151–52
extinction, 24, 26–27, 55, 232, 378
extractivism, 104, 417
    in Chile, 272, 273–74, 281–85, 296–99
    in Uruguay, 291–96
    use of term, 104n

Facebook, 11, 15, 16, 51–52, 105, 154, 159, 162, 192, 209, 230, 321, 334
facial recognition, 57, 103, 104, 115, 161, 435n
Fact Factory, 261
fair use, 91
Fairwork, 202, 206, 416
Federal Trade Commission (FTC), 239, 308, 358
Fedus, Liam, 247, 406
"Feel the AGI," 120, 255
Feynman, Richard, 121–22
firefighting, 237, 260
first mover's advantage, 103
Flamingo Generation, 220–21
Floyd, George, 152–53
"fluid data territory," 299
Formula One, 1, 231
Founders Fund, 38
Foursquare, 32
fraud, 25, 250, 267
free speech, 368–69
Friar, Sarah, 404
Fridman, Lex, 383
Friedman, Milton, 272–73
friendly AI, 57, 319–20
Friend, Tad, 26–27, 31
frontier model, 305–11
Frontier Model Forum, 305–6, 309
FTX, 231–32, 233
    bankruptcy, 257–58, 322, 380
FTX Future Fund, 231–32
Fuentes Anaya, Oskarina Veronica, 197–202, 415–17
Future Perfect, 388
Futures of Artificial Intelligence Research, 273–74

Gates, Bill, 68
    congressional testimony of, 311
    GPT-4, 245–48
    OpenAI demo, 71–72, 132–33, 246

Gates Demo, 71–72, 132–33, 246
Gawker Media, 38
GDPR (General Data Protection Regulation), 136
Gebru, Timnit, 24, 52–53, 108, 160–70, 171–73, 414
Generative Pre-Trained Transformers. *See* GPT
Genius Makers (Metz), 80
Geometric Intelligence, 110
Ghost Work (Gray and Suri), 193–94
Gibstine, Connie, 29–31, 44, 327–28, 331–32, 333, 337
Gibstine, Marvin, 29
GitHub, 135–36, 182–84, 237, 243, 336
    Codex, 184, 243, 247, 269, 318
    Copilot, 184, 237, 336
GiveWell, 230–31, 322
Global South, 16, 89, 165, 186, 190, 193, 222, 278, 291, 416. *See also specific countries*
Gmail, 100
Go (game), 59
Gobi, 269, 348
Godfather, The (movie), 369
Goldman Sachs, 18, 275
Good Ventures, 230–31
Google, 15, 132
    AI research, 64, 70, 72, 100–101, 106, 178
    AI scraping, 136
    Amodei at, 55, 57
    Android, 100, 239
    captchas, 98
    data centers, 274–75, 285–91, 295–96
    DeepMind. *See* DeepMind
    DNNresearch, 47, 50, 98–99, 100
    Frontier Model Forum, 305–6, 309
    GPT-4 and, 249
    Imagen model, 240, 242
    LaMDA, 153, 253–54
    neural networks, 100–101
    Project Maven, 52
    speech recognition, 100
    Sutskever and, 50, 100–101
    techlash, 51
    Transformers, 120–22, 158–59, 160, 165–66, 169, 171–73, 235
    valuation, 70
    Waymo, 100
Google Brain, 72, 159, 162, 166, 167
Google I/O, 379, 380, 383
Google Research, 53, 158, 163
Google Translate, 100, 121–22, 197, 410
Gordon-Levitt, Joseph, 323
government regulations. *See* regulations
GPT-1, 178
    release, 16, 122
    training and capabilities, 122, 123, 124, 235

INDEX

GPT-2, 71–72, 253
  errors, 146
  Gates Demo, 71–72, 132–33
  potential risks, 125–28
  "pure language" hypothesis, 129–30
  release, 75, 128, 314
  scaling, 130–32
  training and capabilities, 124–25, 135, 150, 153, 410
  withholding research, 125, 128, 131, 166
GPT-3, 132–36, 260, 278–79
  API, 150–51, 154–56, 158–59, 162, 163, 213–14, 314
  chatbot imitation, 112
  InstructGPT, 214–17, 246–47
  release, 133–34, 158–59, 160
  training and capabilities, 109, 134–35, 136, 153–56, 179, 242–43, 244, 253
GPT-3.5, 135, 183–84, 189, 217–18, 247, 258, 259–60, 264, 269, 378
GPT-3.75, 378
GPT-4, 189, 244–53
  Bing, 112, 113, 247
  capabilities, 16, 119, 135–36, 245–53, 410
  development, 242, 244–53
  release, 258–62, 323–24
  Superassistant, 247–49, 258–59, 381
GPT-4o, 383–84, 386, 390–91
GPT-4 Turbo, 346, 363
GPT-5, 279, 325
  Orion, 374–75, 379, 380, 405
GPUs (graphics processing units), 61–62, 134, 265–68. *See also* Nvidia
  shortage of, 261
Graham, Paul, 28, 32, 36–39, 40, 69
Gray, Mary L., 193–94
Groom, Lachy, 41
grounding hypothesis, 129–30, 318
Groves, Leslie R., 317–18
Guo, Eileen, 186

Hacker News, 70
hallucinations, 113–14, 217, 268, 358
Hanna, Alex, 414
"hardware overhang," 177, 232, 377
Harris, Kamala, 302
Hassabis, Demis, 24–26, 48, 309–10
"hate scaling laws," 137–38
hate speech, 18, 192, 208
health care and medicine, 12, 19, 76, 77–78, 114, 229, 257, 304, 333
Helion Energy, 186–87, 280
Hendrycks, Dan, 322–23
Hepburn, Audrey, 96
Her (movie), 246, 378, 382, 390–92, 393
Herbert-Voss, Ari, 179, 180–81
Hernández, Andrea Paola, 203–5
Hernandez, Danny, 60
Herzberg, Elaine, 107, 113

Hinton, Geoffrey, 105, 110
  DNNresearch, 47, 50, 98–99, 100
  ImageNet, 47, 59–60, 100–101, 101, 117–18, 259
  neural networks and deep learning, 97–99, 100–101, 109, 183
  Sutskever and, 47, 100–101, 109, 117–18, 121, 254
Hoffman, Reid, 50, 63, 320, 324, 367, 384–85
Hogan, Mél, 274–75
Ho, Jonathan, 235–36
Hollywood, 302–3
Hood, Amy, 72
Hooker, Sara, 306–7, 310, 311
Huffman, Steve, 34
Huggines, Ricardo, 204–5
Hugging Face, 276–77, 420
human brain, 60, 73, 90, 91, 109
human consciousness, 111, 119–20
human control, AI evasion of, 152, 310, 314, 380
human extinction, 24, 26–27, 55, 232, 378
human intelligence. *See* intelligence
human longevity, 186–87
human rights, 19–20, 197, 294
Hurd, Will, 321
Huyen, Chip, 52
Hydrazine Capital, 35–36, 38, 41, 69
hyperscalers, 274–75, 277, 279–80, 285, 294, 296

IBM, 100, 161
  Watson, 99
Imagen, 242
ImageNet, 47, 59–60, 100, 101, 117–18, 259
Imitation Game, The (movie), 81–82, 91
Index Ventures, 203
India, 133, 191, 202, 242, 276, 324
industrialization, 39, 272
industrial revolution, 88–89, 93
inequality, 15, 16, 190, 207, 228, 273, 291
inferencing, 98, 236, 373, 374, 378
Inflection AI, 320, 384–85
"information hazard," 125
Information, The, 33, 213, 280, 371, 403
Inglewood, 279
insider threats, 148
Instacart, 362, 376
Instagram, 334, 404
InstructGPT, 214–17, 246–47
intelligence, 109, 111
  definition of, 90–94
"Intelligence Age," 19, 405
International Energy Agency, 275
IQ tests, 91–92
Irving, Geoffrey, 158–59, 370
Isaac, William, 104

Israel, 47, 207, 337
iterative development, 142, 150, 314–15, 379, 401

Janah, Leila, 191–92, 206
Jeopardy! (TV series), 99
Jernite, Yacine, 276–77, 309
Jobs, Laurene Powell, 2
Jobs, Steve, 2, 34, 35, 37
Johansson, Scarlett, 382, 390–92, 393
John Burroughs School, 30–31, 329
Johnson, Josh, 381
Johnson, Simon, 88–89
Jones, Peter-Lucas, 410–13
Jones, Shane, 238–39
Jonze, Spike, 246
Jordan, Michael, 34

Kacholia, Megan, 166–68, 170
kaitiakitanga, 412
Kalluri, Ria, 102, 106, 418–19
Kaplan, Jared, 156–57
Karnofsky, Holden, 56, 57–58, 230, 321–22
Karpathy, Andrej, 64
Kay, Alan, 321
Kelton, Fraser, 150, 236–37, 241, 247
Kennedy, John F., 54
Kennedy, John Neely, 302
Kenya, 137, 179, 190–92
 data annotation, 15, 18, 190–92, 206–13, 415–17
 RLHF projects, 218–23
ketamine, 35, 42
Khan Academy, 246
Khan, Sal, 246
Khlaaf, Heidy, 179–80
Khosla Ventures, 70
Klein, Ezra, 115
Klein, Naomi, 272
Knight, Will, 126
Koko (gorilla), 254
Kokotajlo, Daniel, 388–90, 394, 403
Kolln, Ryan, 137, 189
Krisiloff, Matt, 41
Krizhevsky, Alex, 47, 100–101, 117–18, 259
Kwon, Jason, 7, 8, 346, 365, 373, 392–96

labor exploitation, 16, 17, 19–20, 89, 133, 190, 194, 295, 414–16, 418. *See also* data annotation
LAION, 137
LaMDA, 153, 253–54
language loss, 409–13
large language models, 15, 71, 115, 133, 153, 156, 158–60
 language loss, 410
 "On the Dangers of Stochastic Parrots," 164–73, 254, 276, 414
Latitude, 180–81, 189

Lattice, 36
Leap Motion, 69, 150, 344
LeCun, Yann, 105, 159, 235, 305
Leike, Jan, 387–88
 alignment and safety, 248, 250, 314, 316, 387–88, 403
 departure of, 387–88, 401
Lemoine, Blake, 253–54
Lessin, Jessica, 33
Library Genesis, 135
Lightcap, Brad, 4–5, 7, 69, 373, 393–94
limited partnerships (LPs), 66–67, 69–71
LinkedIn, 50, 218
LISTSERV, 26, 162, 167, 168
lithium, 272, 283–84
Liu Cixin, 83
Livingston, Jessica, 32, 37–38, 50, 69
Llama, 305
location tracking, 33
Loopt, 32–37, 43, 68
Loopt Star, 33
Lourd, Bryan, 382, 390–91
Lovelace, Ada, 150
Luccioni, Sasha, 276–77, 309, 420
Luka, Inc., 180
Luo people, 207
Lydic, Desi, 381
Lyft, 202, 331

MacAskill, William, 229, 231
machine learning, 77–78, 94–95, 98
Machine Learning for Health, 78
Mądry, Aleksander, 6, 8, 366, 380, 393, 398, 404
Maduro, Nicolás, 195–96
Mahelona, Keoni, 410–13
Makanju, Anna, 7, 154, 256–57, 302, 365
Mallery, Rob, 263
Manhattan Project, 27, 146–47, 315–18
Mannequin Challenge, 103
Māori people, 409–13
Marcus, Gary, 109–10, 118, 183, 252, 302, 307–8, 392
market capitalization, 18, 80, 84, 293
Mars, 23–24, 285
Martin, George R. R., 135
Mathenge, Richard, 416
Mayer, Katie, 150
Mayo office, 74, 316, 434n
McCarthy, John, 89–90, 92, 400
McCauley, Tasha, 321–24, 375
 Altman's firing, 7, 11
 Altman's leadership behavior, 324, 352, 357, 359–60, 361–62
McGrew, Bob, 69, 156, 236–37, 244, 373, 404, 405–6
Mechanical Turk (MTurk), 194–95, 202–3
Meena, 153
megacampuses, 275–76, 283–84

mega-hyperscale, 276
Mejias, Ulises A., 104
meritocracy, 36
Messerschmidt, Neily, 334–35
Meta, 51
  AI investments, 105
  compute, 305
  content moderation, 190, 192, 209
  data centers, 274–75, 281, 285
  Llama, 305
  OpenAI and, 159, 406–7
  open-source, 304–5
  techlash, 51
  Threads, 260
Metz, Cade, 80, 90
Metz, Luke, 247, 406
Miceli, Milagros, 414–15
Michelangelo, 81
Microsoft
  Altman and, 355–56
  firing, 4, 6, 9, 10, 13, 367
  Azure AI, 68, 72, 75, 156, 266, 279
  Bing, 112, 113, 247, 264, 355
  Copilot, 238–39, 247–48, 264
  data centers, 256, 274–75, 277, 278–81, 285, 287, 296–99
  Frontier Model Forum, 305–6, 309
  GitHub, 135–36, 182–84, 237, 243, 336
  Helion Energy, 187, 280
  Inflection AI, 320, 384–85
  market capitalization, 18, 80, 84, 293
  Max Altman at, 36
  ResNet, 309–10
  speech recognition, 100
  Tay, 153
Microsoft Office, 264
Microsoft, OpenAI partnership, 18, 67–68, 71–72, 234, 264–67, 269–70, 402
  ChatGPT, 264, 265–66
  compute phases, 278–81
  GPT-3, 156, 278–79
  GPT-4, 245–48, 279, 324
  investments and funding, 13, 17, 72, 75, 80–81, 84–85, 132–33, 143, 145, 156, 248, 331
Microsoft Research, 68
Microsoft Teams, 264
Mighty AI, 195
military, 52, 304, 380
Millicent, 220–23
Minsky, Marvin, 95, 96–97
Mishra, Nikhil, 58, 254
misinformation, 51–52, 179, 241, 377
Mission District, 57, 73–74
MIT (Massachusetts Institute of Technology), 6, 46, 88, 95, 106, 121, 231, 420–21
Mitchell, Margaret "Meg," 162, 164, 166, 169, 254

MIT Technology Review, 75, 86–87, 126, 169, 186, 370
model weights, 148, 149, 150, 156, 248, 266, 305–9
Moeroa, Raiha, 411
Mohamed, Shakir, 104
monopolies, 39–40, 101, 142, 182, 303
Montgomery, Christina, 307–8
moonshots, 48–49, 51
Moore, Gordon, 60
Moore's Law, 60–61, 116
Morton, Samuel, 91
MOSACAT, 288–92, 294, 297, 300, 417
Moskovitz, Dustin, 230
Mozilla Foundation, 102, 413
multimodal models, 92–93, 158, 175, 176, 234–35, 237, 246, 375
Mundie, Craig, 68
Murati, Mira, 343–51
  Altman and, 244, 345–51, 355–56, 362, 392–93
  firing and reinstatement, 1–5, 9–10, 364–73
  interim CEO, 1–2, 8, 357, 364–65, 366
  leadership behavior, 345–51, 362, 363–64
  background of, 69, 343–44
  chief technology officer, 343, 345–46
  DALL-E and, 241
  departure of, 404, 405–6
  hiring of, 69, 344
  Johansson and equity crises, 392–93
  Microsoft and, 182, 184, 270
  Omnicrisis, 396–98
  Scallion, 381
  Superalignment, 387
  at Tesla, 69, 344, 362
  Toner and, 348–51, 355–56
  VP of Applied, 150, 344–45
Murphy, Cillian, 317
Musk, Elon
  Altman and, 23–24, 26–28, 62–63, 64–66, 147, 316–17, 382
  firing, 368–70, 372, 375
  leadership behavior, 362, 368
  congressional testimony of, 311
  departure from OpenAI, 64–66
  founding of OpenAI, 12–13, 26–28, 47, 49–51, 53–54
  funding, 61–62, 63–64, 66–68
  governance structure of OpenAI, 13–14, 61–63
  Manhattan Project, 316–17
  MIT Technology Review story and, 86
  Neuralink, 63, 73, 147, 320
  Page and, 24, 25–26, 51
  Radford and, 122
  risks of AI, 23–27
  SpaceX, 23–24, 25, 28, 50, 368

Musk, Elon (cont.)
   xAI, 321, 322, 397, 403, 404–5
   Zilis and, 320–21, 324–25
   Zuckerberg and, 406–7
Mutemi, Mercy, 212, 291

Nadella, Satya, 113
   Altman's firing, 4, 6, 10, 367
   congressional testimony of, 311
   GPT-4, 247–48, 346
   OpenAI partnership, 67–68, 71, 72, 248, 265, 270
Nairobi, Kenya, 190–91, 193, 207, 208, 212, 219, 416
Napoleon Bonaparte, 399–400
National Highway Traffic Safety Administration, 107–8
Nectome, 186–87
Nepal, 206
Nest, 134–35, 144–44, 150, 151, 156, 244–45
Netflix, 59, 70
"network effects," 39, 40, 187
Neural Architecture Search, 160, 171, 173
Neuralink, 63, 73, 147, 320
neural networks, 95, 97, 98–101
   hallucinations, 113–14, 217, 268, 358
   limitations and risks, 106–10, 112–15
NeurIPS (Neural Information Processing Systems), 418
   Climate Change AI, 77
   Gebru and racism, 52–53, 161–62
   OpenAI at, 50, 154, 259, 374
   Test of Time Award, 259, 374
neurosymbolic AI, 109–10, 116
New Enterprise Associates, 32
Newsom, Gavin, 311
New York (magazine), 326–27, 328–29, 332–33, 336–40, 343, 352
New Yorker, The, 25, 26–27, 31, 57
New York Times, The, 80, 90, 95, 112, 115, 143, 221, 244, 264, 270, 272, 302, 313, 368, 371, 384, 400–401, 403
New York University, 105, 109, 235
New Zealand, 409–13
next-word prediction, 122, 124, 130
Nkosi, Thami, 104
Noah, Trevor, 11
Noble, Safiya Umoja, 162
noise pollution, 275
nondisparagement agreements, 389–90
North Africa, 205–6
North Korea, 146
nuclear fusion, 141, 186, 187, 280
nuclear-powered submarines, 144
Nvidia, 61–62, 278, 304, 412
   A100s, 175–76, 236, 242
   B100s, 279–80
   H100s, 279
   V100s, 133, 175

Obama, Barack, 25, 43, 154, 207
Odysseus, 279
Okinyi, Albert, 209, 211–12
Okinyi, Cynthia, 208, 209, 210–11
Okinyi, Mophat, 193, 207, 211–12, 291, 415–17
Olin College of Engineering, 121, 411
Olson, Parmy, 18
Ommer, Björn, 236
Omni, 380, 381
Omnicrisis, 390–92, 395–98, 400, 401, 403, 404
"On the Dangers of Stochastic Parrots" (Bender), 164–73, 254, 276, 414
OpenAI. *See also specific persons and products*
   Altman's firing and reinstatement, 1–12, 14, 364–73
   author's reporting, 12, 370–73
   the investigation, 369–70, 375–76, 377, 392
   Altman's vision for, 9, 83, 142–43, 262
   Applied division, 150–52, 154–56, 178–79, 213–14, 236–37, 239–40, 241, 247–51, 253, 267–68, 313, 314, 344–45, 375, 379–80
   The Blip, 375, 377, 384, 386, 396, 397–98
   board of directors. *See* board of directors, of OpenAI
   buildings and office design, 73–74, 316
   business structure and governance, 13–14, 61–67, 369–70
      Altman's restructurings, 86, 402–3, 407
      "capped-profit," 70, 72, 75, 322, 370–71, 401
      for-profit, 13, 14, 61–64, 69–70, 233, 369, 407
      limited partnerships, 66–67, 69–71
      nonprofit, 6, 13, 14, 27, 28, 49, 50, 61, 63–64, 65, 67, 233, 267, 402–3, 407
   charter of, 67, 70, 239, 401
   commercialization, 13, 14, 66–67, 72, 75, 101, 110, 143, 150–51, 154–55, 175, 267, 402
   company conflicts and rifts, 144–47, 149, 155–56, 233–34, 239–42, 267–68, 313–16, 345, 351–52, 387, 396, 402, 403–4
   company culture, 53–54, 127, 146–47, 157, 262–64, 267–68
   company mission, 5, 28, 66–67, 72, 76, 83, 84–85, 240, 385, 400–402, 418
   compensation, 50, 63–64, 69–70
   compute phases, plan, 278–81
   data bottlenecks, 244–45, 280, 309
   The Divorce, 156–57, 181, 213, 230, 233, 242
   employees, 256, 262–63, 385
   equity and equity crisis, 69–70, 388–90, 392–96, 463–64n
   Exploratory Research, 149, 151–52
   founding of, 12–13, 26–28, 46, 47–51
      Rosewood Hotel dinner, 28, 46, 47, 48, 55

Frontier Model Forum, 305–6, 309
funding, 61–62, 65–68, 71–72, 132, 141, 156, 262, 320–21, 331, 367, 377, 405
generative AI and, 110–15, 121–22
Johansson crisis, 382, 390–92, 393
launch of, 50–51, 52–53
logo, 4, 82, 385
Microsoft partnership. *See* Microsoft, OpenAI partnership
Musk's departure, 64–66
naming of, 28
"paradigm shift," 137, 189, 212
recruitment efforts, 53–54, 57–59, 63–64
Research division, 150, 151–52, 156, 177–78, 181–84, 240, 247, 260–61, 268–69, 313, 314, 347–48
  AI Scientist, 183, 318–19, 325, 347, 375
research road maps, 59–61, 175–78, 242
retreat of October 2022, 256–57
Safety, 145–46, 147, 149–59, 179–81, 213–15, 228, 239–41, 248–50, 254–55, 258, 261, 267–68, 305, 314, 317, 351–52, 372–73, 377–78, 380, 387, 388–89, 392–93, 403
scaling, 66, 117–20, 123, 130–32, 146, 159–60, 175–78, 213–14, 242, 278–79, 307, 373–74, 405
tender offer, 2, 4–5, 6, 11, 367
valuation, 2, 11, 14, 18, 49–50, 70, 84–85, 320–21, 406
OpenAI's Law, 60–61, 116, 123–24
OpenAI Startup Fund, 187–88, 324–25, 362
Open Philanthropy, 56, 57–58, 230–32, 322
OpenResearch, 185
open source, 49, 304–5, 308–11, 309, 401
Oppenheimer (movie), 316–18
Oppenheimer, J. Robert, 316–18
Orion, 374–75, 379, 380, 405
Ortiz, Karla, 303
Ostrich, 221
Otero Verzier, Marina, 297–98, 299
Oxford Internet Institute, 202, 416
Oxford University, 26, 55–56, 104, 229

p(doom) (probability of doom), 232, 250, 317, 319–20, 377
Pachocki, Jakub, 145
  AI Scientist, 318–19, 347
  AI security, 145, 148–49
  Altman and, 312, 386–87
    firing and reinstatement, 6, 8, 365–66, 366, 373
  leadership behavior, 347–48, 353, 355–56
  Dota 2, 145, 244–25
  GPT-3, 244–45
  GPT-4, 312
  new chief scientist, 386–87, 406
  Omnicrisis, 396–98
Page, Larry, 24, 25–26, 51, 249
Pakistan, 222

Pang, Wilson, 199
paper clips, 26, 56–57
"paradigm shifts," 137, 189, 212
Parakhin, Mikhail, 355
Park, Matt, 204
Parque de las Ciencias, 292
Patterson, Dave, 172–73
PayPal, 38, 50, 142, 198
PBJ1/PBJ2/PBJ3/PBJ4, 192
peer review, 15n, 170, 374
Pena, Daniel, 294–95, 297, 417
Perceptron, 90, 94–95
Perceptrons (Minsky), 95, 96–97
Perrigo, Billy, 137, 192, 210
Phillips Exeter School, 321
Phoenix, 279
Pichai, Sundar, 169, 311
Picoult, Jodi, 135
Pinochet, Augusto, 273, 296
Pioneer Building, 73–74, 316, 397
Piper, Kelsey, 388–90, 394, 403
plundered earth. *See* extractivism
Png, Marie-Therese, 104
Poe, 324
pornographic content, 108, 162, 189, 237–38. *See also* child sex abuse material
Posada, Julian, 196, 197, 291
poverty, 191, 201, 207, 282, 293, 333–34
Preparedness Framework, 379–80, 404
privacy concerns. *See* data privacy
productivity, 16, 18, 114–15, 222, 265–66
Project Maven, 52
psychological counseling, 191, 209–10, 211
public policy, 19, 43, 54, 75, 81, 125–28, 154, 276, 302–8, 311–12. *See also* regulations
pure language hypothesis, 129–30, 131, 158–59, 234, 318

Q*, 373–74
quantum computing, 141
Queer in AI, 161, 418
Quilicura, Chile, 285–88, 290, 296–99
Quora, 7, 183, 321, 324

racism, 52–53, 56, 91, 108–9, 114, 161–64
Radford, Alec, 121–24, 126, 137
  CLIP, 235
  departure of, 406
  GPT-1, 123, 124, 235
  GPT-2, 135
Raji, Deborah, 56–57, 108, 161, 238, 306–7, 310–12, 419–20
Ralston, Geoff, 34, 36, 142
Ramesh, Aditya, 235, 236
Ramos, Sonia, 281–82, 284–85, 295
rapid generalization, 154
Raven, 279
reality distortion field, 34
Reddit, 34, 151, 163

redistribution of power, 418–21
"red teaming," 179–80, 380
Regalado, Antonio, 186, 187
regulations (regulatory policy), 25, 27, 84, 86, 134, 136, 265, 272, 301, 303–4, 306–7, 311–12, 357, 358, 384
reinforcement learning from human feedback (RLHF), 123, 137, 146, 155, 176, 213–23, 245, 248, 315, 381, 387
Remotasks, 203–4, 218–23, 416
Renaldi, Adi, 186
renewable energy, 77, 275, 277
resiliency screening, 208
ResNet, 309–10
Retro Biosciences, 186–87
Rick and Morty (cartoon), 68
Rickover, Hyman G., 144
Rihanna, 1
Roberts Companies, 29
Robinson, David, 358
Roble, James, 329
robotics, 66, 69, 71, 130, 150, 156, 321
Rodríguez, Tania, 289–90
rogue AI, 55, 56, 145, 230, 231–32, 250, 306, 314, 319–20, 419
Roose, Kevin, 112, 264
Rose, Charlie, 40
Rosenblatt, Frank, 90, 94–95, 97
Rosewood Hotel dinner, 28, 46, 47, 48, 55
Rubik's Cube, 71
Russia, 146
  Ukraine war, 52, 191
Rwanda, 102, 260

Safe Superintelligence, 405
safety. *See* AI safety
Salinas, Alejandra, 290–91
Sama AI, 190–92, 206–13, 218–19, 242, 416
Santiago, Chile, 271–74, 285, 287–88, 295–96, 299–300
Scale AI, 195
  data annotation, 202–6, 213–14
  payment systems, 204–5
  RLHF projects, 218–23
scaling, 115–16, 117–20, 130–32, 146, 160–61
  "hate scaling laws," 137–38
scaling laws, 116, 123, 150, 156–57, 175, 177–78, 306
Scallion, 375, 378–82
Schmidt, Florian Alexander, 196–97
Schulman, John, 258, 387, 404
  InstructGPT, 214–17, 246–47
Schumer, Chuck, 43, 69, 311–12, 419
Scoble, Robert, 33
Scott, Kevin, 4, 68, 71, 72, 182, 247, 266–67, 270, 344
Sears, Mark, 206, 212–13
Securities and Exchange Commission (SEC), 384, 385, 403

Sedol, Lee, 59
self-driving cars, 100, 107–8, 141
  data annotation, 193–95, 202–6, 214–15
Seligman, Nicole, 376
SemiAnalysis, 268, 285
Senate Judiciary Hearing, 301–3, 307–9, 314–15
Sequoia Capital, 32
servers. *See also* data centers
  cooling, 274–75, 277–78, 288–90, 294
  Microsoft, 149
  OpenAI, 257, 260–61, 267
sex bots, 179
sexism, 162, 344–45
Shear, Emmett, 9–10, 34, 367, 369–70
Shopify, 46
Sidor, Szymon, 6, 8, 145, 148–49, 244–45, 318–19, 366
Sierra, 375
sign-up incentive, 267
Silicon Valley Bank crisis of 2023, 41–42
Silverman, Carolyn, 18
Simo, Fidji, 376
Sky, 391
Slack, 3, 9, 81, 156, 240, 263–64, 319, 358, 374, 389, 402–3
slavery, 89, 208, 400
Slowe, Chris, 34
Solon, Olivia, 103
Song, Dawn, 108, 114
Sora, 375
source code, 57–58
South Africa, 104–5, 115
South Korea, 59
SpaceX, 23–24, 25, 28, 50, 368
Spanish conquest of Chile, 271, 272
sparse models, 177–78
specism, 24
speech recognition, 78, 92, 100, 102, 118, 244, 309, 411
  Whisper, 244, 247, 267, 413
Stable Diffusion, 114, 137, 236, 242, 284
Stack Overflow, 183
standardized tests, 91–92, 245–46
Stanford University, 52, 74, 102, 137, 173, 235, 418
  AI Index, 105
  AI Salon, 24
  Altman at, 31–32, 39, 142
StarCraft II, 66
Starlink, 154
Steyerl, Hito, 137–38
Strawberry, 374, 375, 404
stress testing, 179–80
Stripe, 41, 46, 55, 58, 73, 82
Strubell, Emma, 159–60, 171–73, 309
Suleyman, Mustafa, 320, 384–85
SummerSafe, 68
Summers, Lawrence "Larry," 11, 375

INDEX

Superalignment, 316–17, 353, 387–88
Superassistant, 247–49, 258–59, 381
superintelligence, 19, 24, 27, 55
Superintelligence (Bostrom), 26–27, 55, 122–23
Suri, Siddharth, 193–94
surveillance capitalism, 101–2, 103–4, 111, 133, 138
surveillance drones, 52
Sutskever, Ilya
   Alignment Manhattan Project, 315–18
   Altman and, 347–48, 349, 386–87, 397, 401, 406–7
      firing and reinstatement, 1–6, 7, 9–12, 365–66, 368, 373–74
      leadership behavior, 340, 353–59, 363–64
   author's interview, 78–81, 159–60
   background of, 47
   board of directors and oversight, 322–23
   code generation, 152
   culture of OpenAI, 53–54
   deep learning and neural networks, 100–101, 109, 110
   departure of, 386–87, 398, 401
   DNNresearch, 47, 50, 98, 100
   "Feel the AGI," 120, 254–55
   founding of OpenAI, 28, 46, 47–51
   at Google, 50, 100–101
   governance structure of OpenAI, 61–63
   Hinton and, 47, 100–101, 109, 117–18, 121, 254
   ImageNet, 47, 59–60, 100–101, 101, 117–18, 259
   leadership of, 53–54, 58–59, 61–62, 63–65, 69
   Murati and, 343, 344, 347–48, 349
   Omnicrisis and, 396–98, 401
   paranoia of, 148, 149, 441*n*
   personality of, 3–4, 119–20
   Q* (Strawberry), 373–74, 404
   research road map, 59–61
   Safe Superintelligence, 405
   scaling, 117–20, 133, 159–60, 373–74
   Superalignment, 316–17, 353, 387
   Toner and, 325, 343, 351–52, 353–55, 359–60
   Transformers, 121–22
Swift, Taylor, 2
symbolists (symbolism), 94–95, 97, 99–100, 109–10, 116, 217
Syrian refugees, 137–38

Tay, 153
Taylor, Bret, 11, 375
technological revolutions, 16, 88–89, 93
   empires of AI, 16–19, 197, 222–23, 270, 414, 418, 420
technological unemployment, 78–81
techno-nationalism, 308–11
Techworker Community Africa (TCA), 416–17

Te Hiku Media, 411–14
Telemachus, 279
Tenaya Lodge, 255
"10x engineer," 82, 83, 142–43, 175, 177–78, 242
te reo Māori, 409–13
Tesla, 63, 64, 86, 194
   Autopilot, 64, 107–8, 109
   Model X, 69, 344
   Murati at, 69, 344
Test of Time Award, 259, 374
text generation, 112, 113, 121, 124
text-to-image, 176–77, 234–38. *See also* DALL-E
Thiel, Peter
   Altman and, 26–27, 36, 38–39, 39–42
   Founders Fund, 38
   founding of OpenAI, 12–13, 50
   "monopoly" strategy of, 39–40, 142
   Palantir, 38, 69
   PayPal, 38, 40, 142
   Trump and, 38, 42
Threads, 260
Three-Body Problem, The (Liu), 83
Three Mile Island, reopening, 275
TikTok, 304
Tiku, Nitasha, 253–54
Time (magazine), 137, 192, 210, 416–17
Tironi Rodó, Martín, 273–74, 297–98, 300
Toner, Helen, 58
   Altman and board, 7, 11, 253, 321–22, 375, 376
      leadership behavior, 324, 348–51, 353–55, 356–59, 361–62, 364
   "costly signals" paper, 357–59, 364
   Murati and, 348–51, 356–57
   Sutskever and, 325, 343, 351–52, 353–55, 359–60
Tools for Humanity, 185–86
TPUs (tensor processing units), 171
transcription, 220–21
Transformers, 120–22, 158–59, 160, 165–66, 169, 235
transparency, 5, 9, 14, 19–20, 81, 82, 86, 119, 134, 143, 166, 167, 172, 173–74, 230, 301, 384, 403, 406, 419–20
Trump, Donald, 38, 42, 51, 195, 321, 406
Tuna, Cari, 230
Turing, Alan, 81–82, 89, 91, 93, 373
Turing Award, 105, 162
Turing machine, 81–82, 91
Twitch, 9, 34, 367

Uber, 106, 107, 110, 136, 194, 228
Ukraine war, 52, 191
"United Slate, The" (Altman), 42
universal basic income (UBI), 85, 185–86
University of Applied Sciences Dresden, 196
University of California, Berkeley, 49, 56, 108, 118, 217, 235, 419

University of California, Los Angeles, 162
University of California, San Diego, 97
University of Chicago, 272–73, 296
University of Massachusetts Amherst, 79–80, 159–60
University of Toronto, 47, 105, 117
"unknown unknowns," 249
Upwork Research Institute, 18
Uruguay, 272, 417
  data centers, 291–96
  water crisis, 292–95
Utawala, Kenya, 190, 209, 211

Vallejos, Rodrigo, 296–99
veil of ignorance, 3
Venezuela crisis, 195–97, 203
Venezuela, data annotation, 195–96, 198–202, 203–4, 218
Victoria, Lake, 207
Villagra, Julia, 389–90
Vincent, James, 119
Virginia, data centers, 278
Volpi, Mike, 203
Volta, Alessandro, 133

WALL-E (movie), 234
Wall Street Journal, The, 33, 35, 41, 69, 102, 188, 193, 212, 280, 367, 384, 390–91, 416
Wang, Alexandr, 202–3, 213, 218
Warzel, Charlie, 370
Washington Post, The, 69, 114, 253, 371, 400, 403
water pollution, 293
water resources, 15, 17, 271, 273, 275, 277–78, 280–84, 287–96, 297, 299
Watson Health, 99
Waymo, 100
Weil, Elizabeth, 326–27, 328–29, 332–33, 336–40, 343

Weil, Kevin, 404
Weinstein, Emily, 307, 309
Weizenbaum, Joseph, 95–97, 420–21
Welinder, Peter, 150, 155, 250–51
Weng, Lilian, 267, 406
West, Kanye, 221
West, Sarah Myers, 308
Whale, 279–80
Whisper, 244, 247, 267, 413
whistleblower protections, 403
white hats, 107–8
Wikipedia, 57, 125, 135, 221
Willner, Dave, 238, 249–52, 267, 406
WilmerHale, 375
Wong, Hannah, 256, 326–28, 338–40
workplace impacts, 78–81, 114–15, 222, 265–66
World Bank, 207
Worldcoin, 185–86
World War II, 29, 91

X (formerly Twitter), 3, 257, 260, 312, 328, 368–69
xAI, 321, 322, 397, 403, 404–5
Xerox PARC, 54–55

Yale University, 196, 291
Y Combinator, 23, 27–28, 32, 34, 36–38, 39, 43
Yom Kippur, 326–27
YouTube, 34, 51–52, 102–3, 136, 244–45, 334
Yudkowsky, Eliezer, 319–20

Zaremba, Wojciech, 59, 152, 181–82
Zenefits, 36
Zilis, Shivon, 63, 320–21, 324, 384
Zoloft, 329, 330
Zoph, Barret, 247, 381–82, 387, 404, 406
Zuboff, Shoshana, 101
Zuckerberg, Mark, 38, 42, 159, 311, 406–7